# WATER RESOURCES

# WATER RESOURCES
## Distribution, Use, and Management

JOHN R. MATHER

Department of Geography
University of Delaware

JOHN WILEY & SONS, INC.

New York • Chichester • Brisbane • Toronto • Singapore

and

V.H. WINSTON & SONS

Silver Spring, Maryland

**Library of Congress Cataloging in Publication Data:**

Mather, John Russell, 1923-
    Water resources.

    (Environmental science and technology, ISSN 0194-0287)
    Includes bibliographical references and indexes.
    1. Water resources development.  I. Title.  II. Series.
TC405.M38    1984        333.91         83-21795
ISBN 0-471-89401-X

Printed in the United States of America

10 9 8 7 6 5 4 3 2

To my students in Water Resources
at the University of Delaware who have
contributed so much to the organization
and refinement of this material.

# SERIES PREFACE

Environmental Science and Technology

The Environmental Science and Technology Series of Monographs, Textbooks, and Advances is devoted to the study of the quality of the environment and to the technology of its conservation. Environmental science therefore relates to the chemical, physical, and biological changes in the environment through contamination or modification, to the physical nature and biological behavior of air, water, soil, food, and waste as they are affected by man's agricultural, industrial, and social activities, and to the application of science and technology to the control and improvement of environmental quality.

The deterioration of environmental quality, which began when man first collected into villages and utilized fire, has existed as a serious problem under the ever-increasing impacts of exponentially increasing population and of industrializing society. Environmental contamination of air, water, soil, and food has become a threat to the continued existence of many plant and animal communities of the ecosystem and may ultimately threaten the very survival of the human race.

It seems clear that if we are to preserve for future generations some semblance of the biological order of the world of the past and hope to improve on the deteriorating standards of urban public health, environmental science and technology must quickly come to play a dominant role in designing our social and industrial structure for tomorrow. Scientifically rigorous criteria of environmental quality must be developed. Based in part on these criteria, realistic standards must be established and our technological progress must be tailored to meet them. It is obvious that civilization will continue to require increasing amounts of fuel, transportation, industrial chemicals, fertilizers, pesticides, and countless other products and that it will continue to produce waste products of all descriptions. What is urgently needed is a total systems approach to modern civilization through which the pooled talents of scientists and engineers, in cooperation with social scientists and the medical profession, can be focused on the development of order and equilibrium in the presently disparate segments of the human environment. Most of the skills and tools that are needed are already in existence. We surely have a right to hope a technology that has created such manifold environmental problems is also capable of solving them. It is our hope that this Series in Environmental Sciences and Technology will not only serve to make this challenge more explicit to the established professionals, but that it also will help to stimulate the student toward the career opportunities in this vital area.

*Robert L. Metcalf*
*Werner Stumm*

# INTRODUCTION:

## The Study of Water Resources

Water is a unique substance. It has many physical properties unlike those possessed by other liquid, gaseous, or solid materials existing on the earth's surface. Water also possesses important economic, legal, and political aspects as well. It is not distributed uniformly over the surface of the earth. Some areas are blessed with a fairly uniform and more than adequate supply for human needs, but many other regions have a greater need for water than they have supply. Dry conditions in arid or semiarid regions exist for much, if not all, of the year. Water is in great demand in dry areas and considerable time and effort are expended by the inhabitants of such regions in searching for it, in transporting it to the regions of greatest need, and in storing it for later use. In those areas, water is quite costly and ownership of a supply is something to be prized. Money and labor are required to develop available supplies of water, so that lack of adequate water supplies usually indicates lack of financial resources to develop the supply as well as a lack of an available water source. Even in more humid areas, economics enters into the picture; to purify existing water sources for particular human or industrial purposes, or to keep polluted waste water out of the groundwater or surface water supplies, requires vast sums of money. Since water is vital for most domestic and industrial purposes, the economics of water is an important issue in water resources management.

Control of limited sources of water in arid or semiarid regions creates immense political power and authority. Thus, both the politics and the legal aspects become of vital concern in any study of water resources. Knowledge of the politics and of the legal decisions that have been achieved in the regulation and control of water greatly aids in explaining the way water resources management has developed in a particular nation.

While water is neither really created nor destroyed on the face of the earth (small quantities of new water or connate water may be formed in volcanic eruptions but it is a minor quantity in terms of the vast quantities of water now existing on the earth), it does not mean that we always have a fixed quantity everywhere at all times. Both precipitation supplies and evapotranspiration demands vary markedly from place to place and from season to season. Our treatment of the surface of the earth and of the atmosphere can influence the amount of precipitation, the conservation of water on the land, or the way it is absorbed or runs off at the surface. Thus, the ability to produce more water or, at least, to modify the available supply at a place is under partial human control. What we do at the surface of the earth or in the atmosphere may be quite influential in the resulting supply available for human use.

Possibly more than any other aspect of the physical world, water can be thought of as a synthesizer of significant and wide-ranging disciplines. One cannot really look at water from a physical viewpoint and study only the physical aspects or distribution, for then only a small part of the whole subject is seen. Likewise, one cannot really study the legal or economic aspects without understanding the physical aspects of how it moves through the soil and the interconnections between water at the surface and water in the groundwater table. A complete study of water resources per se must bring together the physical and social studies aspects.

Water is a natural subject for study in geography—one of the few disciplines broad enough in its outlook to bridge the gaps between the physical, social, human, and cultural worlds. If we sometimes feel that our various disciplines are becoming too fragmented—too many subdisciplines are being created by narrower and narrower specialists—we can look with pleasure on the field of water resources where we must seek those interrelations between the physical, the human, and the cultural if we are to achieve a worthwhile grasp of the subject. Water is an ideal subject with which to unite the broad discipline of geography.

# ACKNOWLEDGMENTS

I will never be able to express sufficient appreciation to the many individuals and groups who have helped in the preparation of this book. However, special thanks must go to those who have given permission to use material from other books, journals, or reports. I hope all have been acknowledged appropriately; any oversights are unintentional.

Many individuals have also contributed their special skills to the production of the finished manuscript. With apologies to those who may inadvertently be omitted, I extend my deepest appreciation to Mrs. Bernice Williams, Mrs. Jane Austin, Mrs. Elizabeth Pfeufer, and Mrs. Nancy Bristow, all of whom have typed and retyped portions of the text with skill and cheerfulness. Special thanks to Dr. Thomas Meierding for his cartographic advice and to Jeffery Macintire for supervising the preparation of the art work (and preparing some of it himself) as well as to his associates, Robin Spiller, Don Hallegan, and Thomas Cosgrove, who ably prepared the remaining art work. Wendy Sue Rottenberg, Cornelia Thorne, Glenn Hartmann, and Lynda Hall helped research special topics and deserve my sincere thanks. My daughter Ellen must be accorded very special mention not only for her continuing encouragement but also for her work on many technical problems, especially those connected with permissions for borrowed materials and with the references. Finally, my thanks to my wife, Amy, and to my children for their patience and understanding of my preoccupation with manuscript preparation rather than with them over the past 4 years.

# CONTENTS

# PART A:
# PHYSICAL ASPECTS OF WATER
# AND WATER MANAGEMENT

Water as a physical substance is unique in a number of respects. It can exist easily in all three states—solid, liquid, and gaseous—at temperatures found at the surface of the earth and even at the same temperature. Its density is greatest at an above-freezing temperature; thus, as water cools from about $4°C$ to the freezing point and below, the density of water becomes less. It is a universal solvent so that "pure" water in nature is difficult to find. It has the highest specific heat of all common substances. The list could be enlarged but it is already sufficient to show that water has a special place among physical substances.

Any study of geography must impress us with the fact that the physical world is a fairly rational mechanism—complex, closely interrelated, but still not haphazard or without logic. The earth's physical features include oceans and mountains, plains and forests, deserts and lakes, and their location and existence are basically in accord with the so-called laws of nature. Forests are where they are for perfectly logical reasons, just as deserts are found in places where we would expect them to be. One might almost go so far as to say that it is often only humans who choose, at times, to locate illogically. There are well-defined areas on earth where water is in short supply and others where it is fairly abundant. When human demands exceed natural supply, problems exist.

This section will outline water problems that have developed because of the distribution of supply and demand factors and the operation of the hydrologic cycle that supplies the sources of our water, as well as provide physical facts about the principal terms in the hydrologic cycle—precipitation, evapotranspiration, water storage, and water runoff. Knowledge of the physical distribution of water is basic to an understanding of the social, economic, and political problems that surround water resources management.

# THE WATER PROBLEM

Water is not in short supply over the earth as a whole. After all, some 70% of the globe's surface is water. Even in the United States, the overall water picture is quite favorable (Fig. 1.1). Average precipitation over the conterminous United States is some 30 in./year—a value of 4750 million acre-feet of water per year.* Seventy percent is evaporated back to the atmosphere from the vegetation, forests and pastures, cropland, and noneconomic vegetation, as well as from the lakes and water surfaces. Even though only 30% of the precipitation supply is left for other uses—primarily human, agricultural, and industrial—it amounts to some 1370 million acre-feet annually. Only about 25% of this water is actually withdrawn from the surface water supplies and from the groundwater for use in the various processes of human life, and even less—possibly some 7% of the total water in the surface water supplies or only about 30% of that withdrawn from the surface supplies—is actually consumed or returned to the atmosphere and not to the surface water supplies in some modified form. Ultimately, some 1280 million acre-feet run off to the oceans in our surface streams. Of course, not all of this water can be utilized for further expansion of domestic, agricultural, or industrial purposes since some must be left in the streams for recreation, fish and wildlife, navigation, transportation, and other similar purposes. However, overall supplies are still quite adequate for a considerable expansion of human needs before we have to worry about shortages or the possibility of running out of water.

But we must look at the distribution of water across the nation as well as at total supplies. The mean annual streamflow from more than one-half of the country is less than 5 in. In just 10% of the country, the annual runoff is over a 20-in. depth while almost one-quarter of the country has less than a 0.5-in. depth of

---

*An acre-foot is the volume of water contained in the area of 1 acre when covered to a depth of 1 foot—not quite 326,000 gal.

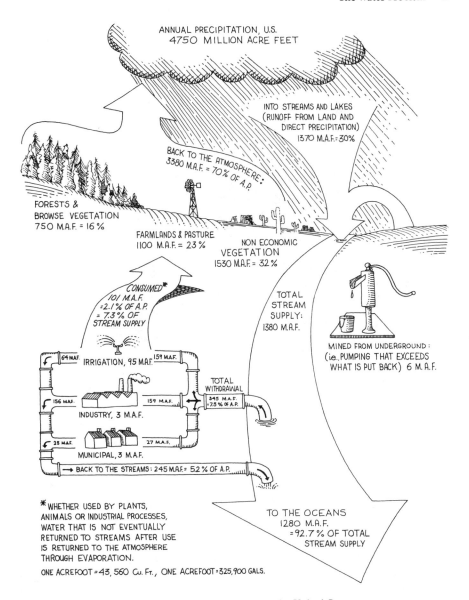

ANNUAL PRECIPITATION, U.S.
4750 MILLION ACRE FEET

INTO STREAMS AND LAKES
(RUNOFF FROM LAND AND
DIRECT PRECIPITATION)
1370 M.A.F. = 30%

BACK TO THE ATMOSPHERE:
3380 M.A.F. = 70% OF A.P.

FORESTS &
BROWSE VEGETATION
750 M.A.F. = 16%

FARMLANDS & PASTURE
1100 M.A.F. = 23%

NON ECONOMIC
VEGETATION
1530 M.A.F. = 32%

CONSUMED*
101 M.A.F.
=2.1% OF A.P.
= 7.3% OF
STREAM SUPPLY

TOTAL
STREAM
SUPPLY:
1380 M.A.F.

MINED FROM UNDERGROUND:
(ie., PUMPING THAT EXCEEDS
WHAT IS PUT BACK) 6 M.A.F.

64 M.A.F.    IRRIGATION, 95 M.A.F.    159 M.A.F.

TOTAL
WITHDRAWAL
345 M.A.F.
=7.3% OF A.P.

156 M.A.F.    INDUSTRY, 3 M.A.F.    159 M.A.F.

25 M.A.F.    MUNICIPAL, 3 M.A.F.    27 M.A.F.

→ BACK TO THE STREAMS: 245 M.A.F.= 5.2% OF A.P.

*WHETHER USED BY PLANTS,
ANIMALS OR INDUSTRIAL PROCESSES,
WATER THAT IS NOT EVENTUALLY
RETURNED TO STREAMS AFTER USE
IS RETURNED TO THE ATMOSPHERE
THROUGH EVAPORATION.

ONE ACREFOOT =43,560 Cu. Ft., ONE ACREFOOT =325,900 GALS.

TO THE OCEANS
1280 M.A.F.
=92.7% OF TOTAL
STREAM SUPPLY

**Figure 1.1** Disposition of the annual precipitation over the United States.
*Source.* © *The Johns Hopkins Magazine* (1966) as adapted from Wolman (1962).

runoff per year. The nationwide average annual runoff is about a 8.6-in. depth (Fig. 1.2).

These figures point out that the distribution of water is quite varied. Some portions of the country are richly endowed and other portions are practically bereft of water. While overall supplies are sufficient, certain areas have considerably

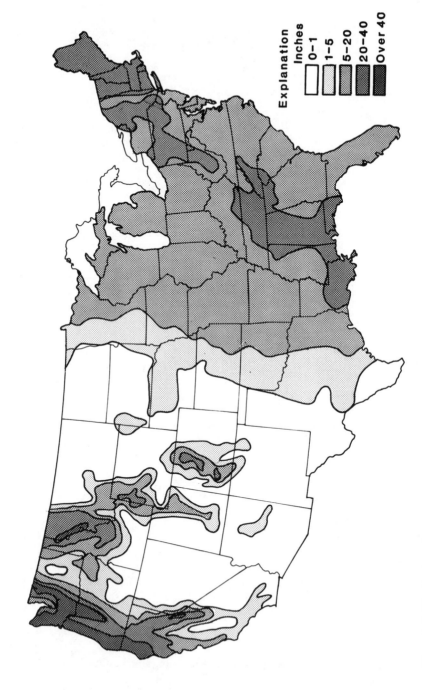

**Figure 1.2** Distribution of average annual runoff across the United States. *Source.* U.S. Water Resources Council (1978).

**Table 1.1** Available Water Resources, Withdrawal, and Consumption in the United States by Water Resources Regions

| Region | Area (1000 sq mi) | Average Runoff | | Estimated Dependable Supply, 1980 (bgd) | Withdrawals 1970[a] (bgd) | Fresh Water Consumed 1970 (bgd) | Annual Flow in bgd exceeded in 90% of Years | Fresh Surface Water Withdrawn, 1970 (bgd) |
|---|---|---|---|---|---|---|---|---|
| | | Inches Per Year | Bgd | | | | | |
| New England | 59 | 24 | 67 | 22 | 9.7 | 0.41 | 49 | 4.1 |
| Middle Atlantic | 102 | 18 | 84 | 36 | 45 | 1.4 | 68 | 25 |
| South Atlantic–Gulf | 270 | 15 | 197 | 75 | 35 | 3.3 | 129 | 20 |
| Great Lakes | 126 | 12 | 75 | 69 | 39 | 1.2 | 54 | 38 |
| Ohio | 163 | 16 | 125 | 48 | 36 | .92 | 75 | 34 |
| Tennessee | 41 | 21 | 41 | 14 | 7.9 | .24 | 28 | 7.7 |
| Upper Mississippi | 190 | 7.2 | 65 | 31 | 16 | .76 | 36 | 14 |
| Lower Mississippi | 96 | 17 | 79 | 25 | 13 | 3.6 | 38 | 8.5 |
| Souris–Red–Rainy | 59 | 2.2 | 6.2 | 3 | .3 | .07 | 2 | .2 |
| Missouri | 515 | 2.2 | 54 | 30 | 24 | 12 | 29 | 18 |
| Arkansas–White–Red | 265 | 6.0 | 73 | 20 | 12 | 6.8 | 36 | 5.2 |
| Texas–Gulf | 175 | 3.9 | 32 | 17 | 21 | 6.2 | 11 | 7.4 |
| Rio Grande | 136 | .8 | 5.0 | 3 | 6.3 | 3.3 | 2 | 3.8 |
| Upper Colorado | 110 | 2.5 | 13 | 13 | 8.1 | 4.1 | 8 | 8.0 |
| Lower Colorado | 137 | .5 | 3.2 | 2 | 7.2 | 5.0 | 1 | 2.8 |
| Great Basin | 185 | 1.0 | 7.5 | 9 | 6.7 | 3.2 | 3 | 5.5 |
| Columbia–North Pacific | 271 | 16 | 210 | 70 | 30 | 11 | 148 | 26 |
| California–South Pacific | 120 | 9.0 | 62 | 28 | 48 | 22 | 30 | 21 |
| United States (conterminous) | 3,020 | 8.3 | 1,200 | 515 | 365 | 87 | 747 | 249 |
| Alaska | 590 | | | | .2 | .02 | | .2 |
| Hawaii | 6.4 | 44 | 13 | | 2.7 | .81 | | .8 |
| Puerto Rico | 3.4 | | | | 3.0 | .17 | | .4 |
| Grand total | 3,620 | | | | 371 | 88 | | 250 |

[a]Including some minor interregional diversions.
*Source.* Murray and Reeves (1972).

5

more than others. The values of runoff withdrawals, and estimated dependable supplies (Table 1.1) given by water resources regions (located on Fig. 1.3), suggest something of the variation in supply and available water that exists in the United States. Hawaii has the greatest annual runoff (44 in.) while, of the regions in the conterminous United States, the New England region has the greatest amount of runoff (24 in.) followed by the Tennessee region (21 in.) and the Middle Atlantic area (18 in.). At the other end of the scale, the Great Basin area of Nevada and Utah has only 1 in. of annual runoff, the Rio Grande area only 0.8 in. and the Lower Colorado area (mainly Arizona) only 0.5 in.

The figures on estimated dependable supply in 1980 suggest that the South Atlantic–Gulf and the Columbia–North Pacific regions will be most favorably situated. This is also true in terms of the volumes of annual flows exceeded in 90% of the years.

Precipitation, of course, provides the basic supply of water for runoff. Figure 1.4, a map of average annual precipitation across the conterminous United States, is closely related to Figure 1.2, the map of runoff. Any discrepancy between the two figures is the result of evapotranspiration losses of water which reduce the available supply and make it unavailable for runoff. The largest values of annual precipitation are found in the southern Appalachian mountains, in Louisiana and along the eastern Gulf coast, and in the mountainous areas of the Pacific Northwest. Most of the western Great Plains and the Rocky Mountains areas have only a limited supply of water. Expressed in terms of volumes of water per square mile, Louisiana is the most favored of all of the conterminous United States with nearly 1000 gal of rainfall per square mile, followed closely by Alabama and Mississippi with nearly 950 gal/mi$^2$. On the dry end, Nevada is by far the driest, receiving only about 130 gal of precipitation a year per square mile.

## DISTRIBUTION OF WATER

Most of us are fortunate to be surrounded by fairly available supplies of water in various forms. Water vapor is present even in the atmosphere of the deserts and it is certainly present in the humid air of tropical and subtropical climates. Water is stored in lakes, ponds, reservoirs, rivers, and streams on the surface of the earth as well as in the great ocean areas of the world. It is also stored in less visible form in the vegetation of the surface and, although many of us may not see them, in the glaciers and ice caps of the polar or high-altitude regions of the world. Finally, water is stored in the upper soil layers to be utilized by plant roots as well as in the deeper soil and rock layers as part of the groundwater. All told, tremendous volumes of water are present on our planet. It has been estimated that if the entire surface of the earth were levelled off and the land areas filled into the ocean depths, we would have a whole planet covered with water to a depth of 10,000 ft. But, again, just as the water distribution in the United States varies appreciably from region to region, so does the amount of water in each of the possible categories in which it can occur in nature.

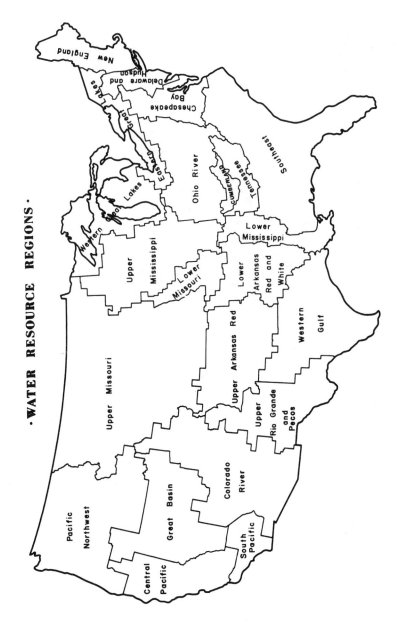

**Figure 1.3** Water resources regions of the United States. *Source.* Murray and Reeves (1972).

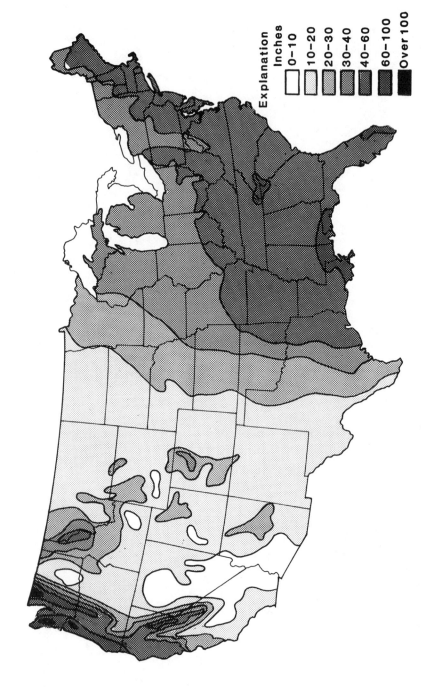

**Figure 1.4** Distribution of average annual precipitation across the United States. *Source.* U.S. Water Resources Council (1978).

(a) *Water in the atmosphere.* Water vapor in the lower atmosphere varies generally from close to 0% to about 4% by volume from region to region. It has been estimated that if all of the water in the atmosphere were precipitated out at one time, there would be enough to cover the entire surface of the globe to a depth of about 1 in. (25 mm). Multiplied by the area of the globe, this gives a water volume of some 13,000 $km^3$ (3200 $mi^3$) in the atmosphere.

(b) *Water in oceans.* A rough figure for the amount stored in the ocean areas of the world can be obtained from an estimate of the average depth of the oceans multiplied by the area of the oceans themselves. While the latter figure can be fairly well determined, the average depth of the oceans is only approximate since bottom contours for all ocean areas are not available. However, based on an estimated average depth of the oceans of 12,500 ft, this results in a total water volume stored in the oceans of the world of 1,350,000,000 $km^3$ (330,000,000 $mi^3$).

(c) *Water in lakes.* An inventory of the amount in the lakes of the world reveals that 80% of all lake storage is found in the 40 largest lakes. The many thousands of smaller lakes provide storage for only a small percentage of the total lake water.

Lakes can be either fresh or salt depending on whether they have outlets or not. Of the salt lakes, the largest, by far, is the Caspian Sea with more than 75% of all the salt water volume in the world—some 80,000 $km^3$ or 19,000 $mi^3$. This volume is about three times the fresh water volume in all five of the Great Lakes between the United States and Canada. Other salt lakes include the Dead Sea (280 $km^3$, 70 $mi^3$) and the Aral Sea (900 $km^3$, 215 $mi^3$). The shallow Great Salt Lake in Utah, while well known in this country, ranks nowhere near these lakes in volume of water stored. Total storage of salt water in lakes with interior drainage approximates 105,000 $km^3$ (25,000 $mi^3$).

Fresh water lakes provide a slightly greater amount of water storage than salt lakes. While none of the fresh water lakes approaches the Caspian Sea in volume of storage, Lake Superior is second in surface area (it has 31,700 $mi^2$ of surface compared with 168,000 $mi^2$ for the Caspian Sea). Lake Baikal is second in volume with some 23,000 $km^3$ (5600 $mi^3$) compared with the 80,000 $km^3$ of water in the Caspian Sea. The Great Lakes have a total water volume of some 22,500 $km^3$ (5400 $mi^3$), approximately equal to that of Lake Baikal. Lake Tanganyika is close behind with a water volume of 19,000 $km^3$ (4600 $mi^3$). All fresh water lakes combined have a total water volume of some 125,000 $km^3$ (30,000 $mi^3$).

(d) *Water in rivers.* While flow volumes of many of the rivers of the world are fairly well known, it is more difficult to estimate the amount of water actually stored in the rivers at any time. The Amazon has a flow equal to about 20% of the total discharge of all of the rivers of the world—more than four times the discharge of the Congo, the river with the next greatest discharge, and more than ten times larger than the discharge from the Mississippi River, which ranks seventh in terms of total discharge among the rivers of the world. To determine the storage of water in a river involves knowledge of its length, width, and depth. Total river storage has been estimated at about 0.7% of all of the water stored in the lake areas of the world—1700 $km^3$ (400 $mi^3$).

Reservoirs constitute a surface storage of water also and so it is necessary to make some estimate of this water volume if we are to obtain a realistic figure for total water stored on the surface. Van der Leeden (1975) has provided a tabulation (based on a Bureau of Reclamation survey) of the storages in the major reservoirs of the world. The list shows a total storage volume of some 2500 km$^3$ (600 mi$^3$) of water. Compared with the water storage in natural lakes, this is still a rather small amount. The total storage of water on the surface in lakes, rivers, and reservoirs is probably very close to 235,000 km$^3$ (56,000 mi$^3$).

(e) *Water stored in glaciers and ice caps.* The amount held in frozen form in polar ice caps and mountain glaciers will vary somewhat through the year with freezing and melting. Again, any estimate of the total volume of water held in this form is quite rough since actual depths of ice thickness in the polar regions—on Greenland and in the Antarctic, especially—are not known exactly. Estimates of the area of polar ice caps total some 6,000,000 mi$^2$—somewhat less than one-half of the total area within the combined Arctic and Antarctic circles. More than 5,000,000 mi$^2$ of this ice cap are found in the South Polar region and some 700,000 mi$^2$ are found in the North Polar region. The rest of the glacier ice occurs in the other continental areas of the globe, mostly at high elevation.

Depth of ice is difficult to estimate without precise soundings of the elevation of the land beneath the ice cap or glacier. However, a thickness of about 2 mi for the polar ice caps and considerably less for the other glaciers has been used to provide an estimated average thickness of 1 mi for the glaciers and ice caps of the world. Using this thickness provides a total water volume of 6,000,000 mi$^3$ (25,000,000 km$^3$) for the storage of water in the ice caps and glaciers of the world.

(f) *Water in vegetation.* Estimates of the total amount of water stored on the surface of the earth in the vegetation are extremely crude, and, of course, the amount varies appreciably from winter to summer. Since it is such a small amount, there is no reason to try to refine our estimate. The total volume in the vegetation of the earth can be considered to be negligible in the overall water budget.

(g) *Water in soil and rocks.* Storage beneath the surface can be separated into several categories—water in the root zone of the soil (top 3 to 4 ft), water from the root zone to the 2500-ft depth (sea level), and water from there to the average depth of the oceans, 12,500 ft. Water below that depth is generally unavailable and will not be considered here.

If we assume that the upper soil layer, the plant root zone depth, has a water-holding capacity of 3 in./ft or 12 in. in a 4-ft root zone, and that, on average, the layer is holding 50% of its available capacity, we obtain a maximum value of some 22,000 km$^3$ (approximately 5000 mi$^3$) of water stored in the root zone.

From the root zone to a depth of 2500 ft, we can assume a porosity of 4% by volume. With a land area of some 57 million mi$^2$ and a depth of 1/2 mi, this provides a water volume of about 1,000,000 mi$^3$ (4,200,000 km$^3$).

Below 2500 ft, we might assume a value of 1% as the porosity of the rock layers. From 2500 ft to a depth of 12,500 ft is approximately 2 mi. Again using a land area of 57 million mi$^2$, we obtain an additional amount of water storage in the deep rock layers of some 1,100,000 mi$^3$ (4,600,000 km$^3$).

The foregoing figures have been summed in Table 1.2 for comparison purposes.

**Table 1.2**  Summary of Water Volumes Stored, by Categories

| Category | Storage (1000 km$^3$) | % of Total |
|---|---|---|
| In atmosphere | 13 | 0.001 |
| In ocean | 1,350,000 | 97.542 |
| In lakes | | |
| salt | 105 | 0.007 |
| fresh | 125 | 0.009 |
| In rivers | 1.7 | – |
| In reservoirs | 2.5 | – |
| In glaciers and ice caps | 25,000 | 1.806 |
| In vegetation | negligible | – |
| In soil and rocks | | |
| root zone | 22 | 0.001 |
| to 2500 ft | 4200 | 0.303 |
| 2500–12,500 ft | 4600 | 0.331 |
| Total (all categories–rounded) | 1,384,000 | 100 |
| Summary:  Total in ocean | 97.5% | |
| in glaciers and ice caps | 1.8% | |
| in soil | .6% | |
| | 99.9% | |

Figures in Table 1.2 are revealing. Vast quantities of water are available on the earth but only tiny amounts are available in fresh water form in areas where a population needs to exist. More than 99.3% of all of our water exists either in saline form (oceans or salt lakes) or in glaciers and ice caps located in regions where there is very little, if any, demand for water and where demand will probably never be very significant. This leaves less than 0.7% as fresh water storage on land and half of this amount is in deep rock layers more than 2500 ft below the surface and so relatively unavailable for use. If the total volume of stored water were our only supply, we would be in critical shape for, in many areas, that would be exhausted in short order. Considering only the fresh water storage on the surface and to a depth of 2500 ft, there is just over $1 \times 10^{18}$ gal of fresh water. Consumption in the United States alone comes to some $0.15 \times 10^{15}$ gal/yr. Worldwide total demand comes to nearly $1 \times 10^{15}$ gal/yr. Since only a small portion of the total stored water is actually available—that occurring in areas where people live and also occurring in large enough supply so that pumping would be economically feasible—the fact that current storage is three orders of magnitude greater than current world needs is hardly reassuring. Water at 2500 ft and held in rocks with a 4% porosity is not available for pumping in large enough quantities to make it worthwhile to develop.

Thus, groundwater or deep storage of water is not going to supply enough

**Table 1.3** Estimated Water Use in the United States: 1900 to 1970 and Projections to 1980 (in billions of gallons daily averages)

| Year | Total Water Use | | Irrigation[a] | | Public Water Utilities | | Rural Domestic[b] | | Self Supplied Uses | | | |
|---|---|---|---|---|---|---|---|---|---|---|---|---|
| | | | | | | | | | Industrial and Miscellaneous[c] | | Steam Electric Utilities | |
| | Total | Ground | Total | Ground | Total | Ground | Total | Ground | Total | Ground | Total | Ground |
| 1900 | 40.19 | 7.28 | 20.19 | 2.22 | 3.00 | 1.05 | 2.00 | 1.60 | 10.00 | 2.40 | 5.00 | 0.01 |
| 1910 | 66.44 | 11.68 | 39.04 | 5.29 | 4.70 | 1.49 | 2.20 | 1.76 | 14.00 | 3.15 | 6.50 | 0.01 |
| 1920 | 91.54 | 15.78 | 55.94 | 8.17 | 6.00 | 1.79 | 2.40 | 1.94 | 18.00 | 3.87 | 9.20 | 0.01 |
| 1930 | 110.50 | 18.18 | 60.20 | 9.09 | 8.00 | 2.30 | 2.90 | 2.40 | 21.00 | 4.37 | 18.40 | 0.02 |
| 1940 | 136.43 | 22.56 | 71.03 | 11.22 | 10.10 | 2.82 | 3.10 | 2.64 | 29.00 | 5.86 | 23.20 | 0.02 |
| 1950 | 202.70 | 35.19 | 100.00 | 19.80 | 14.10 | 3.78 | 4.60 | 4.09 | 38.10 | 7.47 | 45.90 | 0.05 |
| 1960 | 322.90 | 58.17 | 135.00 | 35.24 | 22.00 | 5.68 | 6.00 | 5.58 | 61.20 | 11.57 | 98.70 | 0.10 |
| 1970 | 327.30 | 54.27 | 119.18 | 33.13 | 27.03 | 6.65 | 4.34 | 4.13 | 55.95 | 10.24 | 120.80 | 0.12 |
| 1980 | 442.63 | 63.98 | 135.85 | 38.17 | 33.60 | 7.73 | 4.85 | 4.61 | 75.03 | 13.28 | 193.30 | 0.19 |

[a]Total take including delivery losses but not including reservoir evaporation.
[b]Rural farm and nonfarm household and garden use, and water for farm stock and dairies.
[c]For 1900–1960, includes manufacturing and mineral industries, rural commercial industries, air conditioning, resorts, hotels, motels, military and other State and Federal agencies, and other miscellaneous uses; thereafter includes manufacturing, mining and mineral processing, ordnance and construction.
*Source.* U.S. Bureau of the Census (1975).

fresh water for all our many needs. If it were not for the operation of the hydrologic cycle—the unending flow of water in all of its various states from ocean, to atmosphere, to land, and back to ocean again—we would shortly be without water. The hydrologic cycle, which will be described in more detail in the following chapter, provides a constant replenishment of fresh water in the lakes, reservoirs, and streams as well as in the more porous upper layers of the soil and rock mantle of the earth. The constant evaporation of water from the ocean, made fresh by the process of solar distillation, its movement in great moist air masses from the ocean areas to the land where it falls as precipitation, and its subsequent infiltration and runoff where opportunities exist for use—the hydrologic cycle—is the mechanism that supplies the daily and annual volumes of water needed. Storage in rivers, lakes, reservoirs, and soil is constantly being replenished. In arid areas or where human demands are excessive, the annual renewal of supplies may not be able to keep up with the demand. Here, other sources of fresh water, interbasin transfers, or "mining" of the water stored in the water table must occur in order to meet the demand.

## CHANGING TRENDS IN THE USE OF WATER

The use of water has increased rapidly and significantly over the past few decades around the world. Table 1.3 gives values of estimated water use in the United States, by categories of use from 1900 to 1980 (in two parts)—water demand satisfied from groundwater sources and the total demand for water by category.

Total water demand is estimated to have increased about elevenfold in the period 1900 to 1980. Increased demand cannot continue indefinitely into the future, although some predictions of future use seem to be based on such an extrapolation.

Before considering future water requirements, we should study the pattern of past water demand. Overall demand increased slowly in the first part of the 20th century—by 26 billion gallons/day (bgd) in the period from 1900 to 1910, by 25 bgd from 1910 to 1920, by 19 bgd from 1920 to 1930, and by 26 bgd from 1930 to 1940. Strangely, the increase was only 5 bgd from 1960 to 1970.* Thus the periods of excessive growth in demand were confined to just 30 of the 80 years of this century—from 1940 to 1950, an increase of 66 bgd; from 1950 to 1960, an increase of some 120 bgd; and from 1970 to 1980, an increase of some 115 bgd. The pattern is hardly one of a smooth increase in demand that would give credence to continued future extrapolations of water demands.

The great increase in demand during the 1940 to 1960 period as compared

---

*Murray and Reeves (1972, Table 1.3a) estimate the increase in water use between 1960 and 1970 at 100 bgd. While they agreed with total withdrawals of about 200 bgd for 1950, they had a figure of only 270 bgd for 1960 (some 50 bgd less than that shown in Table 1.3) and withdrawals of 370 bgd in 1970 (some 50 bgd greater than shown in Table 1.3). The lack of agreement suggests some of the problems inherent in water resources management.

**Table 1.3a**   Trends in Water Use (in bgd) for the United States, 1950-1970

|  | 1950 | 1955 | 1960 | 1965 | 1970 | Percent Increase or Decrease 1965-70 |
|---|---|---|---|---|---|---|
| Total population (millions) | 150.7 | 164 | 179.3 | 193.8 | 203.2 | 5 |
| Withdrawals |  |  |  |  |  |  |
| Public supplies | 14 | 17 | 21 | 24 | 27 | 13 |
| Rural domestic and livestock | 3.6 | 3.6 | 3.6 | 4.0 | 4.5 | 13 |
| Irrigation | 110[a] | 110 | 110 | 120 | 130 | 8 |
| Thermoelectric power | 40 | 72 | 100 | 130 | 170 | 33 |
| Other self-supplied industrial use | 37 | 39 | 38 | 46 | 47 | 2 |
| Total withdrawal | 200 | 240 | 270 | 310 | 370 | 19 |
| Sources of water |  |  |  |  |  |  |
| Fresh groundwater | 34 | 47 | 50 | 60 | 68 | 13 |
| Saline groundwater |  | .65 | .38 | .47 | 1.0 | 113 |
| Fresh surface water | 160 | 180 | 190 | 210 | 250 | 19 |
| Saline surface water | 10 | 18 | 31 | 43 | 53 | 23 |
| Reclaimed sewage | – | .2 | .1 | .7 | .5 | 29 |
| Water consumed | – | – | 61 | 77 | 87[b] | 13 |
| Water used for hydro-electric power | 1100 | 1500 | 2000 | 2300 | 2800 | 22 |

[a]Including an estimated 30 bgd in irrigation conveyance losses.
[b]Fresh water only.
*Source.* Murray and Reeves (1972).

with the previous 10-year period might be easily explained on the basis of political-economic factors. The depression years of the 1930s resulted in a relatively modest increase in water demand for industry, power, and irrigation. Money for expansion of demand was simply not available. The years of World War II in the first half of the 1940s and the great industrial-irrigation surge from the later 1940s through 1960 are clearly revealed in the tremendous increase in water demand during those 20 years.

But why (if Table 1.3 is accurate) should there be such a slowdown in demand between 1960 and 1970? True, industrial expansion could not continue at the same rate; but we did not experience a recession of any great consequence and certainly one would have expected a continued increase in irrigation demand—yet the figures actually show a decrease in irrigation demand from 135 bgd in 1960 to 119 bgd in 1970. Several things may have happened in the 1960s to cause some slowdown in water demand but possibly not as great as indicated in Table 1.3.

First, the early part of that decade witnessed a prolonged drought period in the eastern United States. There was serious talk of running out of water, restaurants did not serve water to customers unless they requested it, rainmakers were hired by various groups to try to break the drought and to refill reservoirs. There was a great upsurge in the national consciousness. We began to realize that water was no longer an unlimited resource and that we had no "right" to waste fresh water.

Second, realization of the limited nature of fresh water supplies, coupled with the growing environmental movement, brought to light what harm had been done to streams and lakes through uncontrolled discharge of polluted waters. The whole question of use and abuse of water became more clearly understood and undoubtedly led to a modification of earlier perceptions of water as unlimited, essentially free, always available.

Third, the industrial and agricultural expansion of the previous two decades could not continue without some slowdown. A mild economic recession, a significant increase in agricultural surpluses, and increased foreign competition for products from a rebuilding foreign industry worked toward a slowdown in our own demand for water. These and possibly other reasons may have led to the modest increase in the rate of demand for water in the period 1960–1970.

One might well wonder, if those were actually the reasons for the slowdown in demand during the 1960s, why should they not also be active in the 1970s to limit the increase in demand? The estimated figures included in Table 1.3 show, however, a return to the rapid rise in demand experienced in the 1940–1960 period. Will the lack of an eastern drought or a changing perception of environmental problems, or a new economic boom, lead to a substantial increase once more in water demand? To forecast such a change for the 1980–1990 period from the previous decade, one must forecast significant economic, population, technological, or social changes. The whole aspect of forecasting future demands for water is considered in the next section.

Consideration of the detailed figures included in Table 1.3 reveals that as of 1980 only about 14% of our total water demand will be met from groundwater sources. This figure varies appreciably by category of use, however, since 95% of our rural-domestic use will come from groundwater, 28% of irrigation need will be satisfied by groundwater, 23% of public water supplies will be from groundwater, and 18% of industrial demands will be met from groundwater sources. The steam electric utilities, whose overall demand for water constitutes some 44% of total national demand, will have only 0.1% of their water demand met from groundwater sources.

In 1900, irrigation water use constituted 50% of total national demand. Industrial use equalled 25% of total water withdrawals while steam electric utilities used only 12%. By 1940, those figures had changed only slightly so that irrigation demands were 52% of our total, industrial demands constituted 21%, and steam electric utilities made up 17% of the overall demand. The real change in the requirements for water has come in the last few decades during which steam electric utilities increased their use of water from 17% to 44% of the total U.S.

**Table 1.4**  Water Use Changes (by Actual Volume and Percent) for the United States, 1900–1940, 1940–1980

| Period | 1900–1940 | | 1940–1980 | | 1900–1980 | |
|---|---|---|---|---|---|---|
| | Volume (bgd) | Increase (%) | Volume (bgd) | Increase (%) | Volume (bgd) | Increase (%) |
| Total water use | 96 | 240 | 306 | 225 | 402 | 1005 |
| Irrigation | 51 | 255 | 65 | 92 | 116 | 574 |
| Public water utilities | 7 | 233 | 23 | 230 | 30 | 1000 |
| Rural-domestic | 1 | 0.50 | 2 | 0.67 | 3 | 150 |
| Industrial-miscellaneous | 19 | 190 | 46 | 159 | 65 | 650 |
| Steam electric utilities | 18 | 460 | 170 | 733 | 188 | 3760 |

demand. At the same time, irrigation needs, while doubling since 1940 (1980 irrigation water use alone equals total U.S. water use in 1940), will constitute only 30% of total U.S. water use, down from 52% in 1940. Industrial water use constitutes only 17% of total U.S. water use in 1980, down from 25% in 1900 and 21% in 1940. Thus the pattern of change has been one of explosive growth in overall demand for water and a change in the category of demand. Water use by steam electric utilities has increased more rapidly than all other categories of use (Table 1.4).

Obviously, percentage increases, when the original values are extremely low, give a biased picture, but volume increases can be used to indicate something of the real nature in the changing demand for water. Over the past 80 years, overall water use has gone up by around 400 bgd in the United States. Steam electric utilities' water use has increased by 188 bgd or nearly half of that total increase. Irrigation water use has increased by 116 bgd or just over one-quarter of that increase and industrial demand has accounted for about 16%.

While the increase in demand for water has been tremendous (elevenfold in 80 years), it must be remembered that the use of water to satisfy steam electric utilities is not a consumptive use of water. The water, used once for cooling purposes primarily, can be reused. Its major pollution loading is heat which often does not limit its usefulness for other purposes. However, irrigation water use is primarily a consumptive water use so that any increase in that category may be far more significant in our overall water resources picture than the great increase in nonconsumptive use created by the demand from steam electric utilities.

## THE COST OF WATER

In areas where water is plentiful and unpolluted, it is extremely inexpensive. In fact, most of us take it almost for granted since it constitutes such a small part of the total budget. However, in water-short areas, where water must be transported from some distance or treated to remove salt or other pollutants, costs can rise appreciably. Since domestic demands are still relatively small, cost factors usually do not limit such use, but they may well limit widespread use of the water for industry or agriculture.

The national average cost of water for the typical homeowner is about 35¢ per 1000 gal or 2¢ per day per person. At present, irrigation water varies from about 1¢ per 1000 gal with an adequate streamflow available to about 20¢ per 1000 gal in some citrus orchards in southern California. For most farmers, delivered water falls between 1¢ and 5¢ per 1000 gal.

Small industries in cities usually use local municipal supplies and pay the prevailing rates. However, large industries may require so much water that they can take most of the available municipal supply. To prevent being dependent on such a limited supply that might be subject to rising prices, most large industries develop their own supplies. Often they have to treat this water to obtain desired taste, composition, and temperature. Depending on the degree of treatment, costs for industrial water may vary from 1¢ to 15¢ per 1000 gal.

For comparison purposes, we will see later that desalination techniques often provide water at costs of $1.00 per 1000 gal although there is great hope that these costs may be reduced. However, until they are, such water is not competitive with other sources. If treated seawater is the only available supply, its use for agriculture or industry would be quite limited because of the cost factor.

## ESTIMATING FUTURE DEMANDS FOR WATER

How can one go about the difficult problem of extrapolating water use demands into the future? Estimates are vital, of course, if we are to plan for the future and be ready to meet future problems with satisfactory solutions. But extrapolations based on past performance are frought with problems and are usually no better than uneducated guesses. In a country such as the United States where a reasonable record of past water use is available, the demand has so changed over time that extrapolation is hardly precise. For example, in predicting demand for the 1970-1980 period, should we have been more influenced by the small increase in demand from 1960 to 1970 or the larger increase in demand from 1950 to 1960. Clearly, to provide educated guesses concerning future water use, we must first consider what factors influence demand and then analyze how these factors will change over time.

Since people use water, it is certainly proper to relate water use figures to estimates of population change. But here, of course, we are faced with another problem. How does one forecast population even 5 or 10 years from now. Demographers have proven themselves to be inaccurate in the past and there is little reason to expect greater accuracy in the future. Many factors influence the rate of family formation; we have seen in recent years a significant change in the desire for children and a real decrease in the population explosion of the 1940 and 1950s. Changes in the number of children and the rate of family formation must have an impact on the demand for water in the future but possibly not as great as certain other factors that must be considered. After all (in Table 1.3), the domestic demand for water (public water utilities and rural-domestic) only increased by 33 bgd in the period 1900–1980, less than 10% of the overall increase in water demand during that period. Other aspects of population pressure—indirectly through the need for irrigation, for industry, and most especially for power— have resulted in far greater increases in water demand.

If population changes, directly, are not necessarily the most significant aspect of future water demands, what then might be the factors of greatest concern? In the past 40 years, we have seen a great increase in the need for water for power. Will this need continue in the future or will the changes introduced by the oil pricing policies of the OPEC countries result in significant changes in our own water demands for power? Will the expansion of solar power or the pressure to curb energy use cut the need for water for steam electric utilities or will the expansion of the past few decades continue? While estimates for 5 years hence might be possible, estimates for 20 years ahead are not at all certain in view of the fact that we have little control over oil pricing and cannot fully predict the course of technological developments.

Another question that must be answered is whether development of new labor-saving devices and expansion in the use of those available today will increase or decrease the use of water. In the past, labor-saving devices—dishwashers, washing machines, toilets, air conditioners—as well as devices for leisure-time activities (e.g., swimming pools) have required more water than the devices they replaced. As more material things, which we have equated with a higher standard of living, are required, per capita use of water has shot upward until it is now around 150 gal per person per day, nearly twice the value found in some advanced European countries today. In part, that figure is a reflection of our misunderstanding of the need to conserve water. It also reflects the fact that bathtubs and showers, toilets, washing machines, dishwashers, and air conditioners are part and parcel of almost every modern apartment and suburban home. We might conclude that expansion of the use of these devices in older homes will lead to a further rise in per capita consumption of water. What, however, can be said for the water requirements of a whole new set of labor-saving devices that may be developed in the next 20 years? Will they require more water to operate, just as the last generation of labor-saving devices has required more water than the devices they replaced? Or has the conservation ethic, the higher price for, or shorter supply of, water in recent years been sufficient to encourage a new look at labor-saving devices? Will newer toilets use less water to flush or dishwashers operate on less water for rinsing? These are important questions as we seek to determine possible future demands for water but the answers are hardly straightforward. Answers depend on factors of cost, social acceptability, taste, perception, availability, and many other factors that are hardly even understood by the planners and prognosticators.

Another question concerns not so much the trend of population itself (which we have already indicated is a real unknown) but rather the trend in population relocation. The past several decades have seen a significant migration of people to the so-called "sunbelt" area (Florida and the Gulf Coast, Texas, New Mexico, Arizona, and California). With this migration, of course, goes a change in the need for services and a great change in the location of the demand for water. In the southwestern United States, for example, water is in very short supply and almost all available water is used for agricultural purposes. With an increase in population and a shift to a more suburban way of life, some of the water used consumptively for agriculture will be used nonconsumptively for domestic or municipal purposes. How will that change affect overall demands for water in the future? Will the change be so minor that, as some predict, it will have no real impact on overall national demand, or will, as others suggest, the change from agriculture to suburbanization result in significantly different amounts being used and so influence national water use figures?

An economic question demands an answer as well. What will happen to water demand as the price increases? We are used to adequate supplies of fresh water at very low cost. Not only is water "dirt cheap" but those who have recently bought a load of topsoil know that it is considerably cheaper than dirt. As water costs rise, it is estimated that the demand will decrease. Most planners, however, do not feel that it will be a linear or even smoothly curving relation. Rather, the suggestion is made that water prices could be doubled or even tripled in many places without any real change in demand. In other words, there would have to be a significant increase in the price before much change in water use occurs. Part of this would be due to the great cost of making changes in water-using facilities. For a factory to install water-saving devices, rearrange piping systems within the plant, develop techniques to recycle or to reuse water, the cost of water over a period of time must exceed the cost of the modifications in plant operations. The householder has fewer options to modify water use. People may not wash their cars as often or water the lawns as much if the price of water increases; but generally water is still so inexpensive in the overall cost of living today it will not significantly affect the family budget if it were to change by even 100%.

Significant changes in the cost of water would certainly affect overall demand. If recent legislation to purify wastewater and to increase treatment of water supplies is enforced effectively, and the costs for these activities are passed on to the users, there is no question that the price of water will rise. Certainly this trend toward increasing price should put some brake on increasing use of water and tend to restrict future demand.

Industrial demand for water does not have to continue to increase even though there is a continued increase in industrial production. We have many examples of the same product being made by different factories using quite different amounts of water. For example, production of a ton of steel can require over 60,000 gal of water but it has also been made with as little as 1400 gal. Modification in plant design to reuse or recycle water, or the building of new plants with water conservation as one of the design requirements, can result in significant changes in the amount of water used without any great changes in industrial production.

One final imponderable relates to public attitude toward water and water conservation, or the perception of the public concerning the relative shortage of water, water pollution, and the possibility of reuse of waste water. Future water demands are going to be influenced by public attitudes toward water use and conservation as much as anything else. We have seen the same phenomenon in the use of oil and gasoline in recent years. While there has been a move afoot to try to limit use of oil by increasing the price, the public has generally resisted this action because, in large part, they are unconvinced of a real oil shortage. Efforts to sell smaller, gas-saving cars or to initiate pooling have somewhat failed, again because public attitude is one of skepticism about the question of an oil shortage. When that question is resolved, the public will probably act with dispatch to meet the problem.

Water is certainly more abundant than oil. It will be an even harder battle to convince people in humid areas that water is a limited resource and that they must conserve or change their life styles to adjust to a reduced water situation. We have seen several recent examples in drought-stricken areas of California, in response to a recognized crisis situation, where voluntary restrictions on water use were instituted successfully. The drought was apparent, the reservoirs were obviously without water, and it was clearly a temporary measure. A return to the former way of using water could be foreseen at the end of the drought. That situation was, thus, quite different from trying to convince industry and households in a humid region with essentially adequate rainfall that they must begin to conserve water or recycle it, that they must irrigate the lawn or wash the car with bath water, or that they can irrigate only at certain times of the week. If unconvinced of the real nature of the shortage, if no crises can be seen to exist, cooperation is likely to be half-hearted and conservation unsuccessful. Thus, public attitude and perception of need for conservation, which can so greatly change demand for water, cannot really be predicted for any period in the future. But it is certainly an important unknown.

The previous paragraphs have not been too helpful in providing guidance concerning possible future trends in water use. Forecasting future demand is certainly germane to any attempt at rational planning. But we are left with a wide range of possible scenarios. Straight extrapolation of past conditions would provide us with the possibility of a great expansion of water demand. A great increase in demand would probably result in lack of sufficient water in many places and a great increase in price for water or possibly even an increasing number of crisis situations where water must be rationed. Thus, it might be argued that pure extrapolation into the future, while possibly the best expedient for the near term, will lead to reduction in water use in the more distant future as price factors and public attitude force reductions in the demand for water. It might seem that any logical consideration of future trends in water use would forecast an increase in water use but at a slowing rate. The pattern will probably not be smooth but rather, following the pattern of the past 30 years, quite variable as swings in weather,

**Table 1.5**  Projected Water Use (in bgd) in the United States, by Purpose

| Type of Use | Projected Withdrawals | | | Projected Consumptive Use | | |
|---|---|---|---|---|---|---|
| | 1980 | 2000 | 2020 | 1980 | 2000 | 2020 |
| Rural domestic | 2.5 | 2.9 | 3.3 | 1.8 | 2.1 | 2.5 |
| Municipal (public supplied) | 33.6 | 50.7 | 74.3 | 10.6 | 16.5 | 24.6 |
| Industrial (self-supplied) | 75 | 127.4 | 210.8 | 6.1 | 10 | 15.6 |
| Steam-electric power | | | | | | |
| Fresh | 134 | 259.2 | 410.6 | 1.7 | 4.6 | 8 |
| Saline | 59.3 | 211.2 | 503.5 | .5 | 2 | 5.2 |
| Agriculture | | | | | | |
| Irrigation | 135.9 | 149.8 | 161 | 81.6 | 90 | 96.9 |
| Livestock | 2.4 | 3.4 | 4.7 | 2.2 | 3.1 | 4.2 |
| U.S. total | 442.6 | 804.6 | 1368.1 | 104.4 | 128.2 | 157.1 |

*Source.* U.S. Water Resources Council (1968).

in crisis situations, and in public attitude influence water demand. As more and more areas face real shortages of fresh water, per capita consumption should decrease even without any real change in our standard of living so that overall growth rates in water use should not equal those found during the past 40 years. But forecasters must realize that predicting future water needs is still much more of an art than a science.

The U.S. Water Resources Council in 1968 presented a detailed survey of future demands for water in a report commonly called the First National Assessment. The results are included in Table 1.5 not because of their accuracy (developments in the past few years have suggested certain shortcomings in their projections), but because they give some insight into the problems of forecasting. The projections of the First National Assessment, based generally on continuation of past trends, showed an increase in water demand from 443 bgd to 805 bgd from 1980 to 2000. That change of 360 bgd is three times greater than the change in water use from 1960 to 1980 and might seem to be out of line with other suggestions. The increase to some 1368 bgd in 2020 represents an even greater increase in water use—some 564 bgd in the last 20 years of the forecast period. This daily *increase* in water demand is considerably larger than the total daily water demand in the whole United States in 1980!

The National Water Commission in its Final Report to the President and to the Congress in 1973 reviewed the problems of forecasting future demand and suggested that at least eight variables needed to be considered: population; growth of national income; energy consumption; changes in consumption and export of foodstuffs and fibers; environmental and resource development programs of the government; technological developments; change in demand for recreational water use; and price of water.

Rather than make a single best estimate of water demand at specific times in the future, the Commission suggested the evaluation of "alternative futures" to provide a range of possible future outcomes. The Commission suggested several different scenarios and provided values of future water demand given different changes in the factors that affect water use. For example, four different levels of population for the year 2000 were assumed along with different levels of productivity of the labor force; two different assumptions were made with regard to waste heat disposal and thus the need for cooling water; two different assumptions were made on the dissolved oxygen content of fresh water streams; and two different assumptions were made about the treatment of waste effluent. Other current information on the reuse of water by industries, on water-use coefficients for steam electric power generation, and on water use in the major water-using industries was also incorporated into the prediction models.

The results showed a wide range of possible futures. Water withdrawals for the year 2020 ranged from 570 bgd up to 2280 bgd as opposed to the value of 1368 bgd proposed by the First National Assessment. Actual water consumption as opposed to withdrawals were found to range from 150 to 250 bgd. The First National Assessment predicted a figure of 157 bgd.

Significant changes in the amount of water withdrawals were found to occur depending on the amount of recycling or reuse of water by industry and the amount of heat pollution permitted in the cooling waters used by the steam electric utilities. For example, allowing the water temperature at the point of discharge to increase up to $5.4°F$ could reduce withdrawals for cooling purposes by about 75% from the levels anticipated if current trends persisted. The Commission also noted that requiring an increase in the dissolved oxygen content of stream water from 4 to 6 mg/liter would increase treatment costs by about 50%.

In 1978, the summary volume of the so-called Second National Water Assessment by the U.S. Water Resources Council appeared. While primarily concerned with identifying supply and demand problems, it provided some new figures for total fresh water withdrawals for 1975, 1985, and 2000 by categories of use for the United States (Table 1.6). The figures are unique in that they actually forecast a decrease in total fresh water withdrawals for the nation from 1975 to 2000 from 338 bgd to some 306 bgd. They do anticipate a doubling of saline water withdrawals during the 1975–2000 period so that total withdrawals for all purposes will increase slightly from 398 bgd in 1975 to 425 bgd in 2000. The total estimated water withdrawals in the year 2000, according to the Second Assessment, will be less than the 1980 withdrawals estimated by the First Assessment.

The reports of the National Water Commission serve very useful purposes. While they do not provide a single value of possible future water demand, they show how water demand might be influenced by different human actions and how demand can be modified through the choice of different alternatives for future development. They also show clearly the need for caution in any interpretation of future projections.

**Table 1.6**   Total Estimated Withdrawals and Consumption of Water, by Functional Use, for the United States—1975, 1985, 2000 (million gallons per day)

| Functional Use | Total Withdrawals | | | Total Consumption | | |
|---|---|---|---|---|---|---|
| | 1975 | 1985 | 2000 | 1975 | 1985 | 2000 |
| Fresh water | | | | | | |
| Domestic: | | | | | | |
| Municipal | 21,164 | 23,983 | 27,918 | 4,976 | 5,665 | 6,638 |
| Rural | 2,092 | 2,320 | 2,400 | 1,292 | 1,408 | 1,436 |
| Commercial | 5,530 | 6,048 | 6,732 | 1,109 | 1,216 | 1,369 |
| Manufacturing | 51,222 | 23,687 | 19,669 | 6,059 | 8,903 | 14,699 |
| Agriculture: | | | | | | |
| Irrigation | 158,743 | 166,252 | 153,846 | 86,391 | 92,820 | 92,506 |
| Livestock | 1,912 | 2,233 | 2,551 | 1,912 | 2,233 | 2,551 |
| Steam electric generation | 88,916 | 94,858 | 79,492 | 1,419 | 4,062 | 10,541 |
| Minerals industry | 7,055 | 8,832 | 11,328 | 2,196 | 2,777 | 3,609 |
| Public lands and others[a] | 1,866 | 2,162 | 2,461 | 1,236 | 1,461 | 1,731 |
| Total fresh water | 338,500 | 330,375 | 306,397 | 106,590 | 120,545 | 135,080 |
| Saline water,[b] total | 59,737 | 91,236 | 118,815 | | | |
| Total withdrawals | 398,237 | 421,611 | 425,212 | | | |

[a]Includes water for fish hatcheries and miscellaneous uses.
[b]Saline water used mainly in manufacturing and steam electric generation.
*Source.* U.S. Water Resources Council (1978).

## WATER RESOURCES MANAGEMENT

The foregoing describes the ways in which water is found in nature, the volumes of water available for use, as well as trends in the demand for water. It has led to the general conclusion that water per se is not in short supply but that in certain areas of the world, due to a poor distribution of precipitation or to too great a concentration of people and industry, it is indeed in short supply. As a planet, we will not run out of water although some areas are dangerously close to that condition even now. The problem, then, is not one of insufficient water resources but rather one of water resources management.

Water resources management can be described in simplest terms as having the right amount of water available for a particular use at the right time and with the right quality. Having too much, too little, or the wrong quality of water is poor management because either there is too much for the particular use and it is wasted or there is not enough for the purpose at hand. Water resources management has been practiced in one fashion or another ever since the first cave dweller stored a little water in a gourd or skin container for use at a later time, but, in recent years, water management has become imperative because of the growing demand for water and the need for planning in the use of water.

Water resources management involves both quality and quantity management. While there is a tendency for the public to think in terms of only one use of water—if it is polluted through use, it should be discarded or returned to the river—actually we do reuse our water over and over again. Water returned to the stream is both cleansed by natural and human processes so that it can be reused several times in its passage downstream. This must be a fundamental part of water resources management since if we were not able to reuse water there would be severe shortages in many more areas than at present. Water quantity management involves rational planning for use of water as well as planning for ways to conserve available supplies of water (as shall be seen in later chapters).

Two basic techniques in water resources management are multipurpose use of water and integrated basin development. Multipurpose use or reuse of water has been a technique actively employed by planners for most of this century. In building dams and reservoirs or water diversion and transportation networks, it is not usually feasible to think of only a single use of the water. Water is needed for irrigation purposes during only one season of the year and so another compatible use in other seasons of the year must be developed to maximize the benefit from management of the water resource.

Water is a single resource whether it appears in surface streams, in groundwater storage, or in lakes and reservoirs and its management to satisfy a variety of functions (including water supply, power, navigation, recreation, flood protection, and the like) is a complicated undertaking. Clearly, water management is linked to land management; what we do to our land resource, how it is zoned for use, or the purposes to which it is dedicated will determine, in large measure, the quantity and quality of water available for use. Thus multiple-purpose use and reuse must be closely tied to the second technique for management—integrated basin development.

Some basins are developed for a single purpose, such as flood control or power, but more and more basins are being developed for multiple use. Since land use influences water use, it is necessary to think in terms of the original precipitation falling on the basin, the use of soil conservation techniques to improve the infiltration and percolation of water into and through the soil, and the use of vegetation management to produce the largest amount of water compatible with the most efficient economic management of the basin. Then the downstream development of the basin must be considered in terms of reservoirs and their particular uses. If water is to be stored for power, can it also be used for irrigation, and can the reservoir be fairly empty for flood control purposes at the proper time? If uses are to be made of the water in the reservoir for irrigation and power, will they be compatible with the use of the reservoir for recreation or for fish and wildlife management? Integrated basin development becomes even more complex when the basin covers more than one state or more than one country. Here, interstate compacts or international treaties must be achieved for the best use of available water resources. Without adequate records of quantities of use—or more especially, quantities and qualities of waters discharged into the stream—regulation is extremely difficult to achieve.

Allocation of water in a water resources management system is usually guided by economic and social benefits. Planners must face many constraints in trying to achieve the most rational use of any resource and especially water. O'Riordan and More (1969) suggest that at least eight constraints are placed on decision-makers in the process of water resources development:

(a) *Physical*—there is a physical limit to possible water development.
(b) *Economic*—development will be limited by the total amount of money available and other demands for that money.
(c) *Policy*—certain water demands (such as for domestic purposes) must not be allowed to suffer regardless of what happens to other uses.
(d) *Legal*—court decisions and legislation are often the result of compromise between the demands of different groups with biased interests.
(e) *Administrative*—willing cooperation between different local or state authorities is needed to achieve an optimum water resources management program.
(f) *Ownership*—private land ownership, if opposed to water development plans, may limit such water development.
(g) *Quantification*—inability to quantify benefits and costs, especially those relating to social or aesthetic factors, places greater emphasis on the previous six constraints.
(h) *Perception*—limited understanding of the range of choices available or the range of alternate uses by the decision maker may place some constraint on water resources developments.

In the beginning, the individual water hole, spring, or well was the unit for planning water resources development. Water was not transported any great distance and what went on elsewhere seemed to have little influence on local supplies. Population pressure was low and often great distances existed between towns or industries. As urbanization increased, as more factories and farms filled in the areas between the settled places, it was clear that water resources management had to consider a larger region and the basin or watershed was accepted as the unit for management purposes. In time, even the basin became too small inasmuch as water could be stored in one basin for later transfer to a neighboring basin. Interbasin transfers both of surface water and by means of groundwater flow occur frequently now so that the region rather than the basin is more the unit to be considered in management of water resources. The future may well see the development of more grandiose schemes, such as the North American Water and Power Alliance which suggests the transfer of water from Alaska and northwestern Canada all the way south to northern Mexico and east to the Great Lakes system and thence to the Atlantic Ocean. The Russians have suggested plans for the southward redirection of streams flowing into the Arctic Ocean to the Caspian Sea. Truly such plans will make us think in terms of continents as the proper areas for water resources management.

Many problems in water resources management have stemmed from our scale of thinking and planning. Water in all of its aspects is clearly interrelated. What we do with it in one basin or in one section of the country can now, with modern technology, influence the water situation in another basin or part of the country.

Thus, the recent trend toward thinking of water resources management on larger and larger scales, and more especially in thinking of it as a subsystem of overall regional development, comparable to the management of other physical and human resources, augurs well for the more rational development of water resources in the future (Nanda, 1977). Understanding of the interrelation of water to land, to politics, to economics, to human and cultural factors are all keys to the proper development of the subject and a major purpose of the following chapters.

## 2

## THE HYDROLOGIC CYCLE

Water is continually moving, over countless miles, during the course of a year. It changes state a number of times from solid to liquid, to gas, to appear in different forms on the surface of the earth or in its atmosphere. When looking out at the ocean, only one aspect of the mobility of water is sensed: the surface of the sea being whipped into waves by the wind and the surf crashing on the shore. But once this motion is finished, the ocean is still there and one can imagine that all that motion has occurred without any real movement. Even as one looks at a large river or lake, little real movement is seen; hence the water's great mobility may not strike us. A fast-flowing mountain stream or the clouds scudding across the sky does give an impression of mobility but as we look at clouds, we often do not think in terms of water. The white, fleecy cumulus of fair weather remind us more of puffs of cotton or woolly sheep than of volumes of liquid water moving from one place to another.

Thus, the relative ease with which water moves from one place to another is often underestimated. Yet, if it were not for this mobility of moisture, we would be living in a far different world. Without significant atmospheric transport of moisture, vast areas of our land would be rainless, with only areas adjacent to oceans having sufficient moisture for life. Furthermore, without evaporation of moisture, which removes heat from the evaporating surface and makes possible condensation in some distant place, temperature contrasts between tropics and polar regions would be much stronger.

## CHANGING IDEAS ABOUT THE HYDROLOGIC CYCLE

While most scientists are now aware of the great mobility of water in the hydrologic cycle—the unending flow of water in all of its various states from the ocean to atmosphere, to land, and back to the ocean again—such has not always been the case. The literature reveals a history of confusion and misunderstanding about the mobility of water and even the basic operation of the hydrologic cycle.

All the rivers run into the sea; yet the sea is not full; unto the place from whence the rivers come, thither they return again.

So wrote the biblical author of *Ecclesiastes* of the mobility of water in the hydrologic cycle. And Xenophanes of Colophon (who lived about 500 B.C.) wrote that the sea was the source of water. He was convinced that the ocean was the originator of clouds, precipitation, winds, and rivers (Freeman, 1952).

Plato (about 400 B.C.) provided two possible explanations for the formation of rivers and springs, the aspect of the hydrologic cycle that seemed to be of greatest concern to ancient scholars. He suggested, along with many others, that there were many interconnected passageways in the interior of the earth as well as a large subterranean reservoir called Tartarus. All rivers or streams (and presumably the ocean) flowed back to Tartarus and the rivers and springs were all fed by water from the great underground reservoir.

A second explanation, which is not as well known, appeared in *Critias*. In writing of conditions near Athens thousands of years earlier, Plato argued that the deep soil at that time could store good quantities of water in soil that was in Plato's time used for pottery clay. It would draw water down from the higher ground and so make available numerous sources of springs and rivers (Biswas, 1970). Thus, Plato, while he might be criticized for his use of the subterranean reservoir concept, must also be credited with the suggestion that precipitation might also be sufficient to influence the flow of rivers and springs.

Aristotle (about 350 B.C.) felt there were two types of evaporation, both resulting from the sun's heat. One was the evaporation, as we know it, from moist surfaces; the other, more like smoke, occurred by means of a sort of windy exhalation from dry earth. As the water vapor rose, it lost heat—partly by rising into the cooler regions of the upper air and partly by expansion—and this resulted in condensation of the vapor back into water. The condensed air first formed clouds. If the vapor had not risen far (the condensation of a single day's evaporation), dew or—if the temperature was below freezing—hoarfrost was formed. Rain itself resulted from the condensation of the vapor from a larger area over a considerable period of time.

Aristotle criticized his teacher Plato for use of the subterranean reservoir origin of springs and rivers although he replaced that idea with one no less questionable. He suggested that since cold changes vapor (air) into water above the earth, it would also produce the same result within the earth. Thus, air moving through the earth is changed to water in the cool places of the earth which are often associated with the mountainous areas. While this explanation was more reasonable than Plato's, which would have involved water from Tartarus running up hill to be able to gush forth as mountain springs, it did not suggest how the air (vapor) was to move in large enough quantities through the earth to the mountain areas to be cooled so that streams of water would flow forth as springs. Aristotle, as did Plato, suggested that springs and rivers would receive a great deal of water from precipitation.

The Roman scholar, Vitruvius, writing just a few decades before the time of Christ, had a more clearly defined explanation of the hydrologic cycle. For example, he recognized that valleys between mountains often had large amounts of rainfall and that snow would remain on the mountain sides protected by the thick forests for considerable periods. Snowmelt would be absorbed into the soil, would percolate through the soil to the lowest portions of the mountains or in the valleys where it would come forth as springs or streams.

Vitruvius, as had others before him, recognized that in evaporation only the fresh water was removed and the heavier (or saltier) portion of the water was left behind. He suggested that the rising vapor drives the air before it and also is driven by the air which rushes in behind it. With the movement of the wind, moisture from springs, rivers, the oceans, and marshes is also added to the air through the process of evaporation. The condensed air forms clouds that result in precipitation when they come into contact with mountains because of the force of the impact and the fullness and weight of the vapor in the cloud. This would explain, he felt, why there was always more precipitation near mountains than over plains.

Vitruvius' concept of the hydrologic cycle depended on the heat of the earth, the formation of strong winds, the presence of great amounts of water for evaporation, and coldness aloft to produce condensation. He likened it to a hot bath when water is vaporized from the bath. The rising vapor is condensed on the cooler ceiling, falling as water droplets on the heads of the bathers. Since there is no source of water on the ceiling, the droplets must have come from the hot bath. The analogy was used many times for the next 10 centuries.

Few new ideas about the hydrologic cycle were added until the Renaissance and Reformation brought a growing acceptance of new ideas and a questioning of old "truths." Earlier philosophers had pointed out the similarities between humans and the earth. Leonardo da Vinci (1452-1519), in preparing material for a *Treatise on Water*, which he never completed, commented in detail on the similarities. As humans have a pool of blood, so the earth has its water. As humans breathe and pump blood against gravity even to the head area, so the earth has its tides and can pump water through "veins" in the earth to the highest mountain tops. As blood will pour forth from a broken vein in the forehead, so will water emerge from a broken vein on a mountain top as a spring.

While we may now disagree with Leonardo's description of water issuing forth from broken veins, we can also find comments in his writings to the effect that moisture from evaporation, condensed in the cold air to form clouds, and moved by the winds to various places, will also fall again as precipitation to augment the flow of streams. Thus, in a sense, he was trying to reconcile the two ideas of springs and streams being fed by water moving under pressure from within the earth's surface and springs being fed solely by infiltration from precipitation. To Leonardo, both avenues seemed to be available.

Palissy, a French potter and naturalist (about 1510-1590), may have been the first to state, without qualification, that rivers and springs can have no other source than precipitation. He refuted the age-old arguments of subterranean reservoirs or the changing of vapor to water in cold regions beneath the surface and showed that precipitation must infiltrate the earth (even in mountainous areas), flow downward toward lower areas, come into contact with impervious rock layers, and finally issue forth from the side of the mountain as a spring. As a result, even mountain springs must have a catchment area above them sufficient to supply the volume of water flowing from the spring. A contemporary of Palissy's, Jacques Besson, also stated, in a short book published in 1569, that the source of moisture for precipitation was evaporation from the surface. Precipitation by itself would be sufficient to maintain the flow of rivers and springs, he felt.

In spite of the rather clear explanation of the operation of the hydrologic cycle by the middle of the 16th century, not everyone agreed. Rejecting Aristotle's ideas of the transformation of air to water, the Jesuit professor, Kircher suggested that there were large hydrophylocia (caverns with water in them) within many of the mountain ranges of the world. Rivers flowed out of the caverns. However, to keep water in the caverns, it was necessary for Kircher to postulate openings in the sea floor through which water would flow as rivers or subterranean channels back to the mountain caverns. Kircher never was able to explain satisfactorily how water moved upward from the sea bottom to the mountain caverns against gravity, although several possible explanations were given (Biswas, 1970).

The introduction of a more quantitative hydrology toward the end of the 17th century generally put to rest once and for all the old ideas of Plato, Aristotle, and the others who followed after them about subterranean reservoirs and air changing to water in the cold regions of the earth. Three outstanding scientists of that period—the Frenchmen, Pierre Perrault and Ediné Mariotte, and the English astronomer, Edmund Halley—undertook fundamental quantitative investigations which established some of the basic principles of hydrology. In a work on *The Origin of Springs*, published anonymously in 1674 and later attributed to Perrault, the French naturalist reviewed and fairly well demolished most older views on the origins of springs and rivers; he suggested the need for a quantitative investigation to show that precipitation alone is sufficient to support the flow of rivers and springs. Based on his own computations of how much water might fall on the catchment of the Seine River from its source to Aynay le Duc and comparing this with his estimate of how much water actually flowed in the river over the course of a year, he concluded that only about one-sixth of the annual precipitation is needed to supply all the water flowing in the stream during the year. He quickly generalized from this one example to all of the rivers in the world and concluded that precipitation alone was sufficient to supply the flow in all rivers. The difference between the value of precipitation and the streamflow he assigned to evapotranspiration, deep seepage, storage, and other losses. In spite of his strong statement, Perrault was not a believer in the infiltration of significant amounts of water to recharge the groundwater table.

Mariotte, a close associate of Perrault, published (posthumously in 1686) a further corroboration of the work of Perrault, showing again that precipitation was more than enough to supply the annual flow in rivers. Using a much larger catchment area than used by Perrault (the Seine above Paris), Mariotte came to the conclusion, based on measurements of precipitation, the cross-sectional area of the river, and its velocity of flow, that less than one-sixth of the total amount of precipitation in the catchment area flowed off in the river. Similarly, he concluded that the great spring at Mont-Martre discharged only about one-fourth of the yearly precipitation falling on the catchment area above it.

In the three decades following the publication of Mariotte's work, the English astronomer, Halley, published four significant papers on evaporation; he showed conclusively that the volume of evaporation from water bodies was more than enough to supply the water for streamflow. Water heated and raised from the sea as vapor would be blown over land to be cooled in passing over mountains and would give up precipitation in amounts sufficient to supply the needs of the streams. Halley made actual measurements of the volumes of evaporation from which he estimated the water loss from the Mediterranean Sea per summer day would be some 5.3 billion tons. He further estimated the total inflow of river water to the Mediterranean per day would be about 1.8 billion tons or only slightly more than one-third of the total water loss by evaporation. Enough water evaporated to supply all of the streams and rivers in the area. The quantitative work of Perrault, Mariotte, and Halley did not actually settle the question of the basic operation of the hydrologic cycle, but it did provide a firm basis on which many reputable scientists, in the years following, based their suggestions for modifications of specific details of the overall process. The nature of these arguments over details can be seen by considering the literature of just the past century.

## CHANGING IDEAS ABOUT THE HYDROLOGIC CYCLE IN THE PAST CENTURY

George Perkins Marsh (1864), one of the most articulate geographers of the 19th century, suggested that the precipitation at a particular place might be influenced by local cultural conditions. Quoting the refrain "Afric's barren sand, where nought can grow, because it raineth not, and where no rain can fall to bless the land, because naught grows there," Marsh implied that one can cause an area to become more moist merely by having a vegetation cover that will replenish atmospheric moisture locally by transpiration.

Thornthwaite (1937) suggested that the idea of the close relation between local evaporation and local precipitation was so strongly fixed that many environmental programs had been based upon it. The Timber Culture Act of 1873, for example, was passed in the expectation that the planting of trees by western farmers would not only reduce the danger of wind erosion but would also increase evaporation as well as local rainfall and so eliminate most of the climatic hazards to agriculture in the Great Plains. Just 7 years later, Aughey (1880) suggested that increased rainfall would follow the spread of cultivation rather than reforestation! Cultivated soil would act like a sponge and absorb rain rather than allowing it to flow off. The evaporation of this larger amount of absorbed rainfall would provide the moisture for increased precipitation year by year as cultivation spread. While the particular thrust of this argument was blunted by a serious drought in the

Great Plains shortly thereafter, the implication that local cultural practices play an important role in the local precipitation economy was hardly questioned.

In 1923, Weaver and Clements conducted an investigation of the relative moisture contributions to the atmosphere from both native grassland and field crops in the Great Plains. They found that there was no marked difference in evapotranspiration from either of the two types of vegetation cover. They concluded, therefore (as reported by Clements and Chaney in 1937), that the type of cultivation in the Great Plains area could have no particular influence on the amount of local precipitation. This conclusion, of course, gives implicit support to the idea that if one cultivation practice really did introduce more moisture into the air locally than another, there would indeed be a corresponding change in the local precipitation.

The dramatic dust bowl conditions of the 1930s did more to stimulate studies of the hydrologic cycle and the mobility of moisture than any event for decades. President Roosevelt, acting on advice from many leading scientists, suggested planting vast belts of trees from North Dakota to Texas not only to reduce wind speed and to hold soil in place but also to increase evaporation and thus local precipitation. Caught up by the public's imagination, the idea was converted into a grandiose plan to transform the climate of the entire Plains region as a result of the expected local consequences of tree planting.

Possibly the most significant contributions to the recent literature on the hydrologic cycle were an article by Holzman (1937) and another by Thornthwaite (1937). These articles and several other important ones that followed (Benton and Blackburn, 1950; Benton, Blackburn and Snead, 1950; Budyko, 1956) showed conclusively that local precipitation could not be greatly changed by local cultural practices.

Holzman (1937) determined the changing moisture content of moving air masses as these large horizontally homogeneous bodies of air traveled from land areas to the oceans or vice versa. He concluded that maritime air masses, obtaining most of their moisture from ocean evaporation, were the principal sources of moisture for precipitation in the United States. Most of the moisture added to continental air masses by evaporation from the land would move off the continent in these air masses without falling as precipitation. Cool or cold continental polar air masses warm as they move equatorward across the continent. The warming air, even though it is gaining moisture by evaporation, may never achieve saturation and so there will be little opportunity to squeeze moisture out of the air. Air masses that are cooling, such as maritime tropical air flowing poleward, will have far better opportunities to become saturated by cooling and to lose moisture by condensation and precipitation. Both Holzman and Thornthwaite emphasized the role of continental and maritime air masses in the precipitation process and the significance of mobility of atmospheric water vapor in the hydrologic cycle.

The hydrologic cycle is usually illustrated by means of a simple diagram with arrows indicating directions of moisture flows (Fig. 2.1), to which may be added estimates of the annual evaporation and precipitation on land and ocean areas.

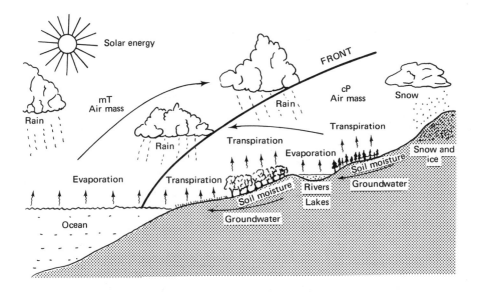

**Figure 2.1**   The hydrologic cycle.

Usually these diagrams are so simplified that they fail to acknowledge the significance of air masses in the process of atmospheric moisture flux, stressed by both
Thornthwaite and Holzman.

Benton and Blackburn (1950) and Benton et al. (1950) estimated, on the
basis of atmospheric moisture profiles and trajectories of air masses, the amount
of precipitation that fell from both maritime and continental air masses at
Huntington, West Virginia as well as over the entire Mississippi River basin. They
concluded that at Huntington, for the year 1946, some 85 to 90% of the precipitation fell from maritime air masses so that only 10 to 15% of the precipitation
could have come from land-based evaporation. Also, only 12 to 14% of the annual
precipitation on the Mississippi watershed came from evaporation from some
land source either within or outside the watershed while 86 to 88% of the precipitation was oceanic in origin. These results were later reinforced by Budyko
(1956) who found that land-based evaporation contributed from 4.1% (October)
to 18.4% (May) of the total monthly precipitation over European Russia. For
the year as a whole, he estimated that 53 mm out of a total precipitation of
487 mm (10.9%) was due to land-based evaporation.

Sutcliffe (1956) made possible the estimation of the actual mobility of atmospheric moisture by calculating the average atmospheric residence time of a given
molecule of water. He estimated this to be about 10 days (from time of entry
by evaporation to time of removal by precipitation). If we assume a mean wind
speed of 15 mi/hr or 360 mi/day of atmospheric movement by one molecule
of water, this would mean an average movement of 3600 mi between points of
evaporation and precipitation for moisture in the atmosphere. Since mean winds

aloft are often much stronger than the wind speed assumed here, atmospheric movement may be considerably greater than this suggested value.

Despite knowledge about the relatively small influence of land evaporation on nearby land precipitation and the significant role of mobility of water vapor, there have still been recent suggestions that local cultural practices can indeed modify local climates significantly. Part of this confusion lies in the geographic scales and orders of magnitude of changes being considered. There is no question that building large urban complexes significantly alters aspects of local heat and water budgets. The creation of large reservoirs or the drainage of extensive swamps modifies local evaporation aspects of the water budget and even possibly, to a small degree, some very local precipitation conditions. The construction of cooling towers to remove heat from industrial effluents probably increases evaporation, the opportunities for fog or icing conditions, and may even cause some additional cloudiness or convective showers, but their role in climatic change over scales of thousands rather than hundreds of square miles is open to much more question.

McDonald (1962) illustrated the small effect on local precipitation of significant local evaporation by detailing the problems involved in trying to increase July water vapor influx to Arizona (and hopefully its July precipitation) by just 10%. Based on determinations of the present water vapor flux into the area and the July rate of evaporation from a water surface, he calculated that a lake some 19,000 mi$^2$ in area (16% of the area of Arizona and larger than the combined areas of Lakes Erie and Ontario) would be necessary. The current water vapor flux into the state is $24 \times 10^9$ grams per second. Summed for just one week, this is equivalent to the entire annual flow of the Colorado River at Hoover Dam, approximately 15 million acre-feet. Therefore, increasing this inflow by 10% would involve creation of a truly magnificent man-made lake. McDonald's further development exposed the fallacy of ever considering such a project feasible or of anticipating any real modification of local climate as a result of the creation of large reservoirs or lakes. The dry conditions that virtually surround the Caspian Sea verify the conclusion that the evaporation of large volumes of water from that sea has negligible effect on local climates. Only in the south, where high mountains exist, is there appreciable precipitation; there it is the orographic uplift and resulting cooling of the air rather than the evaporation from the sea that causes the significant precipitation.

## QUANTITATIVE ESTIMATES OF FACTORS OF THE HYDROLOGIC CYCLE

Estimated volumes of water involved in the precipitation, evaporation, and runoff phases of the annual world water budget have changed over time as a result of our increasing knowledge about the factors involved. For example, L'vovich (1973) tabulates the results of various estimates of world surface runoff (Table 2.1), all but his own 1964 and 1969 estimates being based on the familiar water budget relation P = R + E (precipitation equals runoff plus evaporation). His 1964 and

**Table 2.1**  World Surface Runoff Estimated by Various Investigators

| Author | Year | Surface Runoff (km$^3$) | Author | Year | Surface Runoff (km$^3$) |
|---|---|---|---|---|---|
| E. Brikner | 1905 | 25,000[a] | M. I. L'vovich | 1964 | 37,320 |
| R. Fritsche | 1906 | 30,640[a] | I. Marcinec | 1964 | 30,600[a] |
| G. Wüst | 1922 | 37,100 | R. Nace | 1968 | 42,600 |
| M. I. L'vovich | 1945[b] | 37,000 | M. I. L'vovich | 1969 | 38,150 |
| M. I. Budyko | 1956 | 37,000 | J. R. Mather | 1969 | 37,560 |
| F. Albrecht | 1961 | 33,600[c] | L. I. Zubenok | 1970 | 46,200[c] |
| | | | Baumgartner and Reichel | 1973 | 36,600 |

[a]Without runoff from polar ice caps.
[b]1940 study, published in 1945.
[c]Estimated by L'vovich based on values of runoff in mm provided by original author.
*Source.* Adapted, in part, from L'vovich (1973) by courtesy of the American Geographical Society.

1969 results came from the use of what he calls differentiated water-balance equations as follows:

$$R = U + S; P = U + S + E; W = P - S = U + E$$
$$K_U = \frac{U}{W}; K_E = 1 - K_U = \frac{E}{W}$$

where R is total runoff, U is subsurface outflow, S is surface runoff (rivers), P is precipitation, E is evaporation, W is gross wetness of an area, and $K_U$ and $K_E$ are coefficients characterizing subsurface flow to rivers and evaporation, respectively. The gross wetness factor is directly comparable to the renewable soil moisture storage of an area.

The basic problem in any evaluation of factors of the world water budget results from the paucity of reliable information from much of the world's surface. L'vovich has estimated that perhaps 40% of the land area of the earth had either unreliable hydrologic data or no data whatsoever. This figure increases significantly when the vast ocean areas of the globe are included. Precipitation onto ocean areas has always been difficult to estimate since extrapolations based on island or ship observations can be quite unreliable. Precipitation on land areas is much better known, although in remote mountain areas, where large totals may occur, sizeable errors of estimate still exist. Precipitation in polar regions is largely unknown, but since totals are small, errors are probably less significant. The return flow of moisture by evaporation or evapotranspiration can be estimated over oceans provided energy inputs are known; evapotranspiration is less well understood over land where energy factors cannot be used exclusively since, among other things, the surface is not continually moist. Vast areas of the land surface

**Table 2.2**  Estimates of Annual Precipitation and Evapotranspiration from Land and Ocean Areas of the Globe by 10° Latitudinal Belts (all values $\times 10$ km$^3$)

| Latitude | P(Ocean) | E(Ocean) | P(Land) | E(Land) | Runoff (P–E) |
|---|---|---|---|---|---|
| 80–90°N | | | | | |
| 70–80 | 183 | 218 | 62 | 40 | 22 |
| 60–70 | 314 | 271 | 486 | 227 | 259 |
| 50–60 | 1,115 | 604 | 763 | 417 | 346 |
| 40–50 | 1,639 | 1,258 | 851 | 586 | 265 |
| 30–40 | 1,876 | 2,812 | 850 | 606 | 244 |
| 20–30 | 1,897 | 3,846 | 890 | 537 | 353 |
| 10–20 | 3,700 | 4,925 | 939 | 714 | 225 |
| 0–10 | 7,148 | 4,796 | 1,596 | 1,085 | 511 |
| 10–0°S | 4,408 | 5,024 | 2,021 | 1,144 | 877 |
| 20–10 | 3,095 | 5,683 | 1,142 | 782 | 360 |
| 30–20 | 2,306 | 4,692 | 523 | 396 | 127 |
| 40–30 | 3,016 | 3,801 | 235 | 171 | 64 |
| 50–40 | 3,722 | 2,263 | 59 | 38 | 21 |
| 60–50 | 2,749 | 1,184 | 19 | 6 | 13 |
| 70–60 | 879 | 462 | 37 | 21 | 16 |
| 80–70 | 95 | 67 | 152 | 90 | 62 |
| 90–80 | | | | | |
| Total | 38,142 | 41,906 | 10,625 | 6,860 | 3,765 |

*Source.* Mather (1969) with permission of the American Water Resources Association.

of the earth have no gaged streams to provide runoff values; subsurface flow to the ocean areas can only be estimated by reference to other, more easily measured, hydrologic factors.

Using world maps of average annual precipitation and effective evapotranspiration prepared by Geiger (1965), Mather (1969) obtained values of precipitation and evaporation over the surface of the globe by 10° squares of latitude and longitude. The results are summed by latitudinal belts in Table 2.2. Precipitation in each latitudinal belt on land always equals or exceeds evapotranspiration, while evaporation from the ocean areas can often greatly exceed precipitation in the same latitudinal belt because of the continuous availability of water for evaporation.

A quantitative evaluation of many aspects of the water budget is now possible. For example, since positive values of P–E represent a surplus of water and since storage change over a long period of time can be considered zero, total P–E from land areas must represent runoff or return flow of water from land to ocean. Total runoff from land areas of the world has been found to equal 37,650 km$^3$, a value that quite closely approximates those previously listed by L'vovich (Table 2.1).

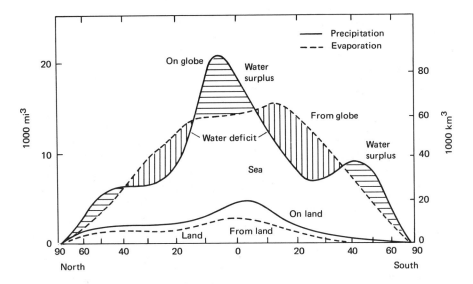

**Figure 2.2** Latitudinal variation of evaporation and precipitation on the globe as well as on the land and sea areas.
*Source.* Mather (1969) with permission of the American Water Resources Association. (Diagram prepared by Prof. Dr. Rudolf Geiger and used with kind permission.)

The latitudinal values of precipitation and evaporation given in Table 2.2 have been plotted graphically in Figure 2.2. The upper two curves represent the latitudinal distribution of precipitation and evaporation over the earth as a whole while the lower two curves represent precipitation and evaporation on the land areas alone. Since the latitude scale is based on total area between latitudinal belts, the area between upper and lower precipitation curves represents graphically the volume of precipitation on ocean areas while the area between the two evaporation curves represents volumes of evaporation from the oceans of the world.

Figure 2.2 clearly shows that a water surplus, especially marked in the belt 0–10°N, exists in the equatorial zone of convergence. In both hemispheres, evaporation exceeds precipitation in the belt from 10 to 40° latitude corresponding generally to subtropical high pressure areas of reduced precipitation. The deficit of moisture is more strongly marked in the Southern Hemisphere than in the Northern. From 40 to 90° latitude, precipitation exceeds evaporation and a water surplus exists. Again, water surplus is greater in the Southern Hemisphere. Precipitation and evaporation are both quite small and nearly in balance in polar regions.

Average values of water budget factors by continents (in both km³ and in mm depth over the continent/year), as determined by several investigators, are given in Table 2.3. The methods used to achieve the continental values are somewhat different since Mather worked from large world maps of precipitation and evaporation,

**Table 2.3** Summary of Continental Values of Precipitation, Evaporation, and Runoff (km³ and mm depth) by Three Investigators

| | Precipitation | | | Evaporation | | | Runoff | | |
|---|---|---|---|---|---|---|---|---|---|
| | Mather 1969 | Budyko 1956 | L'vovich 1973 | Mather 1969 | Budyko 1956 | L'vovich 1973 | Mather 1969 | Budyko 1956 | L'vovich 1973 |
| **Values in km³/yr** | | | | | | | | | |
| Africa | 21,450 | 20,190 | 20,780 | 17,590 | 15,370 | 16,555 | 3,860 | 4,820 | 4,225 |
| Asia | 30,390 | 26,920 | 32,690 | 18,440 | 17,210 | 19,500 | 11,950 | 9,710 | 13,190 |
| Australia | 3,560 | 3,620 | 6,405 | 2,830 | 3,160 | 4,440 | 730 | 460 | 1,965 |
| Europe | 6,380 | 5,980 | 7,165 | 3,730 | 3,590 | 4,055 | 2,650 | 2,390 | 3,110 |
| North America | 15,550 | 16,370 | 13,910 | 9,430 | 9,780 | 7,950 | 6,120 | 6,590 | 5,960 |
| South America | 27,030 | 23,990 | 29,355 | 15,470 | 15,280 | 18,975 | 11,560 | 8,710 | 10,380 |
| Antarctica | 1,890 | | | 1,110 | | | 780 | | |
| Total | 106,250 | 97,070 | 110,305 | 68,600 | 64,390 | 71,475 | 37,650 | 32,680 | 38,830 |
| **Values in mm depth/yr** | | | | | | | | | |
| Africa | 704 | 670 | 686 | 577 | 510 | 547 | 127 | 160 | 139 |
| Asia | 691 | 613 | 726 | 421 | 392 | 433 | 270 | 221 | 293 |
| Australia | 457 | 480 | 736 | 363 | 419 | 510 | 94 | 61 | 226 |
| Europe | 638 | 603 | 734 | 373 | 362 | 415 | 265 | 241 | 319 |
| North America | 636 | 668 | 670 | -386 | 399 | 383 | 250 | 269 | 287 |
| South America | 1,522 | 1,350 | 1,648 | 871 | 860 | 1,065 | 651 | 490 | 583 |
| Total | 775 | 725 | 834 | 500 | 481 | 540 | 275 | 244 | 294 |

**Table 2.4**  Factors of the Annual World Water Balance

| Factor | Amount, $km^3$ |
|---|---|
| Precipitation on ocean areas ($P_0$) | 381,410 |
| Evaporation from ocean areas ($E_0$) | 419,060 |
| Precipitation on land areas ($P_1$) | 106,250 |
| Evapotranspiration from land areas ($E_1$) | 68,600 |
| Runoff from land to ocean ($P_1-E_1$) | 37,650 |
| Land precipitation from land evaporation ($11\%P_1$) | 12,000 |
| Land precipitation from ocean evaporation | 94,000 |
| Atmospheric moisture flow, land to ocean areas | 57,000 |

Budyko from energy balance considerations, and L'vovich from his previously mentioned differentiated water-balance equations.

Mather's values of the three water budget factors for the whole globe fall between those achieved by Budyko and L'vovich, although Mather found more precipitation and evaporation over Africa and less over Australia (in $km^3$) than did the others. Largest discrepancies in the three sets of results are found in Australia and South America where L'vovich has much higher values of precipitation and evaporation in both continents and nearly three times the runoff from Australia than the next highest estimate.

Some of the discrepancy can be explained on the basis of actual areas of continental land masses included. Essentially similar continental areas have been used by Mather and Budyko in achieving their values of precipitation, evaporation, and runoff both in $km^3$ and in mm. L'vovich, however, considered somewhat different continental areas, especially in Australia and North America. Tasmania, New Guinea, and New Zealand were included in the former total while the Canadian Archipelago and Greenland were excluded from the latter total.

Based on the results in Table 2.3, South America would seem to be the best endowed continent in terms of potential water resources, with either Africa or Australia as the least well endowed. Europe and Asia follow considerably behind South America, while North America is fourth on the list according to L'vovich and Mather but second according to Budyko who has relatively low values for Asia and Europe.

Accepting the values for precipitation, evaporation, and runoff given in Table 2.2 as well as the suggestions by Benton et al. (1950) and Budyko (1956) that approximately 11% of the precipitation on the land comes from land evaporation, we may now quantify the factors of the world water budget. Actual values are suggested in Table 2.4. Expressed in diagrammatic form, the annual hydrologic cycle can be represented by a series of tubes whose widths are proportional to the volumes of water involved (Fig. 2.3).

The figures in Table 2.4, shown schematically in Figure 2.3, emphasize the mobility of water since both atmospheric transports of moisture (from land to water and from water to land) exceed the total amount of surface transport of

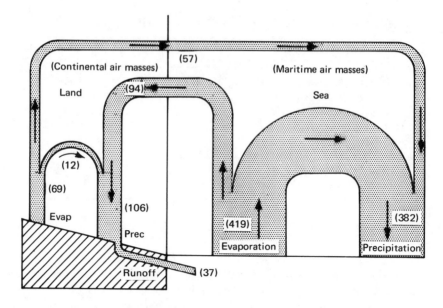

**Figure 2.3** Schematic representation of the hydrologic cycle. Width of each tube proportional to volume of water involved in that phase of hydrologic cycle. (Values in parentheses are water volumes in 1000 $km^3$.)
(Diagram prepared by Prof. Dr. Rudolf Geiger and used with kind permission.)

water from land to oceans by means of runoff. Holzman's earlier qualitative conclusions (1937) are fully justified. Even significant changes in evapotranspiration from the land will have little local influence on precipitation because such a small amount of local evaporation is returned as local precipitation. Unless local orographic influences are involved (upslope motion by air that has received a considerable input of moisture by evaporation), downwind precipitation consequences should be minimal.

## SIMULATION OF THE GLOBAL HYDROLOGIC CYCLE

Manabe and Holloway (1975) have provided an interesting new approach to the world water balance through the development of a numerical model of the atmosphere. Solving the primitive equations of motion in a spherical coordinate system, they have obtained estimated values of wind, temperature, pressure, water vapor, rainfall, snowfall, and evaporation at the surface of the globe. Various assumptions concerning such factors as viscosity, radiative heating and cooling, atmospheric ozone and carbon dioxide content, sea surface temperatures and sea ice cover, cloud cover, and atmospheric water vapor content have been included in order to allow quantitative results to be achieved. Use of seasonal values of such factors as insolation and sea surface temperatures permits estimation of seasonal changes in some of the factors of the water budget. The model has 11 finite difference levels ranging from 80 m to 31 km above the surface, selected to simulate both the stratosphere and the planetary boundary layer. The horizontal grid system has a nearly uniform resolution of about 265 km. The model assumes continents as on the earth with somewhat smoothed topography.

Manabe and Holloway (1975) utilize the model to provide global maps of the winter and summer distribution of pressure, eddy kinetic energy, mean vertical atmospheric motion, surface ambient temperatures, precipitation, and soil moisture storage as well as mean annual maps of precipitation, evaporation, soil moisture storage, runoff, snowmelt, and the distribution of Köppen climatic types. Pole-to-pole vertical cross-sections of the mean stream function of the meridional circulation, the zonal-mean simulated relative humidity, as well as the latitude-time distribution of the zonal-mean rate of precipitation, runoff, evaporation, snowmelt, and snow cover depth over both land and sea are calculated. Comparisons of simulated distributions with those obtained from observed values suggest that the model, as it is now written, can provide reasonable estimates of many of the major features of the seasonal variations in the factors of the hydrologic cycle or world water balance. Some problems develop in mountainous regions where the simulated values tend to deviate significantly from available observations.

Figure 2.4 shows simulated and observed pole-to-pole distributions of the zonal-mean components of the water budget over land. Because of the very high rainfall simulated in the tropics, the value of runoff in that belt also greatly exceeds the runoff estimated by L'vovich and Ovtchinnikov (1964). These two investigators suggest that evaporation exceeds runoff within the tropics (Fig. 2.4), which is corroborated by the data included in Table 2.2. However, the runoff values of Table 2.2 also differ from those suggested by L'vovich and Ovtchinnikov since they show runoff greater than evaporation from 60 to 90°N and from 40 to 90°S. Table 2.2 indicates that runoff is greater than evaporation only in the belts 60–70°N and 50–60°S. The discrepancies emphasize the general lack of agreement between different methods of estimating these parameters and suggest that while Manabe and Holloway have compared their simulated data with one available estimate, great variations in the data from different estimates exist.

Manabe and Holloway do not fully explain the apparent overestimation of tropical rainfall, especially over land, and the resultant large value of runoff. They note that the regions in the model having high rates of runoff correspond to those basins containing major river systems. They also point out, however, that some mid-latitude areas (e.g., the Mississippi River basin in the U.S. and the Yangtze River basin in China) have too low values of runoff because simulated precipitation is deficient in those areas.

The model seems to be able to simulate the major features of the global distribution of different water budget factors even though there has been great simplification of the hydrologic processes included in the model. The current results certainly justify further development of the model for global water budget work.

In passing, all the previous figures for the world water budget are based on present land-sea-ice distributions. During glacial or interglacial periods with sea

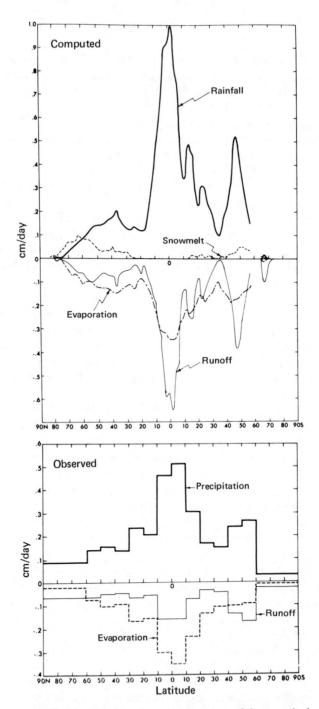

**Figure 2.4** (Top) Zonal-mean components of the water budget over land computed by the model and (bottom) estimated from observed data by L'vovich and Ovtchinnikov (1964).
*Source.* Manabe and Holloway (1975), copyrighted by the American Geophysical Union.

levels much higher or lower than at present, and with either vast quantities of water locked up in ice caps or with essentially no ice present on the earth, values of precipitation and evaporation would be somewhat different from those shown here. Estimates of their magnitude might be attempted, but they would be nothing more than rough guesses. For the present, our estimates are confined to the somewhat more reliable current information on precipitation, evaporation, and runoff. The main concern of the following chapter is to examine the principal components of the water budget so that the current state of our knowledge may be more fully understood.

## 3

ELEMENTS OF THE
HYDROLOGIC CYCLE

The term hydrologic cycle can have several meanings depending on the temporal and spatial scales being considered. On the one hand, it can refer to the annual accounting of the moisture fluxes over the entire globe in all of their various forms. On the other hand, it can refer to daily or monthly accounting of moisture inflows, outflows, and storages over a basin or even at a particular place resulting in day-to-day changes in moisture storage, surplus, or deficit. In the latter use, it is often called a water budget.

## THE BASIN HYDROLOGIC CYCLE

The drainage basin or watershed probably forms the most logical unit with which to start our study of the hydrologic cycle. Most of the interactions we will be concerned about will occur as precipitation falls onto the basin and works its way by one means or another out of the basin (Figs. 3.1a, 3.1b).

Of the precipitation that falls (P), small amounts are evaporated while still in the air or from water intercepted (I) by the vegetation ($e_i$); the rest reaches the surface of the globe. This precipitation is either stored on the surface (S), infiltrates into the surface materials (in), or runs off over the surface ($q_o$) to be stored elsewhere (in lakes, rivers, or ocean). The absorbed moisture will either be stored temporarily in the upper soil layers and be used by the vegetation in transpiration ($e_t$) or evaporated directly from the soil ($e_m$). If the soil is already saturated, the absorbed moisture will seep downward through the upper soil layers (vadose zone V) and possibly through it to the groundwater table, where it passes into groundwater storage (G). Ultimately, this water in the phreatic zone will reappear as baseflow (g) or as springs flowing into distant water courses. Water leaves the basin as basin runoff (Q) as well as by deep flow to other basins, and by evaporation and transpiration.

Four factors in this rather involved set of events—namely, (1) precipitation,

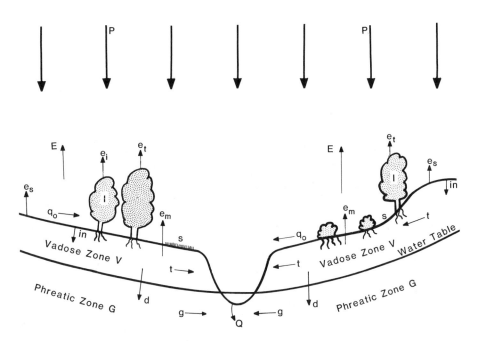

**Figure 3.1a** Cross-section of basin showing relationship of different aspects of the basin hydrologic cycle.

(2) evapotranspiration, (3) soil moisture storage, and (4) runoff (both surface and subsurface)—play primary roles in any consideration of the hydrologic cycle. Because of their importance, the nature and characteristics of each, their measurement or estimation, as well as significant problems or limitations in their use, are briefly reviewed.

## PRECIPITATION

Precipitation represents an available water supply which must be distributed to fulfill various demands for water (storage, evapotranspiration, percolation, runoff). Precipitation occurs in many different forms from drizzle (droplets < 0.5 mm with an intensity less than one mm/hr), to rain (larger, fewer droplets than drizzle, less reduction in visibility), to snow (ice crystals agglomerated into flake form with size depending on moisture content and temperature), to sleet (clear solid grains of ice formed from freezing of rain or melting and refreezing of snow), to hail (ice balls made of concentric layers of clear and milky ice). Many variations of these main types of precipitation exist, including mist, snow pellets, soft hail, graupel, and snow grains, but need not concern us here.

Precipitation may result when large masses of air rise and are cooled by expansion. With cooling, air is brought to saturation since the saturation vapor

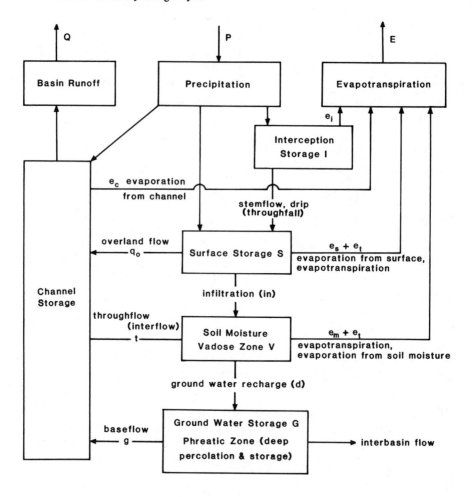

**Figure 3.1b** Schematic relationship of the various components of the basin hydrologic cycle.


pressure* decreases rapidly with decreasing temperature. Thus, air with a vapor pressure below saturation can be brought to saturation (100% humidity) merely by sufficient cooling. Any further cooling results in moisture condensation on available hygroscopic particles (dust, salt, etc.) in the air to form clouds.

This condensation process (which also results in dew or frost formation on


*Vapor pressure is the pressure exerted by the water vapor present in the atmosphere. The pressure of a gas in a chamber is equal to the sum of the partial pressures of each of the gases in the chamber. If there is water vapor, this gas will exert a pressure and it is this pressure that is referred to here. At any given air temperature, there is a maximum value of vapor pressure that can occur. This is the saturated vapor pressure, a measure of the maximum moisture content of the air.

the ground) is quite distinct from the precipitation process. How do cloud droplets that have been formed by the condensation process develop, sometimes within a short period of time, into large raindrops that can produce heavy rains? One process by which this can occur depends on the presence of both ice crystals and water droplets in the cloud. The saturation vapor pressure with respect to ice is less than it is with respect to water at the same below-freezing temperature. With both ice crystals and water droplets present, an average saturation vapor pressure between that for ice and that for water will exist in the cloud. The cloud air surrounding the ice and water droplets will be supersaturated with respect to the ice crystals and unsaturated with respect to the water droplets. Moisture will thus evaporate from water droplets and condense rapidly on available ice crystals. The ice crystals will grow rapidly at the expense of the water droplets; when large enough, they will fall out of the cloud, conceivably melting to raindrops on the way down.

Other possible mechanisms that might cause large raindrops to form rapidly depend on the presence of different sized condensation nuclei in the atmosphere, the influence of electrical charges on cloud droplets, or the occurrence of electrical fields within the atmosphere. Large condensation nuclei are far better collectors of moisture than small condensation nuclei and may actually grow at the expense of small particles. Recent work also suggests that uncharged water droplets in an electrical field or water droplets with unlike charges will coalesce if they come into contact. While the significance of these experimental findings is far from clear when applied to the atmosphere, it is quite likely that they play a real role in precipitation formation.

Many hydrologists and conservationists are concerned with the question of maximum falls of rain during different time periods (e.g., the record one-minute, one-hour, one-day rainfall, etc.). When the value of rainfall amount is divided by the duration of ,the rainfall period, the resulting value of average amount per minute (average intensity) is found to decrease rapidly with increasing duration of the rainfall period (Fig. 3.2).

## Measurement of Precipitation

Precipitation is measured by means of rain gages. The standard National Weather Service gage consists of a 200-mm diameter, funnel-shaped top connected to an inner collecting tube that is just one-tenth the area of the funnel top. This results in a tenfold increase in the depth of water collected in the inner tube and makes it easier to read small amounts of rainfall by means of a dipstick. The gage is 580-mm high (Fig. 3.3a).

To measure rainfall intensity or to undertake time-duration studies, recording gages must be used. A weighing gage has a funnel top that directs the water to a collecting bucket that rests on the platform of a spring scale. The movement of the platform of the scale is transmitted through a series of levers to a pen arm that records on a paper chart mounted on a clock-driven cylinder. A tipping-bucket rain gage substitutes a small tipping collector for the weighing bucket

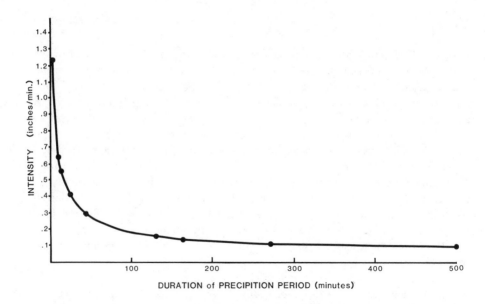

**Figure 3.2** Relation between precipitation intensity (inches/minute) and duration of precipitation period for record short-period rainfalls.

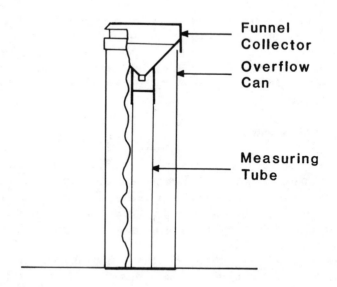

Funnel
Collector

Overflow
Can

Measuring
Tube

**Figure 3.3a** Standard 8-inch National Weather Service rain gage.

48

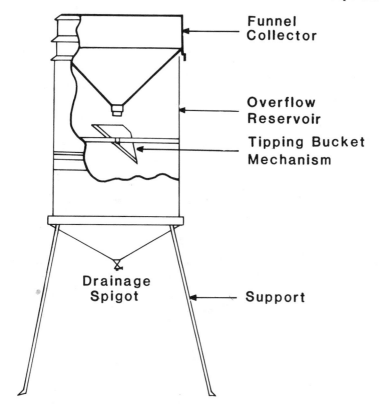

**Figure 3.3b**  Tipping-bucket rain gage.

(Fig. 3.3b). The tipping collector has a capacity of just 0.25 mm and, when full, it will tip rapidly to place an empty collector under the bottom of the funnel. The tipping collector is connected with a recording device that counts the number of tips with time. This type of gage is good for remote operation because the tipping action continually empties the gage and eliminates the need for an observer to service it. However, it is useless during periods of snowfall and may not be able to measure all the rain during times of high intensity due to the time required for the collector to tip back and forth.

Snow serves as an important source of runoff in many areas and thus its measurement is an important aspect of water resources evaluations. Because of the nature of snow, it cannot be caught easily in the standard rain gage or in the conventional weighing or tipping-bucket type gages. The standard gage can be "winterized" by removing the funnel collector and the inside measuring tube and allowing the snow to be caught in the outside 200-mm diameter cylinder. This snow is then melted and the water equivalent of the snow is reported. More frequently, especially in mountainous watersheds, other types of measurements of snow depth and extent are made. Meiman (1976) has summed up a number

**Table 3.1**  Hydrologic Snow Measurements

| Desired Information | Technique | Comments |
|---|---|---|
| Areal extent | Visual observations | Quick, practical method but requires someone thoroughly familiar with area. |
| | Ground and aerial photography | Has advantage of permanent record for checking but expensive and time consuming. |
| | Satellite | Developing technique with good promise; requires sophisticated equipment and highly trained personnel. |
| Depth | Surface probes and snow stakes | Practical method but requires logistical support for field measurements. |
| | Aerial markers | Good operational technique for basin-wide system; requires careful design. |
| | Photogrammetry | Expensive and time consuming; mainly used for research and special operations. |
| | Snowboards | Practical field method for determining new snow; should have frequent (daily) service. |
| Snow water equivalent | Precipitation gauges | Require careful location and design to minimize wind, radiation, and sublimation errors. |
| | Snow pressure pillows | Good technique for linking to radio-telemetry but expensive. |
| | Isotope techniques | Good for large basin measurements in open terrain but requires sophisticated equipment and highly trained technicians. |
| | Surface coring | Mainstay in hydrologic work but requires logistical support for field operations. |
| Snow profile characteristics | Open pits | Direct and practical but time consuming. |
| | Thermistors | Quick, inexpensive way to obtain snow temperature profile. |
| | Isotope profiling techniques | Excellent detailed profile data rapidly available but expensive and requires sophisticated technical support. |

*Source.* Meiman (1976) with permission of the Food and Agriculture Organization of the United Nations.

of different techniques that have been used to obtain information on extent, depth, water equivalent, and snow profile (Table 3.1). Because of the difficulty in making measurements in remote mountainous areas, satellite or aerial reconnaissance is often used. Techniques to transmit data by radio-telemetry are coming into use but the old standard surface probes and snow stakes, surface coring, and open pits still provide a great deal of the information on snow.

Of the 12,000 to 13,000 official rain gages in the conterminous United States, some 3000 are recording gages and the remainder are read daily by cooperative observers. Based on the area of the country, the average density of gages is one for every 650 km$^2$ (250 mi$^2$). This density varies considerably; a density of one gage per 188 km$^2$ is found in New Jersey while there is one gage per 1530 km$^2$ in western Colorado. Density may not be great in some areas of real significance from a hydrologic point of view. Because the rain gage measures precipitation at a single point only, one may question how representative these point observations are of the average conditions over an area. Certainly in areas of marked topographic diversity or during periods of convective storm activity when great local variations in precipitation can occur, point observations may not be representative of conditions in areas only a short distance from the gage.

## Analysis of Precipitation Data

Several techniques to estimate average values of precipitation over an area exist. Rainbird (1970) has listed (a) the use of the arithmetic mean of point precipitation values at all stations in the area; (b) the development of a polygon-weighting method; and (c) the measurement of the area between successive isohyets followed by multiplication of this area by the mid-value of precipitation between isohyets. In the Thiessen polygon method, straight lines are drawn between all adjacent stations. Perpendiculars are constructed through the mid-points of each of these lines. Intersection of these perpendicular bisectors creates a series of polygons, each having a precipitation station near the center. The area of the polygon is then used as a factor for weighting the station precipitation. Figures 3.4a and 3.4b illustrate how monthly precipitation for an area can be calculated by the Thiessen polygon method as well as by the isohyetal and arithmetic mean methods.

Rainbird points out that each technique has certain advantages and disadvantages; the choice of method requires judgment on the part of the analyst. The arithmetic mean and polygon methods are quite objective and are easily applied. No subjective judgments are needed. They might, however, result in the exclusion of other information that could provide additional insight into precipitation distribution. The isohyetal method can be the most accurate provided the analyst approaches his problem with skill and understanding. The location of the isohyets can be adjusted on the basis of other evidence (e.g., topography, vegetation, or wind direction), so that the resulting pattern is able to provide a realistic representation of the actual distribution. The method, however, is time consuming and requires experience on the part of the analyst.

Two limitations in precipitation sampling have been discussed, namely the lack of an adequate sampling network and the difficulty of extrapolating data from point observations to obtain estimates of the areal precipitation volume. A third problem is determining the reliability of any observed value. Both the exposure of the gage and the effect of wind are important in determining the amount of catch. Shading of a gage by trees or nearby houses can limit the amount of precipitation falling in the gage. The gage must be exposed in an open area, at a distance at least several times the height of all nearby tall obstructions away

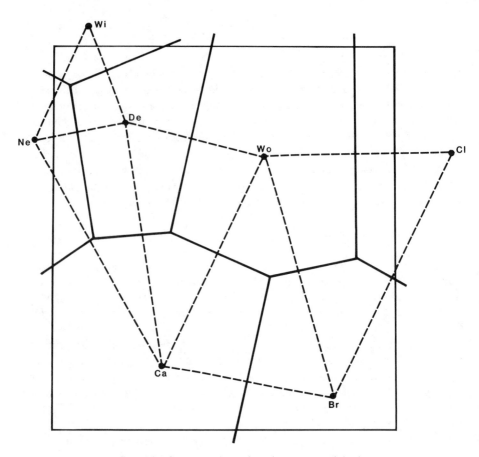

Example of computations of areal average precipitation,
Southern New Jersey, January 1952

| Arithmetic average | | Thiessen polygon | | Isohyetal Method | | | |
|---|---|---|---|---|---|---|---|
| Station | Precip (in) | Area of Polygon as % of Total Area | Weighted Precip (in) (Precip X Area of Polygon) | Area Between Isohyets as % of Total Area | | Mid-value of Precip (in) | Weighted Precip (in) (% X mid-value) |
| | | | | | % | | |
| Wi | 4.73 | .70 | .03 | <4.0 in | 3 | 3.75 | .11 |
| Ne | 5.09 | 1.05 | .05 | 4.0 to 4.5 | 15 | 4.25 | .64 |
| De | 3.98 | 14.30 | .57 | 4.5 to 5.0 | 39 | 4.75 | 1.85 |
| Wo | 5.10 | 31.80 | 1.62 | 5.0 to 5.5 | 34 | 5.25 | 1.78 |
| Ca | 4.97 | 26.65 | 1.32 | >5.5 | 9 | 5.75 | .52 |
| Cl | 5.82 | 7.30 | .42 | Total | 100 | | Total 4.90 |
| Br | 5.12 | 18.20 | .93 | | | | |
| Average | 4.97 | Total 100.00 | Total 4.94 | | | | |

**Figure 3.4a** Construction of Thiessen polygons. Precipitation is measured at stations identified by letters. Dashed lines are drawn between adjacent stations. Solid lines are the perpendicular bisectors of dashed lines. Intersection of solid lines form polygons whose areas determine the weighting factor by which to multiply the value of precipitation at the station within the polygon. Area of polygon is expressed as a percentage of total area.

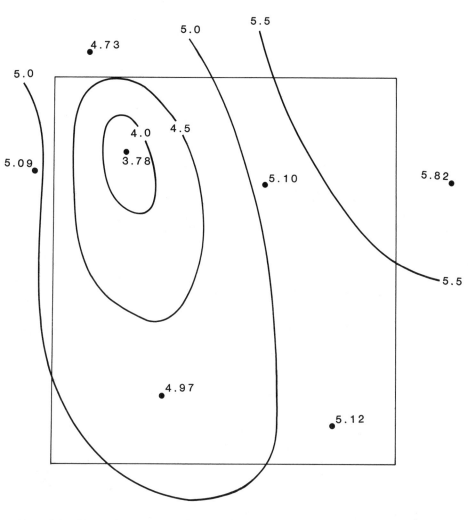

**Figure 3.4b** Lines of equal precipitation (isohyets) over study area. To obtain areal average precipitation, planimeter the area between isohyets. Express this area as a percentage of the total area under study. Multiply this percentage value by the mid-value of precipitation between isohyets and sum the values.

from them. Koschmieder (1934) reported on the effect of wind speed on the resulting catch. As wind speed increased, the amount of catch decreased from the "true" catch which he measured in a carefully constructed sunken rain gage that offered no resistance to the wind (Table 3.2).

Since wind usually increases with height above the surface, more reliable observations should be obtained if the gage is exposed close to the earth's surface where wind speed approaches zero. Rain gages on building roofs will only provide observations within 20 to 30% of the true catch under most ordinary circumstances.

Table 3.2   Effect of Wind Speed on Raingage Catch

| Wind Speed (m/sec mph) | | Deficit from True Catch (%) |
|---|---|---|
| 0 | 0 | 0 |
| 2 | 4.47 | 4 |
| 4 | 8.94 | 10 |
| 6 | 13.41 | 19 |
| 8 | 17.88 | 29 |
| 10 | 22.4 | 40 |
| 12 | 26.8 | 51 |

*Source.* Koschmieder (1934).

Rainbird (1970) has estimated that annual measured precipitation needs to be corrected from 17 to 56% (average between 20 to 30%) due to wind speed and exposure problems based on a Russian study involving several hundred rain gages. Larger corrections are needed with snowfall than with rain and with low intensity or small droplet rain than with more intense or large droplet rain.

Inconsistencies in station records can be evaluated by means of double-mass analysis (graphically plotting the accumulated precipitation over a period of time at several nearby stations and comparing time trends) or by applying simple tests for homogeneity of the data. Various statistical techniques are available to create a homogeneous record if it does not occur naturally.

Missing records in a long data series must be filled by some estimation technique if the long record is to be useful. Long gaps probably should not be filled since estimation techniques are not too reliable. Short gaps may be most conveniently filled by use of a normal-ratio method or by the plotting of isohyets from available data and estimating the missing values from the pattern on the map. The normal-ratio method requires several nearby stations fairly evenly spaced around the station in question. Short-period precipitation (P) at the missing station x is given by

$$P_x = 1/3 \ (P_a N_x/N_a + P_b N_x/N_b + P_c N_x/N_c)$$

where N is the normal precipitation and a, b, and c are three nearby stations with continuous records. Interpolating precipitation values for stations with missing data from an isohyetal pattern is more time consuming but should provide more reliable estimates of missing data if the isohyets are drawn with care.

In recent years, there has been debate over the reliability of observations of extreme precipitation at stations such as La Porte, Indiana. La Porte is located just downwind of extensive steel works around Chicago, Illinois and Gary, Indiana. Over the period 1927 to 1963, the La Porte precipitation observations were considerably greater than those at other nearby stations. With the recent increased interest in man-made weather and climate changes, questions arose as to whether the "La Porte anomaly," as it is now called, is an example of a man-induced climatic change due to the addition of large numbers of hygroscopic nuclei to the atmosphere through industrial activity or whether there has actually been significant observer error (Changnon, 1968; Hidore, 1971; Holzman and Thom, 1970; Mather and Rowe, 1979). The question is still not fully resolved although the precipitation figures in the past few years have failed to reveal the continuation of the very high values of previous years. Even in areas of relatively dense precipitation networks, there are still vast unsampled areas which, together with the various sampling and exposure problems, compound the problem of trying to evaluate the accuracy and reliability of any observation.

## EVAPOTRANSPIRATION AND INTERCEPTION LOSSES

Evaporation or evapotranspiration, the combined water loss from soil and vegetated surface, occurs when there is a change of state of water from liquid to vapor. It is accompanied by an important energy transfer inasmuch as heat is required to effectuate evapotranspiration and this heat is transferred to the air with the vapor as latent heat. While precipitation represents the downward flow of moisture to the earth's surface, natural evaporation is much more than the reverse of this. It is both a return flow of water to the atmosphere and a partial reverse flow to the stream of radiant energy from the sun. Evaporation is a direct function of the vapor pressure gradient between the evaporating surface and the ambient air above. For evaporation to continue, the existence of an external source of energy is required.

### Interception Losses

Interception of precipitation by vegetation plays a role in evapotranspiration. To the climatologist, it matters little whether the water evaporating from a plant comes from the soil via the root system or is merely intercepted rain. Both processes require the same quantity of energy and both constitute evaporation of water. If soil moisture is not actually at the wilting point, the energy consumed in evaporating intercepted water will otherwise be used to evaporate transpired water. Thus, when intercepted rainfall is evaporated, it may be considered as merely relieving the drain on soil moisture.

Amounts of water intercepted vary, of course, with the quantity of rain and with the type of vegetation. Eidmann (1959) studied interception (I) by European beech and Norway spruce in the Sauerland mountains of West Germany for a 5-year period. He found that summer interception by spruce (3 years of data) decreased from 82% to 24% of the rainfall as rainfall increased from 1 mm to more than 20 mm, while it decreased from 72% to 18% of the rainfall for beech in the same rainfall ranges (Fig. 3.5).

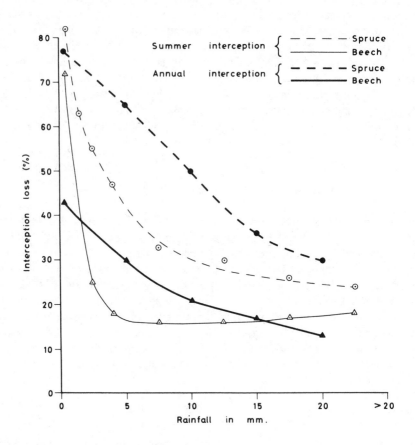

**Figure 3.5** Interception losses from spruce and beech forests.
*Source.* Ward (1975), copyrighted by McGraw-Hill Publishing Company, Ltd., based on data from Eidmann (1959) and Hoppe (1896).

Summing the results of measurements of throughfall (Th), stemflow (Sf), and interception (I) for both the spruce and beech by seasons of the year for 1951–55, Eidmann showed (Table 3.3) the significant influence of both season of the year and vegetation type on interception but not on the other two factors.

Hewlett and Nutter (1969) have shown how the type of tree (coniferous or deciduous) as well as the age of the stand (and therefore the density of cover) affect both the amount of interception and the percentage of total precipitation that is intercepted. The greatest amount of interception is found with coniferous rather than with hardwood trees (because of the greater leaf area and the

**Table 3.3**  Throughfall (Th), Stemflow (Sf), and Interception (I) for Spruce and Beech: Sauerland, 1951–1955

| Period | Rain (mm) | Spruce | | | | Beech | | | |
|---|---|---|---|---|---|---|---|---|---|
| | | Th | Sf | I | I/R% | Th | Sf | I | I/R% |
| Nov.–Apr. | 587 | 465 | 4 | 118 | 20 | 465 | 97 | 25 | 5 |
| May–Oct. | 629 | 428 | 5 | 196 | 31 | 457 | 104 | 68 | 15 |
| Year | 1216 | 893 | 9 | 314 | | 922 | 201 | 93 | |
| Year % | 100 | 73 | 1 | 26 | | 76 | 16 | 8 | |

*Source.* Eidmann (1959) with permission of the International Association of Hydrological Sciences.

nondeciduous character of the pine) and in the more mature stands. In a 60-year old white pine stand, one-fourth of the total precipitation is lost by interception.

In spite of these previous suggestions of differences in interception among vegetation species, Pereira (1973) has summed up observations on cypress, bamboo (both growing in the tropics), and a mid-latitude hardwood forest. He concludes that interception by a forest canopy is governed more by the depth of rain in a particular storm than by the type of tree (Fig. 3.6). The curves relating percent interception to the rainfall depth follow generally the shape found earlier by Eidmann (Fig. 3.5), although the differences at any given value of rainfall are

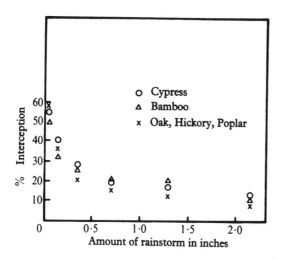

**Figure 3.6**  Interception of rainfall by tropical and temperate forests.
*Source.* Pereira (1973), with permission of the Cambridge University Press.

**Table 3.4**  Percentage Interception[a] of Precipitation by Trees

| Type or Species | Age or Size | Place in Succession | Locality | Interception (%) |
|---|---|---|---|---|
| Hemlock | Mature | Climax | Conn. | 48 |
| Douglas fir | 25 yr | Climax | Wash. | 43 |
| Hemlock | Mature | Climax | N.H. | 38 |
| Spruce fir | Mature | Climax | Maine | 37 |
| Hemlock | Mature | Climax | Adirondacks, N.Y. | 34 |
| Douglas fir | Mature | Climax | Wash. | 34 |
| Hemlock | Mature | Climax | Ithaca, N.Y. | 31 |
| Spruce–fir–paper birch | Mature | Climax | Maine | 26 |
| White pine–hemlock | Mature | Climax | Mass. | 24 |
| Western white pine– western hemlock | Over mature | Climax | Idaho | 21 |
| Maple–beech | Mature | Climax | N.Y. | 43 |
| Mixed | Mature | Climax | N.Y. | 40 |
| Maple–hemlock | Mature, cutover | Climax | Wis. | 25 |
| Beech–birch–maple | Mature | Climax | Ontario | 21 |
| Ponderosa pine | Mature | Preclimax | Ariz. | 40 |
| Lodgepole pine | Mature | Preclimax | Colo. | 32 |
| Ponderosa pine | Mature | Preclimax | Idaho | 27 |
| Jeffrey pine | Mature | Preclimax | S. Calif. | 26 |
| Lodgepole pine | 32 yr | Preclimax | Colo. | 23 |
| Ponderosa pine | Mature | Preclimax | Idaho | 22 |
| Ponderosa pine | Young | Pioneer | Colo. | 18 |
| Calif. scrub oak | 6 ft | – | S. Calif. | 31 |
| Mixed brush | Mature | Preclimax | North Fork, Calif. | 19 |
| White pine–red pine | 40 yr | Preclimax | Ontario | 37 |
| Jack pine | 50 yr | Pioneer | Wis. | 21 |
| Shortleaf pine | 45 yr | Pioneer | N.C. | 16 |
| Quaking aspen | 32 yr | Pioneer | Colo. | 16 |
| Chaparral, mixed | 6 ft | | S. Calif. | 17 |
| Maple–hemlock | Mature (after leaf fall) | Climax (under stocked) | Wis. | 16 |
| Hemlock | Mature | Climax | N.Y. | 13 |
| Oak pine | Open, second growth | Preclimax | N.J. | 13 |
| Ponderosa–lodgepole pine | 25 ft | Preclimax | Idaho | 8 |
| Beech–maple | Mature | Climax | N.Y. | 6 |
| Chamise | 6 ft | Pioneer | S. Calif. | 3 |

[a]Interception includes stemflow and is expressed as a percentage of annual precipitation.
*Source.* Compilation of data from various references from Kittredge (1948).

**Table 3.5** Interception (I) by Agricultural Crops—Summary

| Crop | Plants/m$^2$ | R = 175 mm July 1–31 I% | R = 65 mm Sept. 1–25 I% |
|---|---|---|---|
| Maize | 9 | 17 | 50 |
| | 16 | 20 | 55 |
| | 25 | 19 | 60 |
| | 36 | 32 | 77 |
| Soya beans | 9 | 0 | 28 |
| | 16 | 7 | 31 |
| | 25 | 12 | 56 |
| | 36 | 17 | 47 |
| Oats | 16 | 11 | 5 |
| | 25 | 10 | 6 |
| | 36 | 10 | 13 |
| | 49 | 15 | 26 |
| Vetches | 25 | 12 | 32 |
| Beans | 25 | 25 | 35 |
| Lupins | 25 | 22 | 58 |
| Peas | 25 | 10 | — |
| Clover | broadcast | 40 | — |

*Source.* Wollny (1890).

much smaller. He reports that the same relationship to depth of rainfall also applies to dense stands of Douglas firs in Oregon studied by Rothacher (1963).

Based on an extensive summary of forest interception data, compiled from a large number of sources (Table 3.4), Kittredge (1948) concluded that interception losses varied with density of the crown. Well-stocked stands tended to intercept a greater amount of precipitation than understocked stands. Age also played a role. Stands in the age period between the closing of the canopy and culmination of the current annual increment tended to have a greater interception value than either younger or older trees. Kittredge also felt that interception losses changed with species and forest types as a result of differences in thickness and density of both the foliage and the crowns. Tolerant species were able to intercept more than intolerant species, mesophytic species more than xerophytic species, and climax trees more than those that were preclimax. Even within a stand, interception varied from maximum near the stems to least near the edges of the crowns. Kittredge (1948) reported interception by grasses in Berkeley, California (*Avena, Stipa, Lolium,* and *Bromus*) averaged 26% of the 32.5-in. seasonal rainfall, while Musgrave (1938) showed that bluegrass in Missouri intercepted 17% of the rainfall in the month before harvest.

Although the data are quite old, Wollny (1890) has provided us with a comprehensive set of observations of interception by agricultural plants (Table 3.5).

These data show the effect of increasing size of plants, plant density, and differences in rainfall amounts on interception, although the contribution of each factor cannot be isolated. Wollny's data have been generally corroborated by more recent investigators (Haynes, 1940; Smith, Whitt, Zingg, McCall, and Bell, 1945).

While water dripping from a forest cover will much more readily infiltrate the soil because of the better ground cover under the trees and the slower rate of drop fall, Hoover (1962) points out that interception by the forest canopy will probably be of little significance in reducing flood peaks. In addition, intercepted water is withheld from the storm hydrograph at the beginning of the storm rather than toward the end of the storm when flooding may become a problem.

## Potential Evapotranspiration

In the past several decades, evapotranspiration has become recognized as a fundamental climatic element of concern not only to theoretical climatologists but also to hydrologists, biologists, engineers, and others. However, evapotranspiration has proven to be an extremely difficult element to measure, or estimate, let alone to understand. Part of the problem results from the fact that while evaporation may be thought of as a climatological or meteorological factor, it is also under partial control of vegetation and edaphic factors, antecedent moisture, surface conditions, and cultural factors. Thus, many workers have tried to define a theoretical value, potential evapotranspiration, which will depend almost entirely on the climatic factors of radiation, wind, and humidity. Even these latter two factors have less significance since, under the moist conditions required for potential evapotranspiration, humidity variations should be minimal and wind is less important in removing moist air near the plant to replace it with drier air inasmuch as all air will be fairly similar in moisture content. Potential evapotranspiration depends primarily on energy from the sun, the available energy to evaporate water. Potential evapotranspiration may be defined as the water loss from a vegetation cover that never suffers from a lack of water. The definition should further specify that the soil is covered with a closed, homogeneous layer of vegetation that is everywhere the same in height and type with an albedo (percent of incoming solar radiation that is reflected from a surface) of 20 to 25%.

Penman (1963) has carefully considered the water use of different vegetation with water supply nonlimiting as specified in the definition for potential evapotranspiration. Analyzing data from many different parts of the world that were obtained under a wide variety of experimental conditions, he concludes that there is little consistency in the results. While this might argue for more careful experimental designs, he points out the great difficulties inherent in a homogeneous exposure of the always moist experimental area, the maintenance of satisfactory soil moisture conditions at all times, and problems with adequate buffer areas around the experimental plots to eliminate advection or "oasis" effects.

**Table 3.6**  Peak Water Use (inches/day) in Approximately Homogeneous Climatic Zones of the United States

| Group<br>Frost Free<br>Period (days) | 1<br>250–300 | 2<br>210–250 | 3<br>180–210 | 4<br>180–210 | 5<br>180–210 | 6<br>180–210 | 7<br>210–250 |
|---|---|---|---|---|---|---|---|
| Lucerne | 0.17 | 0.22 | 0.25 | 0.29 | 0.35 | 0.30 | 0.18 |
| Pasture | 0.17 | 0.22 | 0.25 | 0.28 | 0.35 | 0.30 | 0.18 |
| Grain–small | 0.16 | 0.20 | 0.20 | 0.20 | | | |
| Beet sugar | 0.18 | 0.22 | 0.22 | 0.28 | | | |
| Beans–field | 0.16 | 0.18 | 0.20 | 0.20 | (0.30) | | |
| Corn–field | | 0.25 | 0.30 | 0.30 | 0.35 | 0.30 | 0.16 |
| Potatoes | | 0.18 | 0.22 | 0.28 | 0.30 | 0.20 | 0.15 |
| Peas | 0.16 | 0.18 | 0.18 | | 0.25 | 0.20 | |
| Tomatoes | 0.16 | 0.18 | 0.18 | 0.20 | 0.30 | 0.20 | 0.15 |
| Apples | | 0.20 | 0.20 | 0.22 | 0.30 | 0.25 | |
| Cherries | | 0.20 | 0.20 | 0.22 | | | |
| Peaches | | 0.20 | 0.20 | 0.22 | 0.30 | 0.25 | 0.14 |
| Apricots | | 0.20 | 0.20 | 0.20 | | | |
| Tobacco | | | | | | 0.25 | 0.14 |
| Vegetables | | | (0.20) | 0.25 | | 0.20 | 0.12 |
| Strawberries | 0.18 | 0.20 | | 0.25 | | 0.15 | |

| Group | Area |
|---|---|
| 1 | West coast, southern half in fog belt. |
| 2 | West coast, northern half, and southern coastal interior. |
| 3 | Central valley–California and valleys east side of Cascade mountains. |
| 4 | Inter-mountain, desert and high plains. |
| 5 | Mississippi–interior valleys. |
| 6 | Great Lakes. |
| 7 | Atlantic and Gulf coastal zone. |

*Source.* Penman (1963). (Table reproduced by permission of the Commonwealth Agricultural Bureaux.)

Table 3.6, which Penman has adapted from a commercial manual on sprinkler irrigation (Sprinkler Irrigation Association, 1955), includes information on peak water use of a representative group of crops in seven different geographic regions of the United States. He assumed the values of peak use are reasonable estimates of the water demands of complete crop covers exposed to the same weather throughout the period of their active and simultaneous growth. This assumption may be questioned since the active growing seasons of many of these crops is not simultaneous. Lower peak values for peas, for example, than for corn certainly might be anticipated because peas normally grow under cooler climatic conditions than does corn.

Alpatiev (1954) reported on the results of experiments of water use by different crops at the Poltavsky Experiment Station between 1910 and 1924 (Table 3.7). Information on exposure and moisture conditions is unavailable. Therefore,

Table 3.7    Water Use by Different Crops, Poltavsky Experiment Station

| Crop | Vegetative Period (days) | Total Water Use (mm) | Rainfall (mm) | Mean Water Use (mm/day) |
|---|---|---|---|---|
| Oats | 67 | 182 | 92 | 2.7 |
| Buckwheat | 93 | 264 | 184 | 2.8 |
| Barley | 97 | 264 | 187 | 2.7 |
| Spring wheat | 101 | 271 | 187 | 2.7 |
| Maize | 131 | 317 | 265 | 2.4 |
| Sugar beet | 154 | 407 | 305 | 2.6 |
| Mean | | 284 | | 2.65 |

*Source.* Penman (1963). (Table reproduced by permission of the Commonwealth Agricultural Bureaux.)

even though the data show relatively little variation in mean water use per day, the data must be interpreted with care. It does suggest that water use is generally independent of crop type.

Mather (1954) compared measured water use from different crops growing in continuously moist lysimeters with the calculated potential evapotranspiration for the same time periods. In this way, the effect of different lengths of growing season and different seasons of the year of active growth could be eliminated. His data (Table 3.8) showed only about a 10% variation in the ratio of measured to computed water use for a variety of vegetable crops as well as grass. These results have led a number of scientists to conclude that the type of vegetation is of secondary importance in determining evapotranspiration under always moist (potential evapotranspiration) conditions.

**Measurement of Evaporation**

Because of its climatic significance, considerable effort has been directed toward the development of instruments to measure potential evapotranspiration. Soil-filled tanks, covered with a continuous stand of vegetation that is in every way similar to the vegetation that surrounds the tanks and having a water supply fully adequate to the needs of the vegetation, have been used for many years to measure potential evapotranspiration water losses (Fig. 3.7). Water-filled pans such as the standard U.S. Weather Bureau Class A evaporation pan (1.2 m in diameter, 0.25 m deep, exposed 0.15 m above the ground surface) provide only rough estimates of the evaporating power of the air (Fig. 3.7a). Because of its unique exposure with air passing all around it and radiation falling on the walls of the pan, the resulting value of water loss seldom approximates potential evapotranspiration. Inflow-outflow measurements from lakes also provide only estimates of evaporation. The number of observations are quite limited and it is not possible to obtain

**Table 3.8**  Water Use for Different Crops Under Optimum Soil Moisture Conditions—New Jersey

| Crop | Period | | Water Use (mm) Measured | Calculated | Measured Calculated |
|---|---|---|---|---|---|
| Spinach | Aug. 25–Oct. 27, 1947 | | 204 | 183 | 1.11 |
| Corn | June 11–Aug. 23, 1948 | | 335 | 324 | 1.06 |
| Spinach | Sept. 11–Nov. 19, 1948 | | 146 | 126 | 1.16 |
| Peas | April 11–June 1, 1949 | | 178 | 151 | 1.18 |
| Lima beans | July 20–Sept. 25, 1949 | | 296 | 278 | 1.07 |
| Grass | May–Dec., 1952 | A | 778 | 677 | 1.11 |
| | | D | 719 | " | |
| | | B | 771 | " | 1.18 |
| | | C | 818 | " | |
| | | E | 795 | " | |
| | | F | 834 | " | 1.26 |
| | | G | 917 | " | |
| | A, D | Water table at 35 cm. | | | |
| | B, C | Water table at 17 cm. | | | |
| | E, F, G | Watered from above. | | | |

*Source.* Mather (1954).

short-period values of evaporation from such observations. Problems of seepage through the bottom of the lake and unmeasured inflows and diversions make errors likely. Lysimeters (soil-filled tanks buried flush with the ground) have usually been operated to provide drainage estimates rather than to measure potential evapotranspiration. As long as the soil is always kept moist, they can, however, provide reasonable values of potential evapotranspiration. If the soil is allowed to follow normal changes in moisture content due to rainfall, actual evapotranspiration will be evaluated. Watershed studies of precipitation, inflow, soil moisture storage, and surface and subsurface runoff can also be used to provide values of actual evapotranspiration, usually on a longer-term basis. Such natural basins are seldom moist enough to provide values of potential evapotranspiration. The possibility of unmeasured inflows and outflows of water suggests the need for great care in using the watershed approach.

In order to eliminate advection as an additional source of energy for evapotranspiration, evapotranspirometer tanks must be exposed within a field planted to the same vegetation as is on the tanks. This vegetation must receive the same watering treatment as the tanks themselves. The size of the buffer area needed will depend on the climate; in a moist climate, a square 100 m on a side might be sufficient, but in the desert, a square 1000 m on a side might not be large enough. The need for reliable observations in arid areas is very great, but these

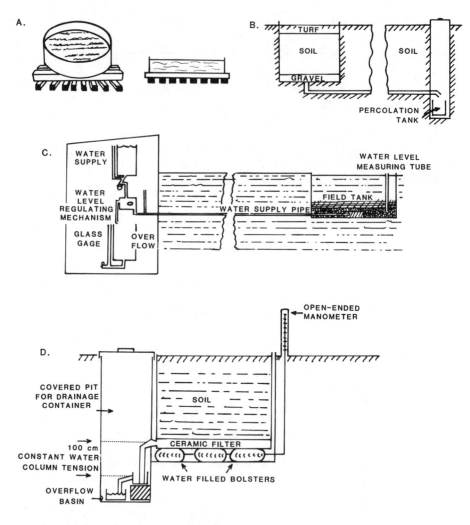

**Figure 3.7**  Different evaporation instruments.
(a)  Weather Bureau Class A evaporation pan.
(b)  Modified evapotranspirometer supplied daily by water through surface irrigation.
(c)  Evapotranspirometer installation utilizing a permanent fixed water table.
(d)  Hydraulic weighing lysimeter with constant drainage tension. (Developed by Forsgate, Hosegood and McCulloch, 1965, and similar to a model used by Dagg, 1970, for study of the water relation in a tea plantation.)

are the areas in which it becomes most difficult to establish and to maintain the exacting conditions required by the definition of potential evapotranspiration.

The question of the advection of sensible heat and its effect on evaporation rates from moist tracts has been studied in some detail (Rosenberg, 1972; Rosenberg and Verma, 1978; Brakke, Verma, and Rosenberg, 1978). Using an irrigated field of alfalfa, 1.9 ha in area, surrounded by fairly dry fields, Rosenberg found that sensible heat advection contributed 15 to 50% of the energy used for evapotranspiration on a daily basis. Values of daily evapotranspiration as measured by weighing lysimeters in the field ranged from 4.75 to 14.22 mm/day and were greater than 10 mm during one-third of the days of observations. Brakke et al. concluded that net radiation, which usually provides an upper limit for evapotranspiration in humid regions, might contribute to less than half of the evapotranspiration from well-watered crops in semiarid areas with the remainder of the energy coming from advection of sensible heat. That these investigators found daily values of water loss well above the potential evapotranspiration is not surprising in view of the oasis-type exposure of their moist field. Their findings are undoubtedly correct, but they relate to the particular size and moisture condition of their field and its exposure rather than to potential evapotranspiration as defined by Thornthwaite.

Failure to maintain proper conditions both within the evapotranspirometer tank and in the field outside results in either too much or too little water loss. When the soil in the tanks is too dry, the evapotranspiration rate drops below the potential. When the soil outside is drier than the soil in the tank, the evapotranspiration rate will exceed the potential because of the advection of dry air. The condition of the vegetation is also important. If the vegetation in the tanks is taller than that outside, the water loss will be excessive. A height difference of just 2 or 3 cm may make a serious difference. If the vegetation in the tanks stands up prominently above the surroundings, the observations of water use are probably worthless and should be discarded. Tall-growing crops will inevitably give excessive values of water use if they are grown in association with lower plants. For that reason, it is almost impossible to compare the potential evapotranspiration of different types of vegetation covers on adjacent tanks at the same time.

The foregoing should make clear why water loss from so-called "standard" evaporation pans can be very different from true potential evapotranspiration. It should also be clear that there is no relation between potential evapotranspiration and expressions relating to the "evaporating power of the air," such as relative humidity or saturation deficit; they are by no means synonymous. The moisture content of air is strongly influenced by the evaporation regime. With dry soils and negligible evaporation, relative humidity is low and saturation deficit high. But if conditions to satisfy potential evapotranspiration exist, evaporation increases and more moisture is added to the air. Since atmospheric moisture is not a conservative property, it is not possible to determine potential evapotranspiration accurately from either relative humidity or saturation deficit. Similarly, pan evaporation is strongly influenced by the moisture content of the air passing over the pan and so it is not possible to determine potential evapotranspiration from pan evaporation. The same criticism, of course, applies to any soil-filled, vegetation-covered tank not operated under the proper conditions. In dry periods, pan evaporation is always higher than potential evapotranspiration. Any formula for determining potential evapotranspiration that contains a humidity term or

**Table 3.9**  Seasonal Consumptive-Use Requirements of Common Irrigated Crops in Relation to the Requirements for Alfalfa in the Western United States

| Crops | Relation of Seasonal Consumptive Use of Crops Listed to the Seasonal Consumptive Use of Alfalfa (Optimum Yield) |
|---|---|
| Alfalfa | 1.00 |
| [a] Almonds | 0.42 |
| [a] Apricots | 0.50 |
| Artichokes | 0.50 |
| [a] Avocados | 0.52 |
| Beans | 0.42 |
| Berries | 0.40 |
| Clover | 0.90 |
| Carrots | 0.40 |
| Corn | 0.65 |
| [a] Citrus | 0.70 |
| Cotton | 0.65 |
| Grain | 0.50 (variable) |
| Grain–Sorghum | 0.35 to 0.40 |
| Grapes | (variable) |
| Hops | 0.50 |
| Lettuce | 0.15 (plant use only)[b] |
| Melons | 0.40 |
| Pasture | 0.90 (variable) |
| [a] Peaches | 0.65 |
| Potatoes–Winter | 0.50 |
| Potatoes–Spring or Seed | 0.30 |
| Seed Peas | 0.30 |
| Sugar Beets | 0.82 |
| Strawberries | 0.60 |
| Tomatoes | 0.45 to 0.50 |
| [a] Walnuts | 0.60 |

[a] Add 20–25% for permanent grass–legume cover.

[b] Water may be applied in addition to that required for growth for the purpose of quality improvement.

*Source.* Hansen, Israelsen and Stringham (1980), after Woodward (1959), with permission of the Irrigation Association, formerly the Sprinkler Irrigation Association.

that is based on pan evaporation will give excessive values in dry areas or dry periods. This discrepancy has resulted in the habitual use of too much water in dry-land irrigation, wasting precious water and damaging valuable land at the same time.

While all crops use about the same amount of water per day when grown under potential evapotranspiration conditions, it is clear that these conditions seldom exist in practice. Thus, the actual seasonal consumptive use of water differs significantly from one crop to another (Table 3.9).

### Alternative Estimates of Evapotranspiration

Reliable measurements are not available at many places so that it is necessary to consider alternative methods to evaluate potential evapotranspiration on the basis of other data. Since size of the evaporating surface influences the rate of moisture loss, it is first necessary to distinguish between small evaporating surfaces surrounded by markedly different surfaces and evaporating surfaces surrounded by essentially similar surfaces. The former case, the *oasis* type, has been investigated by those concerned with point-diffusion or plot-diffusion problems, especially as regards smokes, dust, and pollutants (Sutton, 1953). This case is also of real interest to the climatologist concerned with the representativeness of desert observations of evaporation or the validity of pan evaporation. The latter case, the *mid-ocean* type (Penman, 1956), assumes a similar environment over such a large area that humidities will not vary horizontally. In that situation, it can be shown theoretically that evaporation rate is proportional to the product of the wind speed and vapor pressure gradient.

There has been a great increase in the number of indirect methods for the absolute determination of evapotranspiration. These have mainly involved (a) mass transport techniques, (b) aerodynamic or profile techniques, (c) eddy correlation techniques, (d) energy-budget techniques, and (e) empirical (generally book-keeping) techniques (Rosenberg, Hart and Brown, 1968). Mass transport expressions have generally developed from Dalton's (1802) early expression $E = C(e_o - e_a)$ where C is a constant, empirically determined and usually containing a wind speed term, while $e_o$ and $e_a$ are the saturation vapor pressure at the surface and in the air above. Since there is great difficulty in measuring the vapor pressure at an evaporating surface, many variations of the Dalton expression substitute other vapor pressure measurements (such as the saturation vapor pressure at the temperature of the surface, if that can be determined) with corresponding loss of accuracy. Various methods of expressing the effect of wind speed have been introduced into recent mass transport expressions.

Use of aerodynamic or profile techniques requires making certain assumptions concerning the turbulent diffusion of heat and water vapor in the atmosphere. Vertical diffusion of moisture is assumed to be proportional to the product of the height gradient of moisture content and a turbulent-diffusion coefficient. This latter coefficient is a function of the intensity of air turbulence in the surface layer, and assumed to be dependent on the wind speed profile. When the air temperature is not near neutral stability, atmospheric buoyancy influences the diffusion coefficient.

The more recently introduced eddy correlation technique recognizes that upward diffusion of water vapor can only occur if upward moving turbulent eddies are more moist than downward moving eddies. The magnitude of the flux is determined from simultaneous observations of vertical wind speed and moisture content of the air. The average product of these two terms, when multiplied by air density and specific heat, gives the moisture flux due to turbulent transport plus the flux due to average vertical air movement (the mean flux). The difference between these two fluxes is the flux due to turbulent eddies. Sensitive and fast response instruments are required for these measurements so that reliable observations are difficult to obtain.

Energy-budget techniques involve the partitioning of available net radiation, $R_n$ (incoming long- and short-wave radiation minus outgoing reflected and long-wave radiation) into its different categories of use at the earth's surface. Evaporation is determined as a residual of the other measured terms. If the energy used in photosynthesis and other minor exchanges is neglected, the energy budget can be written as

$$R_n = S + H + LE$$

where S is the soil heat flux, H is the atmospheric heat flux, and LE is the energy going into the evaporation of water. When the soil is fully charged with moisture, Bowen (1926) felt that S would be quite small. Given this condition, B = H/LE. Thornthwaite and Mather (1955) suggested that under optimum moisture conditions, $R_n$ would be primarily used for LE so that both S and H could be neglected as a first approximation. The effect of advection of energy was also considered to be negligible in this case.

Penman (1949, 1956) and others have combined energy-budget and aerodynamic approaches into what may be called combination techniques in an effort to eliminate most unmeasured terms. Penman's expression for evapotranspiration from a moist, short green cover ($E_T$) is

$$E_T = \frac{\dfrac{\Delta R_n}{\gamma L} + f(u)(e_a - e_d)}{(\Delta/\gamma) + 1}$$

where $\Delta$ is the slope of the saturation vapor pressure curve vs. temperature (mb/K), $\gamma$ is the psychometric constant* (mb/K), $R_n$ is net radiation (erg/cm² sec), L is latent heat of vaporization for water (cal/g), f(u) is a wind function equal to $0.35(0.5 + u_0/100)$ mm/day over open water, and $e_a - e_d$ is the saturation vapor

---

*The ratio $\Delta/\gamma$ is dimensionless and is essentially a weighting factor used to assess the relative effects of energy supply and ventilation on evaporation.

pressure difference obtained from the air temperature and the dewpoint temperature both taken at screen height. Inasmuch as observations of duration of bright sunshine, mean air temperature, mean vapor pressure, and mean wind speed are required, Penman agrees that his expression is more difficult to evaluate than other, possibly more empirical, expressions. He would argue that the range of possible applications might, however, be much wider since the inclusion of more factors active in the evaporation process makes the expression more valid under a wider range of meteorological conditions. The expression has been widely tested with very satisfactory results. Many later investigators have suggested ways to simplfy the Penman equation with the loss of some degree of validity. The original expression, or some of the later modifications, have proven to be extremely valuable in a number of different types of water budget studies.

Various empirical or bookkeeping methods to determine potential evapotranspiration exist, the most familiar, possibly, being due to Thornthwaite (1948). Thornthwaite's expression for potential evapotranspiration is useful because it is simple to evaluate, requiring only temperature and daylength. It provides reasonably reliable values of monthly evapotranspiration, especially if marked monthly changes in humidity (as in a monsoon climate) do not occur. The Thornthwaite expression is less effective on a daily basis because daily variations in wind speed and humidity are not included in the expression for potential evapotranspiration. It can be widely evaluated, however, due to its reliance only on routinely measured air temperature.

Thornthwaite, fitting data of evaporation from watersheds and irrigation plots to air temperature, obtained the following expression for unadjusted potential evapotranspiration (in cm/month)

$$e = 1.6 \left(\frac{10\,t}{I}\right)^a$$

where t is monthly temperature ($^\circ$C), I is an annual heat index (determined from the sum of the 12 monthly heat index values, $I = \Sigma i$, where $i = (t/5)^{1.514}$ and t is mean monthly temperature; a is a nonlinear function of the heat index equal to

$$a = 6.75 \times 10^{-7}\,I^3 - 7.71 \times 10^{-5}\,I^2 + 1.79 \times 10^{-2}\,I + 0.49$$

These expressions are complicated and mathematically inelegant. They can only be evaluated readily with the use of tables and nomograms (Thornthwaite and Mather, 1957) or computer programs (Willmott, 1977). Unadjusted potential evapotranspiration is the water loss for a 30-day month with each day 12 hr long. This value is adjusted by a factor which expresses how the actual day and month length differ from these values.

Many other expressions for potential evapotranspiration exist, ranging from the very simple assumption that it can be approximated (in mm) by the mean monthly temperature ($^\circ$C) doubled (Gaussen, 1955), or even tripled in some months (Walter, 1955), to the very involved Penman method, already described,

**Table 3.10** Coefficients in Multiple Regression Equation[a] for Calculating Latent Evaporation ($y$) from Meteorological Variables ($x$)[b]

| Method | $a_0$ | Max. Air Temp, °F $a_1$ | Daily Temp. Range, °F $a_2$ | Daily Solar Energy Q, Top of Atmos. ly $a_3$ | Daily Total Solar Sky Rad. ly $a_4$ | Daily Total Wind Run, Miles $a_5$ | Daily Mean Vapor Pressure Deficit, mb. $a_6$ | Multiple Regression $R^c$ |
|---|---|---|---|---|---|---|---|---|
| I | −87.03 | $9.28 \times 10^{-1}$ | $9.33 \times 10^{-1}$ | $4.86 \times 10^{-2}$ | 0.00 | 0.00 | 0.00 | 0.67 |
| II | −55.60 | $6.87 \times 10^{-1}$ | $2.84 \times 10^{-1}$ | $9.13 \times 10^{-3}$ | $6.85 \times 10^{-2}$ | 0.00 | 0.00 | 0.74 |
| III | −42.28 | $-2.28 \times 10^{-2}$ | 1.09 | $5.06 \times 10^{-2}$ | 0.00 | 0.00 | 2.99 | 0.81 |
| VI | −108.80 | 1.13 | $9.20 \times 10^{-1}$ | $3.59 \times 10^{-2}$ | 0.00 | $1.31 \times 10^{-1}$ | 0.00 | 0.72 |
| V | −26.69 | $-2.32 \times 10^{-2}$ | $5.57 \times 10^{-1}$ | $1.96 \times 10^{-2}$ | $5.31 \times 10^{-2}$ | 0.00 | 2.41 | 0.83 |
| VI | −78.68 | $8.97 \times 10^{-1}$ | $3.40 \times 10^{-1}$ | $1.66 \times 10^{-3}$ | $6.13 \times 10^{-2}$ | $1.18 \times 10^{-1}$ | 0.00 | 0.80 |
| VII | −69.30 | $3.50 \times 10^{-1}$ | 1.04 | $4.03 \times 10^{-2}$ | 0.00 | $1.01 \times 10^{-1}$ | 2.31 | 0.82 |
| VIII | −53.39 | $3.37 \times 10^{-1}$ | $5.31 \times 10^{-1}$ | $1.07 \times 10^{-2}$ | $5.12 \times 10^{-2}$ | $9.77 \times 10^{-2}$ | 1.77 | 0.86 |

[a] Regression equation $Y = a_0 + a_1x_1 + a_2x_2 + a_3x_3 + a_4x_4 + a_5x_5 + a_6x_6$.

[b] Latent evaporation (cc) $\times$ .0034 (ins./cc) = $PE$ (ins.); temperatures in degrees F; energy in ly per day; and vapor pressure deficit $e_w - e_s$ in mbs.

[c] $R$ = Multiple regression coefficient for 900 test cases; $R$ = 0.40 for $P$ = 0.01 and six independent variables.

*Source.* Baier (1967) with permission of the *Annals of Arid Zone*.

or one by Baier and Robertson (1967) that involves a multiple regression equation of the form

$$Y = a_0 + a_1 x_1 + a_2 x_2 + a_3 x_3 + a_4 x_4 + a_5 x_5 + a_6 x_6$$

where Y is the latent evaporation in cc which can be converted into inches of potential evapotranspiration by multiplying by .0034.

Baier (1967) has suggested eight possible variations of this expression using different combinations of available data, as can be seen by the list of $a$ coefficients included in Table 3.10. Multiple regression coefficients based on 900 test cases ranged from R = 0.67 to R = 0.86 using different combinations of the six variables (maximum air temperature in $°F$; temperature range; daily solar energy at the top of the atmosphere, $Q_0$ in 1y; daily total sky and solar energy on a horizontal surface estimated from $Q_s = Q_0 (0.251 + 0.616 \, n/N)$ where n is daily bright sunshine and N is duration of daylight, both in hours; daily totaɪ wind run in miles; and daily mean vapor pressure deficit in mb). The coefficients in Table 3.10 were derived by use of Bellani plate atmometer readings of latent evaporation at six stations across Canada during summers from 1953 to 1957.

In selecting a method for estimating potential evapotranspiration, investigators must decide between ease and simplicity of evaluation, on the one hand, and increased accuracy through the inclusion of more factors that are related to evapotranspiration, on the other. The methods of Penman and Baier, for example, appear to provide reliable values of daily or monthly evapotranspiration under a wide range of environmental conditions. But they require considerable data and so are neither easy to use nor of widespread applicability. Estimates are often used for some of the generally unmeasured terms but, when this is done, some of the advantages of using the more complex methods are lost. The simple Thornthwaite expression can be readily applied worldwide, and while it may be less accurate on a monthly basis (and surely is less accurate on a daily basis), its wide usefulness often outweighs its inability to reflect short-period changes in wind or humidity. Although many studies have sought to demonstrate the advantages of one method over another, there is still no universally agreed upon method for computing potential evapotranspiration.

### Distribution of Potential Evapotranspiration

Thornthwaite (1948) has provided a map of the distribution of potential evapotranspiration in the conterminous United States (Fig. 3.8) which differs appreciably from earlier maps based on annual Class A pan evaporation or lake evaporation. The differences can, in large measure, be explained by the relation that exists between rate of water loss with varying humidity conditions and size of evaporation areas (Fig. 3.9). Small evaporation surfaces (Class A pans, small lakes) are strongly influenced by the moisture conditions of the air passing over them. A continous supply of dry air will result in large values of evaporation from such finite evaporating surfaces. As the air becomes more moist, or as the evaporating

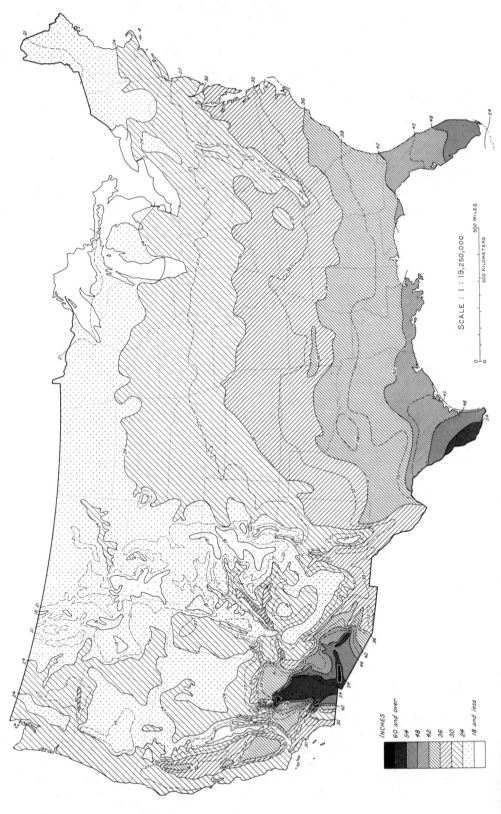

SCALE : 1 : 19,250,000

500 MILES

500 KILOMETERS

INCHES

60 and over
54
48
42
36
30
24
18 and less

72

**Figure 3.8** Average annual potential evapotranspiration over the United States.
*Source.* Reprinted from Thornthwaite (1948), with permission of the American Geographical Society of New York.

surface becomes larger so that the air blowing over such a surface rapidly becomes moist before passing over most of the evaporating surface, the size of evaporating surface has less influence on rate of evaporation. Potential evapotranspiration is almost everywhere less than the values determined by pan or lake evaporation measurements because the influence of air with low to moderate relative humidity has been eliminated.

While numerous studies of the different expressions for potential evapotranspiration have appeared and differences of opinion have occurred concerning which expression to use in water budget analysis, there seems to be little recognition of the fact that the most significant error in water budget input terms may be in precipitation rather than in potential evapotranspiration. While errors of 20% may occur in values of evapotranspiration, such errors are commonly found in precipitation observations due to wind effects, exposure problems, interpolation from point observations, and other instrumental problems. In any evaluation of the water budget, one must look carefully at the problems and limitations inherent in precipitation data before uncritically accepting these data as "correct" and attributing all inconsistencies to evapotranspiration.

## SOIL MOISTURE

A third major factor in the water budget is the movement and storage of moisture in the soil. Not only does the storage capacity of the soil play a role in modifying

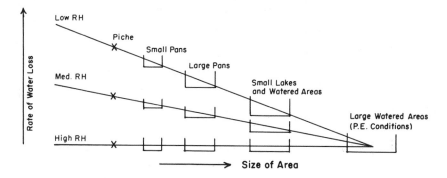

**Figure 3.9** Schematic relation between rate of water loss under different conditions of relative humidity (RH) and evaporation surface areas.
*Source.* Thornthwaite and Mather (1955).

**Table 3.11** Guide for Judging How Much of the Available Moisture Has Been Removed from the Soil

| Soil Moisture Deficiency (%) | Feel or Appearance of Soil and Moisture Deficiency in cm of Water per m of Soil | | | |
| --- | --- | --- | --- | --- |
| | Coarse Texture | Moderately Coarse Texture | Medium Texture | Fine and Very Fine Texture |
| 0 (Field capacity) | Upon squeezing, no free water appears on soil but wet outline of ball is left on hand 0.0 | Upon squeezing, no free water appears on soil but wet outline of ball is left on hand 0.0 | Upon squeezing, no free water appears on soil but wet outline of ball is left on hand 0.0 | Upon squeezing, no free water appears on soil but wet outline of ball is left on hand 0.0 |
| 0–25 | Tends to stick together slightly, sometimes forms a very weak ball under pressure. 0.0 to 1.7 | Forms weak ball, breaks easily, will not slick. 0.0 to 3.4 | Forms a ball, is very pliable, slicks readily if relatively high in clay. 0.0 to 4.2 | Easily ribbons out between fingers, has slick feeling. 0.0 to 5.0 |
| 25–50 | Appears to be dry, will not form a ball with pressure. 1.7 to 4.2 | Tends to ball under pressure but seldom holds together. 3.4 to 6.7 | Forms a ball somewhat plastic, will sometimes slick slightly with pressure. 4.2 to 8.3 | Forms a ball, ribbons out between thumb and forefinger. 5.0 to 10.0 |
| 50–75 | Appears to be dry, will not form a ball with pressure. 4.2 to 6.7 | Appears to be dry, will not form a ball.[a] 6.7 to 10.0 | Somewhat crumbly but holds together from pressure. 8.3 to 12.5 | Somewhat pliable, will ball under pressure 10.0 to 15.8 |
| 75–100 (100% is permanent wilting) | Dry, loose single grained, flows through fingers. 6.7 to 8.3 | Dry, loose, flows through fingers. 10.0 to 12.5 | Powdery dry, sometimes slightly crusted but easily broken down into powdery condition. 12.5 to 16.7 | Hard baked, cracked, sometimes has loose crumbs on surface. 15.8 to 20.8 |

[a]Ball is formed by squeezing a handful of soil very firmly.

*Source.* Hansen et al. (1980) with permission of John Wiley and Sons.

the amount of water from a given rainstorm that will find its way as percolation to the groundwater table or as throughflow (interflow) to the surface stream, but the antecedent soil moisture conditions, which express the state of soil moisture storage at some designated time previous to the time under consideration, are also related to the potential for flooding from a given storm. Because of the significance of soil moisture information, there has been a marked interest in instrumental techniques to measure this quantity. Nevertheless, determination of soil moisture content is still far from an exact or simple process.

## Evaluation of Soil Moisture Content

Most instruments to measure soil moisture are limited in one way or another. For instance, soil moisture tensiometers usually cease to function when the capillary tension in the soil is greater than one atmosphere. Electrical methods, gravimetric plugs, or absorption blocks are generally not too sensitive at very high moisture contents. The presence of electrolytes in the soil can also lead to erroneous determinations of soil moisture with certain instruments. For accurate moisture measurements, good contact between soil and the sensor must be maintained. This is quite difficult in some cases, especially in soils which undergo large volume changes upon wetting or drying.

Soil moisture content, as measured instrumentally or by weighing and drying a soil sample, is representative only of the area from which the sample or measurement is taken. A short distance away, changes in topography or soil character can yield a far different soil moisture value. For example, Staple and Lehane (1962), comparing observations from some 64 locations on Wood Mountain clay loam soil at Swift Current, Canada, taken on stubble fields at seedtime each year for 7 years, found the standard deviation of total moisture content in 3- or 4-ft profiles to vary from 6.9 to 16.3% in different years. The standard deviation averaged about 10%, or 0.9 in. in the profiles sampled. Taylor (1955) found a sample standard error of almost 20% of the mean moisture content in replicated measurements of soil moisture content using three different measurement techniques—resistance blocks, neutron loggers, and gravimetric observations. Thus, to determine moisture content in a field of just a few acres requires averaging a large number of observations. In those few places where a sufficient number of soil moisture observations are being made, the length of these records is often not sufficient to permit their use in studies of the probability of occurrence of different soil moisture contents through the year. For long-range planning, such probability studies are essential.

Lacking instruments to provide actual values of soil moisture content, farmers and engineers have developed indirect techniques to estimate the degree of wetness of a soil (Table 3.11).

To determine the soil water content in an area, a large number of climate, soil, slope, and vegetation factors need to be considered for each level and sloping piece of land (Fig. 3.10). On the ridge or higher-lying land, water is added to the soil only by precipitation (P) and snowmelt (Sm). The precipitation that

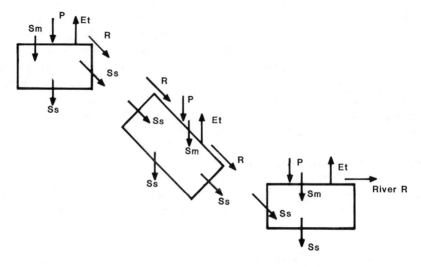

**Figure 3.10** Water additions and removals in the three major topographic sections, the ridge, slope and valley bottom. Soil moisture storage in each section represents a balance between all water inflows and all water outflows (letters identified in text).

falls at rates greater than the infiltration capacity of the soil will run off overland (R) to lower-lying areas. Water can be removed from the ridge area by overland flow (R), subsurface flow (Ss), evapotranspiration (Et) from the vegetation cover or evaporation from the soil or ponded water.

On sloping land, water is also added and lost in the same way as on the ridge land. In addition, water may be added by surface and subsurface flow from higher-lying areas. In the generally level valley or bottom land, water is added to the soil by means of the same methods as on the slope. Water is lost from valley areas, however, mainly by streamflow, deep percolation, and transpiration and evaporation from the soil or ponded water. In level bottom land, even though some over-land and subsurface flow may occur, it is less important than in the other two topographic sections unless a surface stream drains the area. Movement of water in vapor form in the soil has not been considered in this discussion.

A number of meteorologic, pedologic, and botanic factors must be evaluated in order to solve the soil moisture budget for each topographic section. Exact determination of the moisture content at a place requires information on amount and intensity of precipitation, slope, permeability, and infiltration capacity of the soil and its water-holding capacity, depth to any impeding stratum, and rate of transpiration from the vegetation cover. Since some of the needed information is not available in usable form for large portions of the earth, precise evaluations of the soil moisture balance are seldom possible. Expressing some of the unknowns in terms of known meteorologic variables does permit estimates of the balance for many areas, however,

## Infiltration

Infiltration is the process by which water enters the surface of the soil and percolation is the process of movement of water through the soil to the groundwater table or to nearby surface streams. Garstka (1978) has listed the following 10 factors as affecting infiltration: (a) soil texture, (b) soil structure, (c) initial soil moisture content, (d) profile of soil moisture in the root zone, (e) turbidity of the infiltrating water, (f) alkalinity of the soil, (g) soil and water temperature, including the possible occurrence of frost, (h) rainfall intensity and duration, (i) hydrophobic coatings on soil particles, and (j) air trapped in the soil column. He pointed out that many of these factors are interrelated. Land-use treatment can exert a significant influence on infiltration in the basin under average conditions although variations in land-use conditions seem to have less effect under extreme rainfall events.

The infiltration capacity of a soil is the maximum rate at which water can penetrate into the upper layer of a soil. Infiltration rates vary from soil to soil, being greater for more open sandy soils than for tighter clay soils. Wiesner (1970), for example, has shown that infiltration rates for course sand with a good structure range from 0.75 to 1.00 in./hr while they approach 0.50 in./hr for fine sandy loams and to 0.30 in./hr for clay loams. With poor soil structure, the corresponding values are 0.50, 0.30, 0.25 in./hr, respectively, for coarse sand, find sandy loam, and for clay loam.

While cultivation of soil may initially increase infiltration rates, a later heavy application of water will consolidate the surface soil, which will result in a lower infiltration rate. Vegetation, of course, will protect the surface from being sealed by rain splash. Wiesner (1970) suggests that infiltration rates will increase from 0.40 in./hr on bare crusted ground to 1.2 in./hr on a heavily grazed permanent pasture to 2.0 in./hr on a lightly grazed pasture. A good quality grass cover will increase infiltration rates by a factor of two over a poor stand of row crops. Denser stands of vegetation or more luxuriant vegetation will provide higher rates of infiltration than will less vigorous or more open stands (Wiesner, 1970).

Alderfer and Bramble (1942) reported some work on the effect of the ecosystem on infiltration rate (Table 3.12). A forest floor, even if it has lost its litter and

Table 3.12  Effect of Ecosystem on Infiltration Rate

| Ecosystem | Capacity (mm hr$^{-1}$) |
| --- | --- |
| Undisturbed forest floor | 60 |
| Forest floor without litter and humus layers | 49 |
| Forest burned annually | 40 |
| Pasture, unimproved | 24 |

*Source.* Alderfer and Bramble (1942).

**Table 3.13**  Alphabetical Listing of Some Major Soils of the North Central Region of the United States and Their Mean Equilibrium Infiltration Rates, as Measured with a Sprinkling Infiltrometer

| | Rate (in/hr) | | |
| --- | --- | --- | --- |
| Soil | Tilled Surface | Grass Surface | Test Location |
| Bearden silty clay loam | .2 | .9 | N. Dakota |
| Blount silty clay loam | .3 | .3 | Ohio |
| Bodenburg silt loam | .4 | .4 | Alaska |
| Canfield silt loam | .4 | 1.1 | Ohio |
| Cincinnati silt loam | 1.1 | 1.1 | Indiana |
| Cisne silt loam | .6 | .6 | Illinois |
| Clermont silt loam | .8 | .8 | Indiana |
| Dudley clay loam | .4 | 1.0 | S. Dakota |
| Elliott silt loam | 1.5 | 1.5 | Illinois |
| Emmet loamy sand | 1.5 | 1.5 | Michigan |
| Fayette silt loam | .2 | .6 | Wisconsin |
| Flanagan silt loam | 1.9 | 3.6 | Illinois |
| Grundy silt loam | 1.1 | 1.1 | Iowa |
| Holdrege silt loam | 1.2 | 1.2 | Nebraska |
| Hoytville clay | .3 | 1.0 | Ohio |
| Ida silt loam | .7 | 1.5 | Iowa |
| Keith very fine sandy loam | .7 | 1.5 | Nebraska |
| Kenyon silt loam | 1.3 | 1.3 | Minnesota |
| Knik silt loam | .4 | .4 | Alaska |
| Kranzburg silty clay loam | .8 | 1.0 | S. Dakota |
| Miami sandy loam | 1.2 | 1.2 | Michigan |
| Minto silt loam | .4 | .4 | Alaska |
| Moody silt loam | .4 | 1.4 | Iowa |
| Morton loam | .4 | 1.0 | N. Dakota |
| Plainfield loamy sand | 3.7 | 4.4 | Wisconsin |
| Port Byron silt loam | .5 | 2.0 | Minnesota |
| Russell silt loam | .5 | .5 | Indiana |
| Sharpsburg silty clay loam | 1.3 | 1.3 | Nebraska |
| Sims loam | 1.1 | 2.5 | Michigan |
| Sinai silty clay loam | .6 | 1.1 | S. Dakota |
| Webster clay loam | 1.0 | .6 | Minnesota |
| Withee silt loam | .2 | 1.1 | Wisconsin |

*Source.* North Central Regional Committee 40 (1979).

humus layer or been burned over annually, still maintains a fairly high infiltration rate (approaching 2 in./hr). Surface runoff from forested areas is fairly uncommon because of these high values of infiltration. Marked changes in infiltration rate accompany a vegetation succession from pasture to forest.

The North Central Region Committee 40 (1979) has provided experimental results of infiltration rates on tilled and grassed surfaces for a wide range of soils in the North Central region of the United States (Table 3.13). Loamy soils generally show a much higher infiltration rate than do clay soils, but the variation from one silt loam, to another, for example, can be as great as from 0.4 in./hr up to 3.6 in./hr. It is not easy to generalize by soil texture alone.

Infiltration rates decrease as soils become wetter, approaching a lower steady state as infiltration continues. Final infiltration rates differ with soil types, ranging from greater than 20 mm/hr in sand soils to less than 1 mm/hr in heavy clay soils. Vegetation cover and land use are also important in determining infiltration rate, its change with time, as well as the total amount of infiltration over time (Fig. 3.11). Lull and Reinhart (1972) have reported infiltration rates as high as 50 in./hr in hardwood stands in West Virginia. They conclude that forest soils not only have infiltration rates that will exceed essentially all rainfall rates but that they also have the ability to absorb additional overland flow from nearby areas with lower infiltration rates.

In a study of the infiltration of snowmelt and precipitation into different natural and cultivated areas in the forest steppe zone in Russia during the spring breakup period, Grin (1967) found a significant difference between forested areas and fields of winter wheat or stubble. Fall plowing evidently increases infiltration rate and this has been a common practice among Russian farmers for more than 40 years. L'vovich (1973) reports that deep fall plowing has resulted in a significant increase in annual soil moisture storage in the past three decades.

## Movement of Water in Soil

Capillary suction which exists on the menisci of water films in contact with soil particles is one of the most effective mechanisms by which water moves into dry soils (Carson, 1969). The magnitude of the suction force is a function of the radius of the meniscus as well as of the surface tension of water. Menisci in the soil pores have a similar suction force. Interestingly, quite similar results are obtained in evaluating the upward movement of water into a dry soil from a groundwater table or the downward infiltration of water from a surface water supply. In each case, if the soil is initially dry, a wetting front moves vertically away from the water supply. At the wetting front, there is an abrupt change from wet to dry in the soil; behind the front, the moisture content of the soil shows little change with distance from the front. After infiltration ceases, there is a slow redistribution of water in the upper soil layers below the initial wetting front. Some water drains through the initial wetting front under the force of gravity until the soil mass behind the front is only holding moisture in the capillaries between soil grains. Gardner, Hillel and Benyamini (1970) have shown that

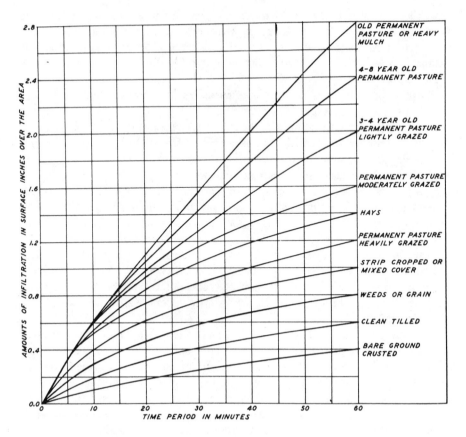

**Figure 3.11** Curves of accumulated water infiltrated into Piedmont soils under different types of vegetation.
*Source.* Holtan and Kirkpatrick (1950) copyrighted by the American Geophysical Union.

drainage past the wetting front can continue for a matter of days after the initial infiltration and movement of the wetting front ceases (Figs. 3.12, 3.13). If the water table is close enough or if large pores or cracks exist in the rock layers, this drainage might ultimately reach the groundwater table.

Field capacity is a term that is frequently used in relation to soil moisture problems even though it is poorly defined. It represents an effort to identify an arbitrary point on a drainage curve marking when rapid drainage from the particular soil layer has ceased. Hewlett and Nutter (1969) have pointed out that it is not a unique value for all soils or even for a single soil at different times because of the influence of hydraulic conductivity on rate of drainage. Gardner (1968) has discussed the concept of field capacity as marking the upper limit of available water. He carefully pointed out problems that can exist when field capacity is defined as (a) the soil water content when downward drainage becomes

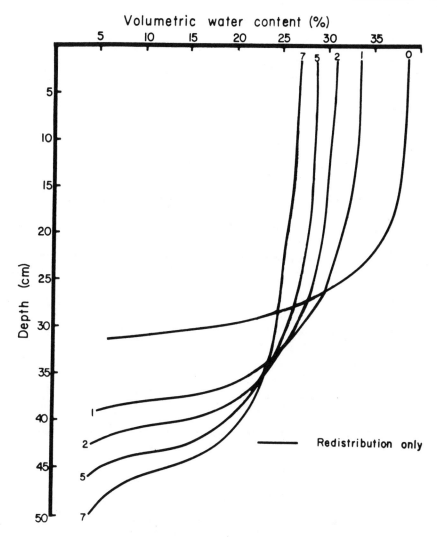

**Figure 3.12** Successive moisture profiles of soil columns following an irrigation of 100 mm. Profiles are designated according to number of days after irrigation.
*Source.* Gardner, Hillel and Benyamini (1970), copyrighted by the American Geophysical Union.

negligible, (b) the moisture content in the soil at a tension of one-third atmosphere (333 millibars), or (c) the water content of the soil 2 to 3 days after a soaking rain. While all three of these definitions have been applied to field capacity, the user must realize that the actual soil moisture content at that point is only approximate, varying with capillary conductivity or soil water diffusivity. It is still a very useful concept for soil moisture work.

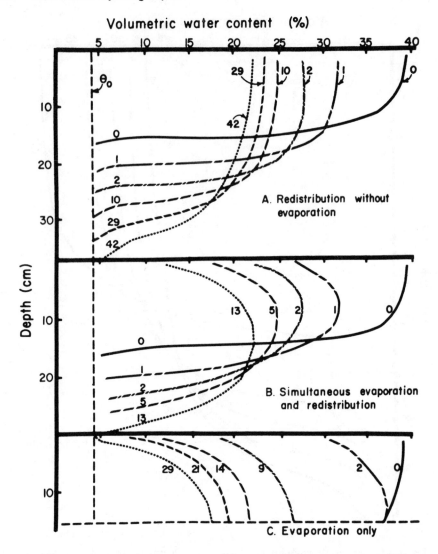

Figure 3.13 Successive moisture profiles of soil columns following an irrigation of 50 mm. Profiles are designated according to number of days after irrigation.
*Source.* Gardner, Hillel and Benyamini (1970), copyrighted by the American Geophysical Union.

Soils differ appreciably in the amount of water that is actually available in the capillaries at a tension of one-third atmosphere (used in this case to define field capacity) because of large variations in the total quantity of capillaries available for storage. Clays have many times the number of capillaries that sands have. Thus, clays may have a moisture content between 30 and 40% by volume at field

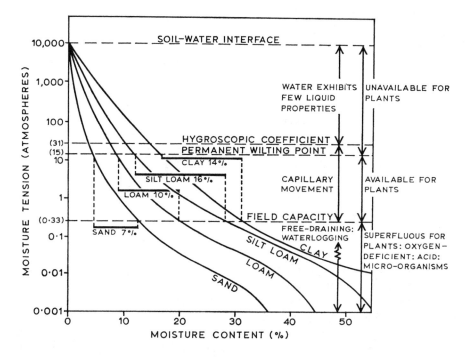

**Figure 3.14** The relationship between moisture content and moisture tension in four different types of soil, indicating the range of moisture available for plants.
*Source.* More (1969) with permission of Methuen and Co., after Buckman and Brady (1960), copyrighted by Macmillan Publishing Co. Inc.

capacity, loams 20 to 30% by volume, and sands only 10 to 15%. The relationship between moisture content and moisture tension for four different types of soils is shown in Figure 3.14.

Evaporation as well as plant roots will be able to remove water from the capillaries. However, as the soils dry, it becomes increasingly difficult for water removal to continue for the remaining water is held in the capillaries with increasing tension. Finally, at about 15 atmospheres tension, plants can no longer remove water from the soil and a point of permanent wilting is achieved. Plants will die if other water is not made available at this time.

The amount of water held in the soil between the field capacity and the permanent wilting point is called the available water inasmuch as this is the amount of water the plants will have available to them from soil moisture storage. Because of small pore spaces and the large number of capillaries, tension in the capillaries of clay soils will reach 15 atmospheres (permanent wilting) at a moisture content of 15 to 20% by volume, while in sands the moisture content may drop to close to 5% by volume before permanent wilting is reached. Thus, the percentage of available water varies from about 14% in clays to 16% in silt loams, to 10%

**Table 3.14**  Variation of Permeability and Soil Type with Depth, in an Ohio Forest Soil at Gradient of 15°

| Soil Depth (cm) | Texture | Saturated Permeability (mm/hr) |
|---|---|---|
| 0–56 | Sandy loam | — |
| 56–90 | Sandy loam | 286 |
| 90–120 | Loam | 17 |
| 120–150 | Clay loam | 2 |

*Source.* Whipkey (1965) with permission of the International Association of Hydrological Sciences.

in loams and only 7% in sands. Sandy soils not only hold less water at field capacity but also dry more rapidly to the wilting point than do loams or clays.

Permeability, which expresses the ease with which water can move through earth materials, is often measured in terms of discharge per unit area under specified hydraulic conditions. Permeability of unconsolidated clays might range from 0.00001 to 0.001 gal/day per square foot while for alluvial sands it may vary from 10 to 10,000 gal/per day per square foot. As soils are modified and become more compacted with depth, permeability changes, often becoming considerably less with depth through the upper soil layers (Table 3.14).

## RUNOFF

Runoff, the fourth of the basic factors of the water budget, is often considered as a residual after evapotranspiration has been removed from the available precipitation. Nevertheless, runoff is often the most important of the water budget factors since it represents a form of surplus water, a resource that can be utilized by humans for new industrial or agricultural development, for power or transportation, for recreation, or for increased domestic use. It represents a kind of climatic potential available for our additional needs—but only if it is not rendered unfit through our misuse.

### The Hydrograph

A hydrograph is a diagram showing, for a given point on a stream or in a drainage system, the discharge, stage, velocity, or other property of water with respect to time. A storm hydrograph relates stream discharge conditions to the available supply of water by precipitation in a particular storm. The hydrograph expresses the sequence of relationships that occur over time between precipitation and the various basin environmental factors such as slope, soil characteristics, vegetation cover, infiltration rate, permeability, and distance. Figure 3.15 is a hypothetical

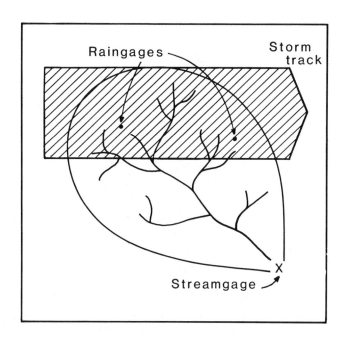

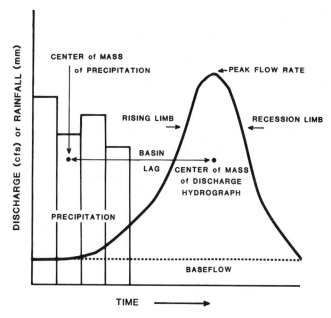

**Figure 3.15** Hypothetical storm and resultant stream hydrograph illustrating significant concepts.

storm hydrograph for a basin to illustrate some of the basic terms and concepts used in hydrograph analysis.

In humid regions, most streams are fed by water from soil and groundwater storage as well as from overland flow. They are effluent streams and their levels represent the level of the water table in the vicinity. When there has been a long dry period, water tables fall and with them the volume of water in the streams. The fairly constant low-level flow found in the streams at this time is called base-flow or sustained flow. After an initial period of rain, there is usually a fairly rapid increase in discharge to the stream until a peak or crest is reached followed by a more slowly declining recession or decrease in the rate of discharge. The basin lag is the time between the center of mass of the precipitation and the center of mass of the discharge hydrograph. The lag is mainly a function of two basin characteristics, the basin slope and the basin length. Lag time can be greatly altered by human action in the basin. The peak discharge usually occurs shortly after rainfall has ceased; the flow after that time is largely determined by the water in storage in the basin and various basin characteristics. Additional rain during the recession period may result in a second peak in discharge before the flow slowly declines to its original baseflow rate.

This generalized description has considered what happens in the case of effluent streams rather than with so-called influent streams that lose water to the water table as they cross dry areas since the water table is located well below the surface. Such streams are fed by melting snow or excessive precipitation falling in certain headwater or source regions; they respond to the availability of water from these more distant source regions. Scattered showers along their courses may add water locally as a result of overland flow of intense precipitation but the amounts of water involved are generally small in the overall budget of the stream.

## Overland Flow and Throughflow

In this section, we are concerned almost entirely with the runoff of water that finds its way to the surface water courses within a few hours or days either overland or through the upper soil and rock layers of the earth's crust. Overland flow is the surface flow that occurs when rainfall intensity exceeds the rate of infiltration of water into the soil. Water which infiltrates the upper soil layers and then moves downhill through the upper soil layers is known as throughflow (or interflow). We need to recognize the contributions of both overland flow and throughflow to the stream regime.

Because of the influence of vegetation cover on infiltration rate, significant overland flow only rarely occurs under humid conditions although it can be fairly common in semiarid areas. Where conditions for overland flow do exist, typical velocities are of the order of 200 to 300 m/hr. With such flow, the rainfall from all parts of small basins will rapidly reach the stream course.

Water that infiltrates into the upper soil layer first fills the soil capillaries before moving downward to recharge the groundwater table or downslope within the upper soil layers as throughflow to the surface water course. Often the greater part of the soil water appears to reach the stream course by means of through-flow rather than by discharge from the groundwater table but the volumes involved are rarely distinguishable.

As moisture moves downward through the soil into regions of decreasing permeability, less water will contribute to deep percolation while more water will be deflected laterally and contribute to downslope throughflow. With continued rain, the upper soil layers become saturated (the more permeable upper soil layers fill from below because the less permeable lower layers are unable to remove the water rapidly enough by throughflow or percolation) and saturated overland flow can occur. This can occur at much lower rainfall rates than will be necessary to produce overland flow but it will only occur after a considerable period of rainfall which has saturated the upper soil layers.

Rates of water movement as throughflow are many times slower than overland flow velocities (possibly as much as 20 to 30 cm/hr as opposed to 200 to 300 m/hr for overland flow) so that the time for throughflow from the more distant parts of even a small basin to reach the water course can be days or even weeks. Much of the baseflow to the surface stream comes from throughflow in the soil. The remainder of the baseflow will come from groundwater table discharge.

Whipkey (1965) has provided a series of discharge hydrographs from different depths within the soil on the basis of a simulated 2-hr rainstorm producing 5.1 cm/hr of water on a 16° slope (Fig. 3.16). The hydrograph shows a negligible contribution from overland flow with very rapid and peaked outflow from the upper soil layers. Deeper soil layers with lower permeability contribute small and constant amounts of water over a much longer period of time.

The actual area of a basin contributing to the stream discharge in a particular storm period will vary appreciably depending on the relative proportion of overland flow, throughflow, and baseflow. Where vegetation is essentially absent and soils are thin, overland flow is dominant and essentially all of the basin area may contribute to streamflow from a particular storm. With well-vegetated, deep soils, overland flow is absent and throughflow is the major contributor to streamflow. In those cases, relatively small portions of a basin may actually contribute the bulk of the water to the hydrograph of a particular storm.

## Effect of Basin Topography and Geology

Topography or surface configuration and basin size and shape influence runoff through the factor of "concentration time," the length of time required for runoff to travel overland from the hydraulically most distant part of the basin to the basin outlet or point of reference. With steep slopes, concentration time is reduced and opportunities for flash flooding as a result of intense precipitation increase. Concentration time is usually longer in broad, flat basins than in narrow, steep ones or longer in narrow basins than in more pear-shaped basins. The Soil Conservation Service (Snider, 1972) provides an excellent illustration (Fig. 3.17) of the effect of basin shape on the basin hydrograph.

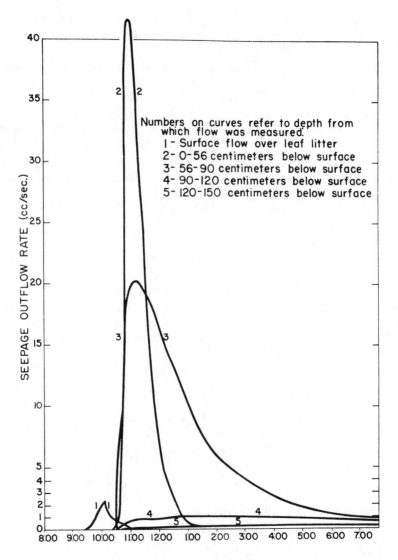

**Figure 3.16** Discharge hydrographs of flow within the soil resulting from a simulated storm of 5.1 cm/hr, lasting 2 hr, on a 16° slope which had previously drained for more than 4 days. The rapid, although small, overland flow results from the initially low permeability of the dry surface soil, which rapidly increases with wetting. The lag before throughflow begins is the time taken for rain to infiltrate vertically to the 90-cm deep, less-permeable interface. *Source.* Whipkey (1965), with permission of the International Association of Hydrological Sciences.

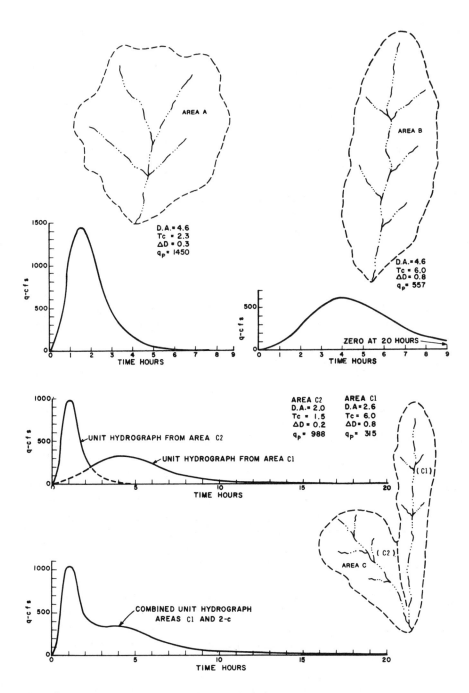

**Figure 3.17** The effect of watershed shape on the peaks of unit hydrographs. *Source.* Snider (1972).

Geology of the stream basin is a very significant factor affecting stream runoff. The physical character of both the underlying rocks and the weathered material is important in determining the rate of movement of water through the upper soil and rock layers. It is also useful in determining storage characteristics and yield of a particular strata. In humid areas, discharge of streams draining areas with permeable sandy soils underlain with gravels or sandstones should be fairly well sustained. Such streams seldom go dry and the range in discharge values from highest to lowest is relatively small. Streams draining areas with clays, shales, or unfractured igneous materials through which water moves rather slowly will be marked with more flash floods, with higher peak flows during rainfall periods and with lower flows during dry periods than will the sandy basins. The range of discharge between wet and dry conditions will be appreciably larger in clay, shale, or unfractured material than in sandy basins, all other factors being equal.

Streams draining areas with unconsolidated materials, especially in arid areas, will carry a great deal of sediment and so make stream gaging difficult; streams draining formations with more consolidated rocks, especially in humid areas, should have small volumes of sediment. Geology will also affect the quality of water depending on the nature of the materials through which the water percolates on its way to the stream. This aspect is beyond our present concern for runoff in the water budget.

### Determination of Streamflow

We have long been concerned with runoff of water because of our dependence on a reliable supply for human use and because of the destructive ability of severe floods. Old Egyptian and Chinese scrolls and tablets abound with references to floods and to irrigation works; several of the laws in the Code of Hammurabi (circa 1760 BC) dealt with problems of control of irrigation supplies. Yet, actual measurements of streamflow are a relatively recent development. Hero of Alexandria, in the 2nd century BC, suggested how cross-sectional area, velocity, and time were all necessary to compute flow rate from a spring. However, his early instructions on how to carry out these measurements were difficult to apply in practice. Problems existed with the measurement of both cross-sectional area and flow velocity. Determining the movement of leaves or chips in the water was a poor substitute for flow rates since the object did not always move with the speed of the surface water and the surface water velocity was not necessarily representative of the average flow in the stream. Though French hydrologists began measuring streamflows and discharges on a regular basis in the later 1600s, it was not until 1732 that Pitot developed an instrument that could measure the velocity of the flow at different places in a river section with some reliability.

The cross-sectional area of a stream is not easily determined, especially in streams that carry considerable sediment and are constantly changing bottom configuration. Streams are seldom constant in height or stage so that cross-sectional area changes with rising or falling of the stream level, as does velocity. Cross-sectional area is normally taken at a place where changes are minimized

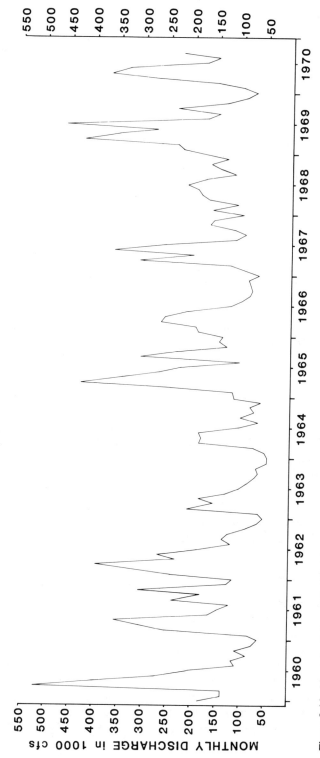

**Figure 3.18** Monthly discharge (in 1000 cfs) of the Mississippi River at St. Louis, 1960–1970. *Source.* Based on data from the U.S. Geological Survey Water Supply Papers.

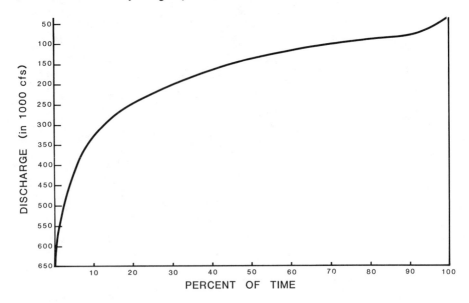

**Figure 3.19** Gage-height-duration curve, Market Street Gage, St. Louis. Curve shows percent of time indicated gage height was equaled or exceeded.
*Source.* Based on daily stages recorded from 1861 to 1960 by U.S. Geological Survey in their Water Supply Papers.

or controlled in every possible way while velocity is measured by a current meter exposed at different depths and in different parts of the particular cross-sectional area and averaged. In view of these difficulties, few reliable records exceeding 70 years in length are available in the United States. Most records are well short of that length. Even those long records that are available often have gaps when instruments failed, gages were blocked by ice, or flooding conditions made it impossible to take observations or destroyed their reliability. Review of discharge measurements from stream gaging stations listed in any Water-Supply Paper of the U.S. Geological Survey raises many questions for which answers are not possible. The brief remarks accompanying the discharge data point out times when observations were less reliable than at other times, when no records were taken, or when estimates had to be made and why. These comments try to provide useful information for interpreting the record but frequently raise more questions than they answer.

The records of streamflow are often of two types: (a) stage or elevation of the water level and (b) river discharge or volume of flow per unit time. Discharge can be determined, of course, from data on stage and flow velocity provided the cross-sectional area of the stream is known and constant over time. Thus, many streams have rating curves which allow conversion from one set of records to the other. Stage and discharge records are used for different purposes. Stage or elevation data provide information on the height of the water in the river channel.

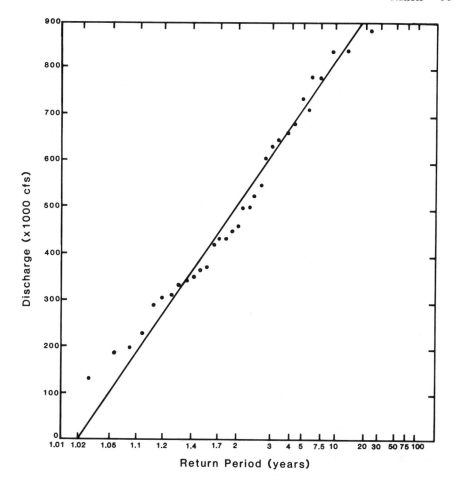

**Figure 3.20**  Recurrence interval for annual flood peaks, Mississippi River at St. Louis, Missouri, 1927–1958.
*Source.* Based on data from the U.S. Geological Survey Water Supply Papers.

Such data are directly applicable to the occurrence of floods or of low water conditions. Discharge or flow volume per unit time is concerned with the amount of water flowing in the channel. These values are useful in determining water supply or the ability of the stream to dilute waste material.

Hydrographs showing the values of stage or discharge over time (Fig. 3.18) do not provide information on the number of consecutive days that river height or volume would be above or below certain values or the frequency of occurrence of certain limiting values. Such information must be obtained from a gage-height-duration curve or a discharge-duration curve (Fig. 3.19). This type of graph summarizes data obtained over a long period of record and provides information on the likelihood or percentage of time the river can be expected to be above or below certain height levels or different values of discharge. These diagrams are useful for planning purposes because they provide statistical information that can be used to determine the likelihood of different water levels or the opportunities for flooding of different amounts provided the river is not undergoing changes in discharge or stage as a result of human activity. That information still does not indicate the number of consecutive days such conditions might last, the recurrence interval between floods, or many other pieces of hydrologic information needed by hydrologists.

Various statistical techniques have been developed to evaluate the frequency and recurrence intervals of flood events. These usually involve plotting the highest flood flow each year for a number of years of record on specially prepared graph paper that automatically gives the information on the recurrence interval between flows of particular amounts or levels of certain heights (Fig. 3.20). These techniques provide sound information for planning purposes only if all limiting conditions are understood. Chief among these is the fact that available records are usually quite short. Also, human activity in the stream basin may rapidly and seriously modify flood peaks so that information obtained from past records may have no future validity. It must also be recognized that just because there is a recurrence interval of 10 years or 100 years between events does not mean that such a period must elapse between such events—only that such events occur on the average that frequently. They may occur twice or three times in successive years and then not again for several times longer than the calculated recurrence interval. Thus, such information must be used with care for planning purposes.

In concluding this section on runoff, it is well to consider briefly a point made by Kazmann (1972) in his book on hydrology. He contrasts "modern" vs. "classical" hydrology by suggesting that modern hydrology is classical hydrology with data made unreliable as a result of human modifications of stream basins. Channel modifications, development of dams and reservoirs, increasing diversions of water for agricultural and industrial purposes, and other activities all result in changing relationships between precipitation, evapotranspiration, and runoff, especially in small or moderate-sized basins. Kazmann points out that classical hydrology assumes that hydrologic measurements are valid statistically and result from a natural set of circumstances. Thus, if those same conditions occur again, the same hydrologic response would be found. This is not the case in modern hydrology. Old relations among intensity and distribution of precipitation, evapotranspiration, and stream response are no longer found since we have changed land use in basins, added impervious surfaces, storm sewers, straightened channels, and made other modifications that influence hydrologic relationships.

## PART B:
## SOCIETAL USES OF
## WATER RESOURCES

Water has many important uses in a modern society. It is not enough to think only of domestic, agricultural, and industrial uses (the subject of the first chapter in this section). The use of water for recreation, power, navigation, and even for fish (the subject of Chapter 5) is also significant. The present section discusses these various demands on our water supplies in order to provide the necessary framework for understanding future needs and the legal, political, and economic problems which have and will continue to surround the use of our water resources.

## 4

MUNICIPAL, INDUSTRIAL, AND AGRICULTURAL
USES OF WATER

## MUNICIPAL WATER USES

If we eliminate the large volume of water used for cooling purposes in power generation, the withdrawals of water for municipal and industrial purposes are about equal. Municipal demand includes water for domestic purposes, commercial uses, fire protection, street washing, and lawn and garden irrigation. In many areas, industrial demand is also met from municipal systems although many industries supply their own water needs.

Domestic water use is low during the night. There is usually a peak use between 8 and 9 a.m. and another peak near supper time between 6 and 8 p.m. Lawn sprinkling also rises to a peak at that time. Commercial and industrial use of municipal water is generally fairly steady all day, dropping off at night (Fig. 4.1). While the United States appears to have the basic water resources to meet foreseeable demand for water for municipal and industrial purposes for the next several decades, clearly some shortages or problems are likely to develop in certain areas because (a) storage or distribution systems lack sufficient capacities to satisfy peak demands, (b) waste water treatment processes to protect existing supplies are being developed slowly, and (c) the quality of the water source has deteriorated (National Water Commission, 1973, p. 166).

### Development of Municipal Water Systems

The last will and testament of Benjamin Franklin left £1000 each to Philadelphia and to Boston, the two cities with which he had been most closely associated during his lifetime. The money was to be kept continuously invested by the municipalities. With interest, Franklin calculated that each fund would grow to £131,000 in 100 years. He stipulated that £100,000 should be spent in useful public works and the remainder would be again invested at interest. Franklin went on to suggest that Philadelphia might expend its £100,000 in piping the water of Wissahickon

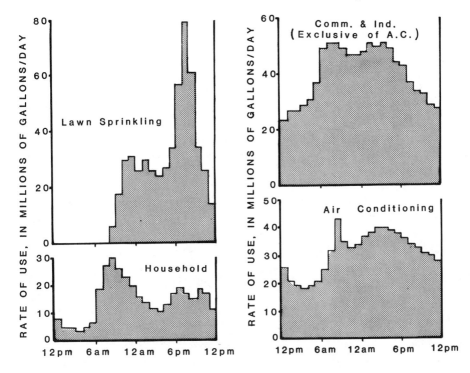

**Figure 4.1** Hourly trends in water use on maximum day (July 12) of use in 1954 of public water supply in Kansas City, Missouri. Based on data prepared by M. P. Hatcher for 1955 conference of American Water Works Association.
*Source.* Savini and Kammerer (1961).

Creek into the city to supply the needs of the inhabitants, if such a move had not already been undertaken. Thus, Franklin, earlier than most others, understood one significant need of the city where he lived and he took steps to answer that need.

The cities of 1790, the year when Franklin died, were not much more than overgrown villages. They were expanding rapidly and many facilities to meet the needs of the people, including water supplies and waste disposal, were grossly inadequate and fire and police protection were minimal. Epidemics of smallpox, yellow fever, and other fatal diseases periodically swept the cities. Over 4000 deaths occurred in Philadelphia in the yellow fever epidemic of 1793—almost 10% of the population. One of the results of the three yellow fever epidemics to sweep Philadelphia in the 1790s was the effort to promote cleanliness within the city. While the cause of the disease was still unknown, it was generally felt that increasing cleanliness and ventilation would somehow help to combat it. To remove the accumulated dirt and filth, large amounts of water were needed. Even in the 1793 epidemic, it was suggested that the fire companies should flush the streets daily. Noah Webster wrote:

Water is perhaps the best purifyer of the homes and streets of cities, as well as of infected clothes. The use of water cannot be too liberal . . . (Webster, 1796, p. 29).

By 1798, there was a growing suspicion that the water of Philadelphia was no longer safe to drink. Each house had its own privy which emptied into a bog-hole sunk into the ground. Each street had its wells with pumps from which water for drinking and cooking was obtained. Many were convinced that the well water was being contaminated by drainage from the bogholes. Thus, springs were highly prized and many individuals transported their water for drinking into the city from some distance to avoid the possibly polluted waters of the urban area.

An early attempt had been made in Boston, in 1652, to bring water to the city from a small reservoir through bored logs but the project did not continue for too long. A much more ambitious project developed in Bethlehem, Pennsylvania, in 1754–water was pumped through pipes to every house in town from a local creek. Some houses had water available in nearly every room (Blake, 1956).

In 1798, Benjamin Latrobe, a young engineer, made a report on the means to supply water of good quality to Philadelphia. He felt that the Delaware River was too polluted for use but that the Schuylkill River had the desired quality. While it would require very large pumping equipment to elevate the water to all parts of the city, he felt that the project was feasible. Others in the city wanted to bring water through a canal from Norristown to Philadelphia, a distance of 16 mi. The canal would also be used for navigation purposes. There was consider-able disagreement among the supporters of each water scheme.

Because of the urgent need for fresh water for the city, it was finally decided, in 1799, to go ahead with the Schuylkill River water project. Construction took 20 months and more money than originally estimated. But on January 27, 1801, water was pumped from the Schuylkill to a reservoir and from there it flowed by gravity through wooden pipes to hydrants in the streets of Philadelphia. Citizens could pay to have their own homes connected to the water mains but it was several years before many were willing to pay for water rather than to obtain it free from the wells still in use on the street corners. There were, of course, frequent break-downs in the large pumps supplying water to the reservoir so that the city was often without Schuylkill water and had to revert to local wells.

By 1800, 17 cities in the country had public water systems. There was not a great need for such systems in smaller towns at that time since water demand was generally low and, in the absence of paved streets, recharge to individual wells was sufficient to supply the requirements of each household. It was only as population pressures increased that significant demand developed to expand public water systems. While the Philadelphia example showed the difficulty and problems in creating a waterworks in a city, it was a step that had to be taken if cities were to grow and waterborne diseases were to be kept under control.

The need for a greater volume of water within limited areas, and the decrease in groundwater recharge as large sections of our cities were paved, made individual wells for each household unfeasible. By the middle of the 19th century, more than 130 cities and towns, usually the larger urban areas, had public water systems while, by 1895, the number of such systems had reached more than 3000 (White, 1969).

## Control of Municipal and Industrial Water Supplies

It has generally been an accepted fact in the United States that cities and industries obtain their own water supplies without Federal aid. The Water Supply Act of 1958 did permit enlarging the capacity of Federal reservoirs to store water for municipal and industrial (M and I) purposes but the cost of the additional capacity had to be borne by the municipal and industrial users. This Act did not extend Federal control into this traditionally non-Federal area of water supply. However, in more recent years, a series of developments have suggested a larger Federal role in supplying M and I water. For instance, the Rural Development Act of 1972 permits the Secretary of Agriculture to assume up to one-half the costs of reservoir storage needed for current M and I purposes and to give grants and loans for building water supply facilities in communities with up to 10,000 inhabitants, a significant increase over the previous limit. Such grants and loans can also be made available to businesses located within or near cities of up to 50,000 inhabitants. The Department of Housing and Urban Development (HUD) can also grant assistance to small communities to enlarge their water supply facilities, further expanding the Federal role in an area of traditionally local responsibility. There seems to be considerable concern that these steps will result in even further Federal control of local water supplies, just as it has in the field of pollution control.

There are about 30,000 water utilities in the United States at present, of which nearly 6000 are investor-owned. These latter systems serve about 30 million people. It is estimated that, in all, about 75 million people receive water from municipal systems, including those that are investor-owned. In most cities, the administration of the waterworks is an integral part of the city government, with the waterworks often operated as a public utility. Waterworks operators are, of course, interested in the quality of their water and this quality is greatly affected by factors of watershed protection, land management, river maintenance, conditions of storage, and pollution control. Clearly, it is less expensive to keep undesirable materials and chemicals out of the water than to remove them once they have entered the water supply. While water in most parts of the country is quite inexpensive, it is still important to keep treatment and distribution costs as low as possible so that no one will be denied needed water on the basis of cost. Figure 4.2 illustrates the sequence of steps followed in the treatment of municipal water for Philadelphia at the large Torresdale intake plant. The various sedimentation, filtration, and chemical treatment steps suggest why the cost of water in municipal systems will have to rise and why waterworks operators seek intake sources of the highest initial quality. The average annual domestic water charge is rather low, being less than $10.00 per person per year. That figure varies appreciably across the country and reflects the ease with which water can be obtained, the type of treatment necessary, and the varying costs of distribution to users (Table 4.1).

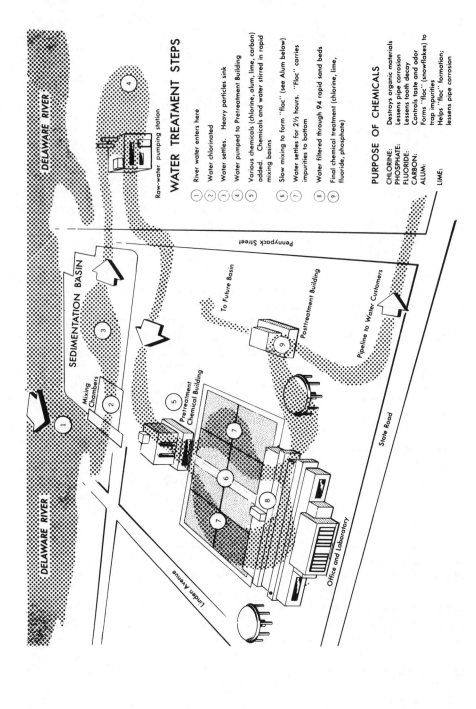

DELAWARE RIVER

DELAWARE RIVER

SEDIMENTATION BASIN

Mixing Chambers

Pretreatment Chemical Building

Pennypack Street

To Future Basin

Posttreatment Building

Pipeline to Water Customers

State Road

Office and Laboratory

Linden Avenue

Raw-water pumping station

## WATER TREATMENT STEPS

1. River water enters here
2. Water chlorinated
3. Water settles. Heavy particles sink
4. Water pumped to Pretreatment Building
5. Various chemicals (chlorine, alum, lime, carbon) added. Chemicals and water stirred in rapid mixing basins
6. Slow mixing to form "floc" (see Alum below)
7. Water settles for 2½ hours. "Floc" carries impurities to bottom
8. Water filtered through 94 rapid sand beds
9. Final chemical treatment (chlorine, lime, fluoride, phosphate)

## PURPOSE OF CHEMICALS

CHLORINE: Destroys organic materials
PHOSPHATE: Lessens pipe corrosion
FLUORIDE: Lessens tooth decay
CARBON: Controls taste and odor
ALUM: Forms "floc" (snowflakes) to trap impurities
LIME: Helps "floc" formation; lessens pipe corrosion

100

**Figure 4.2**  The Torresdale Plant of the city of Philadelphia, Pa. on the Delaware River. Normal capacity 282 million gallons per day; maximum capacity 423 million gallons per day.
*Source.* Swenson and Baldwin (1965).

## Municipal Water Quality

A 1969 survey of public water supply systems by the Bureau of Water Hygiene, including eight large municipal systems and one entire state (Vermont), has given us a sample of the water quality available to about 17.5 million people. It was concluded that the water used by some 15.5 million of these individuals was entirely safe and of good quality. Most of these individuals lived in cities with over 100,000 population. The remaining group of some 2 million used water of inferior quality. It was safe to drink but it had objectionable odor, taste, appearance, or other characteristics that made it less acceptable for human consumption. The results do not necessarily suggest the need for more Federal control of water quality although the Environmental Protection Agency does suggest that local and state programs of quality control are not as effective as they should be.

The late 19th and early 20th centuries witnessed a significant change in the treatment of water supplies. Possibly the first sand filtration plant in the United States was built in Poughkeepsie, New York, in 1871, while a similar sand filtration technique was employed in Lawrence, Massachusetts, in 1886, to control disease. Jersey City, New Jersey began the first chlorination of water supplies to destroy both viruses and bacteria in 1908. Such fairly routine treatment of water supplies has resulted in a significant decrease in the number of water-related diseases over the years. For example, in 1906, before filtration of drinking water was started in Philadelphia, there were more than 9000 cases of typhoid fever. By 1910, with filtration, the number of cases had dropped to under 1000 per year. Chlorination, which began in 1913, dropped the number of typhoid cases to less than 600 in 1914 and to less than 200 by the mid-1920s (Kilbourne and Smillie, 1969).

That typhoid and cholera deaths have been reduced to low levels does not, of course, mean that waterborne diseases have been eliminated entirely. For example, from 1938 to 1945, there were 327 separate outbreaks of waterborne diseases (mainly gastroenteritis and dysentry) affecting over 100,000 individuals (Eliassen and Cummings, 1948). Between 1961 and 1970, some 125 outbreaks occurred, including some 19,000 cases of gastroenteritis (Bardo and Cassell, 1971).

## Per Capita Consumption

Per capita use of water in the United States is very large in comparison with other so-called industrialized countries. A figure of 150 gal/day is often quoted as the

**Table 4.1**  Projected Annual Cost per Person for Water from Municipal Systems (in Dollars) Within Continental United States, by States[a]

|  | 1954 | 1980 | 2000 |
|---|---|---|---|
| U.S. average | 9.79 | 9.79 | 10.05 |
| Alabama | 8.68 | 8.28 | 8.15 |
| Arizona | 9.70 | 8.23 | 8.09 |
| Arkansas | 8.85 | 9.71 | 9.96 |
| California | 11.17 | 10.26 | 10.13 |
| Colorado | 14.54 | 12.69 | 12.64 |
| Connecticut | 10.68 | 10.51 | 10.60 |
| Delaware | 6.48 | 5.90 | 5.81 |
| District of Columbia | 8.22 | 9.43 | 9.85 |
| Florida | 10.88 | 8.85 | 8.63 |
| Georgia | 8.04 | 7.56 | 7.43 |
| Idaho | 14.41 | 13.84 | 13.60 |
| Illinois | 8.47 | 8.40 | 8.45 |
| Indiana | 10.39 | 10.15 | 10.11 |
| Iowa | 9.39 | 9.74 | 9.84 |
| Kansas | 11.70 | 11.37 | 11.38 |
| Kentucky | 6.60 | 6.95 | 5.65 |
| Louisiana | 8.07 | 8.03 | 8.14 |
| Maine | 5.21 | 5.24 | 5.23 |
| Maryland | 6.48 | 5.80 | 5.76 |
| Massachusetts | 8.50 | 9.43 | 9.84 |
| Michigan | 11.76 | 10.63 | 10.30 |
| Minnesota | 6.89 | 7.08 | 7.21 |
| Mississippi | 8.46 | 8.77 | 8.99 |
| Missouri | 8.87 | 10.84 | 10.96 |
| Montana | 12.76 | 12.52 | 12.50 |
| Nebraska | 9.16 | 9.15 | 9.16 |
| Nevada | 13.14 | 11.54 | 11.32 |
| New Hampshire | 9.16 | 9.98 | 10.13 |
| New Jersey | 8.07 | 8.14 | 8.33 |
| New Mexico | 13.23 | 12.00 | 11.79 |
| New York | 9.59 | 10.31 | 10.64 |
| North Carolina | 9.08 | 8.84 | 8.83 |
| North Dakota | 10.74 | 10.85 | 10.99 |
| Ohio | 9.83 | 9.71 | 9.69 |
| Oklahoma | 11.69 | 12.08 | 12.42 |
| Oregon | 10.18 | 9.96 | 9.88 |
| Pennsylvania | 9.66 | 10.14 | 10.12 |
| Rhode Island | 7.82 | 7.85 | 8.15 |
| South Carolina | 6.64 | 6.07 | 5.92 |
| South Dakota | 10.44 | 10.41 | 10.16 |
| Tennessee | 9.91 | 10.11 | 10.04 |

**Table 4.1**  (contd.)

|              | 1954  | 1980  | 2000  |
|--------------|-------|-------|-------|
| Texas        | 10.59 | 9.94  | 9.95  |
| Utah         | 14.93 | 13.42 | 13.03 |
| Vermont      | 10.25 | 11.54 | 11.62 |
| Virginia     | 6.80  | 6.15  | 6.07  |
| Washington   | 12.33 | 11.76 | 11.68 |
| West Virginia| 8.64  | 9.83  | 10.22 |
| Wisconsin    | 8.81  | 8.51  | 8.31  |
| Wyoming      | 15.72 | 15.21 | 15.10 |

[a]Includes operation, maintenance, and amortization.
*Note.* Averages of state figures are not the same as the national averages because of different bases used in computing these two sets of figures.
*Source.* Select Committee on National Water Resources (1960b).

---

per capita average but any consideration of the actual per capita use by city or by state shows great variations from one part of the country to another and helps put into perspective what is meant by such an average figure (Fig. 4.3; Table 4.2). Per capita use is about 75 gal/day in Vermont, 81 gal/day in Arkansas, and about 270 gal/day in Nevada. In six states having annual rainfall of less than 15 in., the median per capita consumption is 210 gal; in 11 states with annual precipitation exceeding 45 in., per capita consumption is just 119 gal/day. This is a simple generalization which does not always apply but it does suggest one reason for the great variation in water use per person.

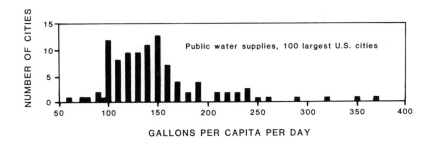

**Figure 4.3**  Range in per capita use of water in 100 largest cities of the United States, 1962. Cities showing high per capita use generally have high industrial usage while cities with low per capita values have industries that are generally self-supplied.
*Source.* Feth (1973).

**Table 4.2**  Projected Per Capita Water Use from Municipal
Systems in Continental United States, by States

| | [Gallons per capita per day] | | |
|---|---|---|---|
| | 1954 | 1980 | 2000 |
| Total United States | 147 | 148 | 152 |
| Alabama | 105 | 100 | 98 |
| Arizona | 170 | 143 | 141 |
| Arkansas | 81 | 88 | 90 |
| California | 147 | 135 | 133 |
| Colorado | 174 | 152 | 151 |
| Connecticut | 137 | 134 | 135 |
| Delaware | 142 | 126 | 125 |
| District of Columbia | 133 | 154 | 161 |
| Florida | 122 | 99 | 97 |
| Georgia | 127 | 119 | 117 |
| Idaho | 170 | 163 | 161 |
| Illinois | 171 | 169 | 170 |
| Indiana | 155 | 151 | 151 |
| Iowa | 110 | 113 | 114 |
| Kansas | 121 | 118 | 118 |
| Kentucky | 125 | 131 | 136 |
| Louisiana | 117 | 117 | 118 |
| Maine | 142 | 139 | 139 |
| Maryland | 131 | 117 | 116 |
| Massachusetts | 121 | 134 | 140 |
| Michigan | 213 | 193 | 187 |
| Minnesota | 108 | 111 | 113 |
| Mississippi | 107 | 110 | 113 |
| Missouri | 135 | 165 | 167 |
| Montana | 191 | 185 | 185 |
| Nebraska | 173 | 173 | 173 |
| Nevada | 270 | 230 | 226 |
| New Hampshire | 124 | 131 | 133 |
| New Jersey | 113 | 114 | 116 |
| New Mexico | 150 | 134 | 132 |
| New York | 138 | 148 | 153 |
| North Carolina | 108 | 106 | 106 |
| North Dakota | 101 | 98 | 100 |
| Ohio | 149 | 147 | 147 |
| Oklahoma | 130 | 134 | 138 |
| Oregon | 134 | 132 | 130 |
| Pennsylvania | 147 | 154 | 154 |
| Rhode Island | 111 | 110 | 114 |
| South Carolina | 108 | 98 | 96 |
| South Dakota | 107 | 101 | 99 |
| Tennessee | 118 | 119 | 118 |

**Table 4.2** (contd.)

|              | 1954 | 1980 | 2000 |
|--------------|------|------|------|
| Texas        | 119  | 112  | 112  |
| Utah         | 223  | 200  | 194  |
| Vermont      | 75   | 83   | 83   |
| Virginia     | 100  | 90   | 89   |
| Washington   | 196  | 187  | 185  |
| West Virginia| 112  | 127  | 132  |
| Wisconsin    | 158  | 153  | 149  |
| Wyoming      | 196  | 186  | 185  |

*Note.* Averages of state figures are not the same as the national averages because of different bases used in computing these two sets of figures.
*Source.* Select Committee on National Water Resources (1960b).

---

Using a value of 150 gallons per person per day, it is estimated that 41% (62 gal) is attributable to purely domestic use (washing, toilets, lawn sprinkling, food preparation, etc.), 18% (27 gal) to use by commercial establishments, 24% (36 gal) to industrial use, and 17% (25 gal) to public use (fire protection, street cleaning, park irrigation, etc.). Consideration of these various uses suggests why there could be such a variation from one part of the country to another.

Water use within the home varies appreciably from one part of the country to another with the life-style of the inhabitants and with the economic situation. Todd (1970) provides the following figures for water use in an average Akron, Ohio, home: washing, 1%; drinking and food preparation, 5%; bathing, 37%; lawn watering, 3%; toilet flushing, 41%; dishwashing, 6%; household cleaning, 3%; clothes washing, 4%.

Residential use was estimated by the California Department of Water Resources (1976) as 44% exterior and 56% interior. Interior use was estimated to be broken down as toilet flushing, 42%; bath, 32%; kitchen, 8%; laundry, 14%; and cooking, 4%. The two lists of estimated residential water use do not differ greatly in percentage figures although actual volume of water used may be somewhat different.

The 150 gal/day per capita figure is usually achieved by dividing total gallons of water pumped by a municipal water system by the number of individuals served by the system. While the number of gallons pumped is a fairly reliable figure, the number of individuals served must be estimated from the number of households in the water service area and the average family size as obtained from census figures. Thus, it is only a rough estimate of the number of individuals served. There could be appreciable errors in resort or tourist areas where a large number of transients swell the population totals for varying periods of time without ever appearing as part of local households. Also, certain parts of the country have a larger percentage of retired families (without children) in which the national average of individuals per household may not apply. Therefore, the figure for number of individuals using water may be somewhat in error in the computations of per capita use.

Industrial, commercial, and public uses of water are included in the per capita figure even though these are not at all related to domestic usages. It might well be argued that commercial uses will be about the same across the country for towns and cities of approximately the same size. For that reason, this use should be fairly uniform from one place to another. Public usages should vary somewhat with climatic factors. Cities and towns in drier areas might use more water for irrigation of public parks or street cleaning than would cities in more moist areas where the normal rainfall might satisfy most irrigation and street cleaning needs.

It is in the matter of industrial usages that significant changes can occur from place to place. Since many industries supply their own water, their demands are not included in the total pumped by the municipal system. Where industries use municipal water, their demands are included. The pattern of industrial usage varies markedly across the country. In predominantly rural or agricultural states, most water requirements are satisfied by private wells or ponds and not from municipal systems. Industrial usage would be quite small and total per capita use might be small. However, in very dry areas, where water tables may be well below the surface and costs for supplying water from a private well are high, there would be a tendency for industries to obtain water from municipal systems and so increase the per capita figure. In that case, a few industries might take a disproportionate amount of water from municipal systems in the drier parts of the country. The location and amount of industry, whether or not they supply their own water, the extent of the municipal system, the pricing policy for water in the area, as well as the climate, all enter into a consideration of the per capita usage. One must be careful in comparing figures from one country to another to be certain that the same terms are included in the usage figures for each country. Also, in evaluating our own wasteful habits, it must be remembered that only about 40% of the per capita figure results from purely domestic use. Savings here would be important but they contribute only in part to the overall figure of per capita use.

## Balancing Municipal Water Supply and Demand

Baumann and Dworkin (1978) have suggested a number of alternatives for bringing municipal water supply and demand into balance (Table 4.3). Several of their suggestions are discussed below; others will be covered in later chapters in this volume.

Efficient management of municipal water systems is necessary to keep costs down and to supply users with high quality water in sufficient amounts and at the right pressures to satisfy needs. As water demand grows and supplies become more costly, municipal systems all introduce certain practices to control wasteful use. Leakage surveys are becoming more important in recent years as it is apparent that large quantities of water are lost in the distribution systems. In a survey of

**Table 4.3**  Alternatives for Balancing Supply and Demand for Municipal Water Supply

| Do Nothing | Increase Supply | Decrease Demand |
|---|---|---|
| (1) Accept shortage<br>Unplanned rationing | (1) Increase system capacity<br>Divert surface water<br>Develop new groundwater<br>supplies<br>Conjunctively use ground<br>and surface water<br>(2) Improve efficiency<br>Reduce evaporation<br>Detect leaks<br>(3) Modify weather<br>(4) Desalinize water<br>(5) Renovate wastewater<br>Use for potable purposes<br>Use for nonpotable purposes | (1) Meters<br>(2) Price elasticity<br>Peak pricing, i.e.<br>Peak summer pricing<br>Marginal cost<br>pricing<br>(3) Restrictions<br>(4) Educational campaign<br>emphasizing water<br>use conservation<br>(5) Technological innovations<br>and application, e.g.,<br>changes from water cool-<br>ing to air cooling |

*Source.* Baumann and Dworkin (1978) with permission of the Association of American Geographers.

379 cities, Seidel and Baumann (1957) found that 10% had unaccounted for water losses that were greater than 25% of the water pumped. True, one-third of the cities had leakage losses of less than 5% of the water pumped but also half of the cities had leakage losses of more than 10% of the water pumped by the municipal system. Reduction of leakage losses could well take the place of developing alternate sources of water.

Metering has been suggested as a way to cut demand for water since it would introduce a rational way for effective price control. It was estimated that complete metering of water in New York City might reduce average daily consumption by about 10% (D'Angelo, 1964). Howe and Linaweaver (1967) showed that the average household water use was 458 gal/day in metered areas and 692 gal/day per household where a flat-rate charge was made for water. They felt that the major reason for this difference was the amount of water used in lawn irrigation. Many studies show a 20 to 25% reduction in demand with complete metering of an area. For instance, the city of Boulder, Colorado experienced a 40% drop in demand as it went from 5% metered to 100% metered. The effect of metering was to give the city water system the capacity to serve an additional 11,000 individuals without any need to increase the water supply (Hanke and Flack, 1968). Though the initial reaction to metering is a great reduction in water usage, there is some indication that once the immediate impact of metering has worn off, usage will edge back up again but not to its former non-metered levels. Metering should be practiced to cut water waste rather than to reduce needed water use for there is still sufficient water in most areas if it is not wasted or rendered unfit by pollution.

There is, of course, some connection between price and demand for water but the actual relation is not clearly defined. Water is vital to our continued existence and at present water is extremely inexpensive in most areas. Thus, even if the price of water were tripled, it would still be inexpensive and domestic demand would not be affected appreciably. Industry is much more sensitive to the price of water. If water were to become more expensive, it is quite possible that industrial demand would be lowered markedly. Industry would find other ways to satisfy their needs at the lowest possible cost. Several techniques will be discussed later in this book.

Education, of course, is one of the major techniques available to bring supply and demand into balance. Education will work on both sides of the balance, for with increased renovation of waste water and its reuse as a fresh water source, a detailed program of consumer education will be necessary to convince users that the water is suitable for domestic purposes. It is only natural for us to shy from renovated water from a sewage treatment plant even though it may be far purer than the treated water from a river or reservoir we drink instead. The idea of reusing sewage water will have to be sold gradually over time on the basis of extensive educational campaigns. Although we may fully admit that the water we use has been used before, it does not seem to be quite the same as reusing water from the local sewage treatment plant.

In developing plans to build a plant to recycle renovated water in Denver, Colorado, some 500 individuals were asked how they felt about the reuse of water (Heaton, Linstedt, Bennett and Suhr, 1974). The initial response was significantly negative. These same individuals then received more information on the quality of the renovated water and the overall implications of water renovation and reuse. With education, the rate of positive response from the sample group increased until, at the end, some 85% of the original group of 500 individuals indicated a willingness to drink the renovated waste water.

Education can contribute to a reduction in demand by showing how water can be conserved around the home or in factory operations. Already information on correct lawn sprinkling techniques, on using less water in toilet flushing, on the use of water in showers and baths, and on not serving water with meals is being circulated. These and other educational campaigns will make us aware of how little acts of conservation can result in sizeable municipal savings.

In evaluating Federal water programs in relation to municipal water supplies, the National Water Commission (1973) pointed out several significant problems that should be considered. These included:

(a) The need to restate the Federal policy with regard to municipal and industrial water systems so that the dominant role of the state and local governments is not compromised. This should be followed by efforts to bring existing laws into conformity with the restated Federal policy.
(b) The need to coordinate plans to guide metropolitan development with plans to guide river basin and region development.
(c) The need to coordinate plans for new suburban water systems with those of the water system for the central metropolitan area. Similarly, plans for water

withdrawals for industries supplying their own water must be coordinated with plans for all other withdrawals.

It is clear that there will continue to be a strain on many existing municipal water systems as more people continue to move into urban centers and our economic activities expand. Planning, followed by effective management, will be necessary if periods of shortages and crises are to be avoided. Without careful planning, real problems can be expected to develop in the next few decades in the more water-short areas. By the year 2000, it would appear that municipal water services will be even more efficient than they are at present, metering will be fairly universal, widespread programs of leak detection and plumbing inspection will have been instituted, and there will have been moderate increases in prices. Such changes should control domestic use within realistic and feasible limits.

## INDUSTRIAL USES

Industries vary in their source of water; some utilize municipal supplies, some have their own wells, while some utilize available surface supplies (such as rivers or lakes). Table 1.3 shows the great change in the volume of water used for industrial purposes over the past 80 years. Total industrial uses, from all sources, has gone from 10 bgd in 1900 to an estimated 75 bgd in 1980. Of this 1980 total, only a little over 13 bgd comes from groundwater supplies. If steam electric utility demand is also included in industrial water demand, the figures are even more striking. Combining both industrial and steam electric utility demand, we find water usage has risen from 15 bgd in 1900 to an estimated 268 bgd in 1980 but, of that latter total, only about 13.5 bgd are obtained from groundwater supplies. The bulk of this industrial water supply comes from various surface sources.

The amount of water used by different types of processing and fabricating industries is shown in Table 4.4. The table also provides interesting information on the number of gallons used per employee per day. The petroleum and coal products industry group, because of their small number of employees, uses the most water per employee by far. The high percentage of water intake that appears as waste water in most industries (last column) suggests that very little water is consumed in the manufacturing process.

Industries differ appreciably in the type of water they can use. For example, dissolved solids concentrations in industrial water supplies have been found to vary from 150 mg/l to over 3000 mg/l from textile to petroleum industrial water supplies while boiler make-up water can have another order of magnitude of dissolved solids and still be used (Table 4.5).

The trend in water reuse and recycling has been strongly upward in many industries. Table 4.6 shows the 20-year trend in seven major industries as well as an average for all industries. In 1973, the national average for all industries was 2.89. This means that each gallon of fresh water withdrawn from the original source was used almost three times before it was discharged from the plant. In petroleum and coal products industries, the recycle factor was over 6. Since a

**Table 4.4**  Quantity of Water Used by Manufacturers (Data Gathered for Manufacturing Establishments with 6 or More Employees)

| Industry Group | Number of Employees | Annual Water Intake (billions of gal) | Gallons Per Employee Per Day (thousands) | Intake Water Appearing as Waste Water (%) |
|---|---|---|---|---|
| Processing industries: | | | | |
| Food and kindred products | 1,589,380 | 812 | 1.400 | 91 |
| Lumber and wood products | 489,354 | 161 | 1.146 | 82 |
| Paper and allied products | 583,234 | 2,078 | 9.762 | 94 |
| Chemicals and allied products | 734,261 | 3,899 | 14.584 | 94 |
| Petroleum and coal products | 152,470 | 1,400 | 25.157 | 94 |
| Rubber and plastic products | 406,777 | 168 | 1.439 | 95 |
| Stone, clay, and glass products | 550,451 | 264 | 1.434 | 88 |
| Primary metal industries | 1,122,911 | 4,587 | <u>11.196</u> | 94 |
| Weighted average | – | – | 6.507 | – |
| Fabricating industries: | | | | |
| Tobacco products | 76,989 | 4 | .168 | 67 |
| Textile mill products | 854,543 | 158 | .644 | 91 |
| Furniture and fixtures | 360,882 | 8 | .079 | 94 |
| Leather and leather products | 322,747 | 20 | .215 | 94 |
| Fabricated metal industries | 1,058,954 | 76 | .249 | 93 |
| Machinery, except electrical | 1,424,432 | 172 | .421 | 95 |
| Electrical machinery | 1,502,324 | 114 | .264 | 87 |
| Transportation equipment | 1,593,285 | 252 | .551 | 95 |
| Instruments and related products | 301,650 | 31 | .363 | 90 |
| Miscellaneous manufacturing | 371,858 | 19 | <u>.175</u> | 93 |
| Weighted average | – | – | .378 | – |

*Source.* Feth (1973) as adapted from Reid (1971).

1960s study suggested a recycle ratio of 3.0 by 1980, it is clear that industry is undertaking recycling of water more rapidly than expected at that time.

In 1965, 58% of 1467 plants reported some in-plant treatment of intake water. Many of these plants supplied their own water and needed some type of treatment to bring the raw water up to their high quality demands for manufacturing. Sixty-nine percent of 1962 plants surveyed reported that they treated all or some of the waste water from the manufacturing process before it was allowed to enter nearby receiving streams. Plants providing no treatment of their waste water vary appreciably from industry to industry (Table 4.7). Of course, some of the plants reporting no in-plant treatment might discharge their effluent to municipal treatment systems.

**Table 4.5** Total Dissolved Solids Concentrations of Surface
Waters that Have Been Used as Sources for Industrial Water
Supplies

| Industry/Use | Maximum Concentration (mg/l) |
|---|---|
| Textile | 150 |
| Pulp and paper | 1,080 |
| Chemical | 2,500 |
| Petroleum | 3,500 |
| Primary metals | 1,500 |
| Copper mining | 2,100 |
| Boiler make-up | 35,000 |

*Source.* Reproduced from *Water Quality Criteria*, National
Academy Press, Washington, D.C., 1972.

Industrial demand for water can usually be categorized into one or more of
the following six areas: (a) cooling, (b) processing or manufacturing operations,
(c) generation of power, (d) cleanup and other sanitary purposes, (e) fire protection, (f) miscellaneous.

While the use of water for cooling purposes far exceeds all other uses, it is
not a consumptive use. Aside from the heat that is added in the process of cooling,
the water is usually available for further reuse. A large electric generating station
may use more than 500,000 gal of water per minute to control temperatures in
condenser operations. The temperature of the water used for cooling purposes,
its volume, and its quality must all be considered. The most important impurities
are, of course, those that can form scale on pipes and other surfaces along with
suspended matter, dissolved gases that might result in corrosion to metal parts,
and acids. With boilers now producing steam at higher pressures and temperatures and operating at higher generating rates, the quality of boiler feedwater
becomes more important. Under high temperatures and pressures, silica may
dissolve or vaporize with the steam and deposit on heating surfaces or turbine
blades, interfering with operations or causing pipe failure.

Most surface water that receives industrial waste is contaminated by oil to
some degree. Free oil can be removed by treatment but oil in a combined or emulsi-
fied form may pass through conventional treatment. The use of streams for waste
disposal makes those waters less usable by others without expensive treatment.
The quality of water used by industry, whether for cooling purposes or for processing, is important because of odor, taste, color, or other pollution problems. Many
industries using surface water supplies must install their own treatment facilities.
Special treatment facilities are particularly important in food and beverage industries where water must be of highest quality. However, there are many operations

**Table 4.6**  Trends in Water Use in Manufacturing Industries (bgd) and Recycle Ratio

| Industry Group | Year | Gross Water Used (Including Recirculated) | Total Water Intake | Recycle Ratio (Gross/Total) |
|---|---|---|---|---|
| All industries | 1973 | 118.9 | 41.2 | 2.89 |
| | 1964 | 81.8 | 38.4 | 2.13 |
| | 1954 | 57.6 | 31.7 | 1.82 |
| Food and related products | 1973 | 4.3 | 2.2 | 1.95 |
| | 1964 | 3.3 | 2.1 | 1.57 |
| | 1954 | 3.6 | 1.7 | 2.12 |
| Stone, clay and glass products | 1973 | 1.3 | 0.6 | 2.17 |
| | 1964 | 1.1 | 0.7 | 1.57 |
| | 1954 | 1.6 | 0.8 | 2.00 |
| Lumber and wood products | 1973 | 0.7 | 0.4 | 1.75 |
| | 1964 | 0.6 | 0.4 | 1.50 |
| | 1954 | 0.4 | 0.4 | 1.00 |
| Paper and allied products | 1973 | 22.3 | 6.6 | 3.38 |
| | 1964 | 15.0 | 5.7 | 2.63 |
| | 1954 | 11.6 | 4.9 | 2.37 |
| Chemicals and allied products | 1973 | 30.4 | 11.4 | 2.67 |
| | 1964 | 21.2 | 10.7 | 1.98 |
| | 1954 | 11.8 | 7.4 | 1.59 |
| Petroleum and coal products | 1973 | 22.4 | 3.5 | 6.40 |
| | 1964 | 16.9 | 3.8 | 4.82 |
| | 1954 | 11.4 | 3.4 | 3.35 |
| Primary metal industries | 1973 | 24.2 | 13.5 | 1.79 |
| | 1964 | 18.4 | 12.6 | 1.46 |
| | 1954 | 13.6 | 10.5 | 1.29 |

*Source.* U.S. Bureau of the Census (1975a).

in factories where water of highest quality is not needed. In these cases, water taken from sewage treatment plants, or recycled within the factory itself, may suffice as long as piping facilities are available to segregate this water from fresh water supplies. The Sparrows Point, Maryland plant of Bethlehem Steel uses treated Baltimore City sewage. They have found it to be superior to the saline Chesapeake Bay water or the available groundwater which they previously used. Some 120 million gal/day of treated sewage effluent are purchased from the city at a bargain price of 1¢ per 1000 gal. The factory is now considering possible changes that will allow it to increase the amount of treated sewage water it can use (Kasperson and Kasperson, 1977).

**Table 4.7**  Variation in In-Plant Waste Treatment
Facilities by Industry

|  | % Without In-Plant Waste Treatment Facilities |
| --- | --- |
| Automobile | 48% |
| Beet sugar | 6 |
| Coal preparation | 12 |
| Corn and wheat milling | 50 |
| Distillery | 62 |
| Food processing | 27 |
| Machinery | 71 |
| Meat | 5 |
| Natural gas compression | 0 |
| Photographic | 50 |
| Poultry processing | 12 |
| Pulp and paper | 32 |
| Salt | 87 |
| Soap and detergents | 56 |
| Sugarcane | 52 |
| Tanning-leather | 38 |
| Textiles | 73 |

*Source.* National Association of Manufacturers
(1965) with permission of the National Association of Manufacturers and the Chamber of Commerce of the United States.

Because special quality water is needed in certain industries, expensive treatment
facilities must often be installed even if water comes from municipal systems.
The treatment may consist of sedimentation to remove silt and slime, sand filters
to remove finer suspended particles, use of chemicals (such as chlorine and activated carbon) to remove taste and odors, and aeration, coagulation, and more filtration to remove other hard-to-remove materials and to improve the oxygen content
of the water. It can add a significant amount to manufactured products in industries that require large volumes of water, even though the cost of water may
be low for domestic uses. Thus, having good water to begin with is a great savings
to those industries.

It is very difficult to generalize about the volumes of water needed to produce
different products because water needs vary appreciably from one part of the
country to another or from company to company depending on the manufacturing

**Table 4.8**　Water Use to Produce One Unit of Product in Different Industries

| Industry | Unit | Gallons Per Unit |
|---|---|---|
| Automobile | Vehicle (including automobiles and trucks) | 10,000 |
| Distilling | One proof gallon of whiskey or spirits | 125 to 167 |
| Meat | 1000 lbs. of finished product | 2750 |
| Petroleum | One barrel of crude oil processed | |
| | based on total use | 1741 |
| | based on recirculated water | 1534 |
| | based on fresh water intake | 204 |
| | based on water consumed | 30 |
| Pulp and paper | Ton | |
| | including cooling water | 57,000 |
| | excluding cooling water | 35,800 |
| Soap and detergents | Case | 3 to 100 |
| | Drum | 33 to 38 |
| Steel | Net ton | |
| | fully integrated mill | 20,790 |
| | rolling and drawing shops | 3544 |
| | blast furnace smelting | 24,798 |
| | electrometallurgical production of ferroalloys | 17,384 |
| Tanning | Square foot of hide tanned | 0.2 to 64 |
| | average | 10 |

*Source.* National Association of Manufacturers (1965) with permission of the National Association of Manufacturers and the Chamber of Commerce of the United States.

process being used and the availability of water. It is also difficult to separate out the exact amount of water actually needed to make a particular product since, in many manufacturing enterprises, a mix of products results from a given use of water. The rough figures that are available indicate anywhere from 0.5 to 170 gal of cooling water for the generation of 1 kw hr of steam power, from 1400 to 65,000 gal of water for a ton of finished steel, from 0.25 to 14 gal for a pound of carbon black, from 13 to 305 gal for a pound of butadiene, and from 0.8 to 45 gal of water for refining 1 gal of oil.

A study by the National Association of Manufacturers (1965) provided information on average volume of water used to produce a unit of product (Table 4.8). They found tremendous variation in certain industries because of special situations. For example, to produce 1 ton of salt required anywhere from 6 to

67, 640 gal of water and to produce 1 ton of sugarcane water requirements ranged from 3000 to 68,300 gal, with most of the responses falling between 20,000 and 40,000 gal. The variations from high to low suggest the amount of water savings that are possible if water becomes limited and its price increases. Some of the large volumes of water used in industry may result from the fact that inexpensive supplies are available rather than from real need for that much water in the processes of manufacture.

### Geographic Distribution of Industry

Early forecasts of future industrial demand for water were based on straight extrapolation of past conditions with no consideration of reuse or recycling of water or the use of other water-saving techniques as the cost for water might increase. One forecast by the President's Materials Policy Commission in the 1950s estimated an industrial demand for some 215 bgd by 1975 not including the water use for steam electric generation. This figure was equal to 63% of the estimated total demand for water at that time. The actual figures for 1975 total no more than 200 bgd including both saline and fresh water uses for manufacturing, steam electric generation, and the water needed in the minerals industry. This total value is about 50% of our overall water use.

Those 1950 forecasts led to all sorts of dire predictions about the future of water supplies and to suggestions concerning the future location of industry. For example, it was felt that the location of industry in the second half of the 20th century might be dictated largely by the availability of water of the proper quality and quantity.

In a study of the geographic distribution of manufacturing, Burrill (1956) found that it was highly localized in the northeastern part of the United States in the 1950s (Figs. 4.4, 4.5). Manufacturing is largely an urban phenomenon for it requires people, markets, and services. Obviously, certain industries, such as mining, agriculture, or fishing, do not fit this mold and must locate where the raw materials are found. However, two-thirds to three-fourths of all industrial water intake was found to occur in the so-called manufacturing belt of the northeast. Forty percent of the industrial water use occurred in just the three states of New York, Pennsylvania, and Ohio.

That industrial demand for water is even more highly localized than the geographical distribution of industry itself is due to the very large water demands of just a few selected industries. Burrill found that just four industry groups took some 84% of all water used by industry in 1953. These four groups and their water demands were primary metal industries (blast furnaces and steel mills), 32%; chemical and allied products, 25%; petroleum and coal products (coke production and oil refining), 17%; and pulp and paper manufacture, 10%. All four industry groups are well represented in the manufacturing complex of the northeastern United States. The location of those industries will, in large measure, determine the location of the industrial water demand. Since those industries are, at present, located in areas of relative water availability, there is little reason to

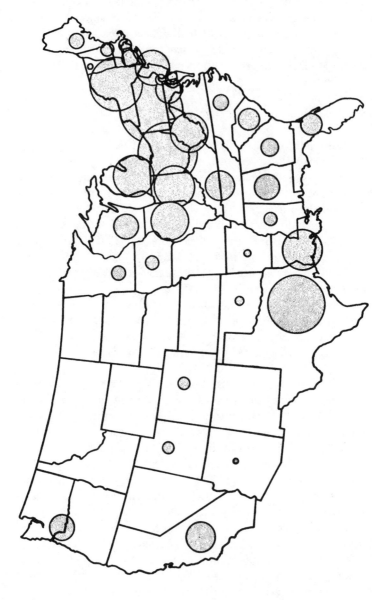

**Figure 4.4** Water use in manufacturing, 1953. Amounts represented by symbols for selected states are in billions of gallons for the year: Pennsylvania, 1397; Texas, 1159; Indiana, 629; Alabama, 207; Arkansas, 20. U.S. total 11,430. The figure for Kentucky-Tennessee is the East South Central states total minus Alabama and Mississippi; that for Washington-Oregon is the Pacific states total minus California; that for Oklahoma is the West South Central states total minus Arkansas, Louisiana, and Texas. Amounts in states with no symbol could not be derived from regional totals. Central electric stations are not included in the tabulation. *Source.* Bureau of the Census estimates; Burrill (1956), copyrighted by the American Association for the Advancement of Science.

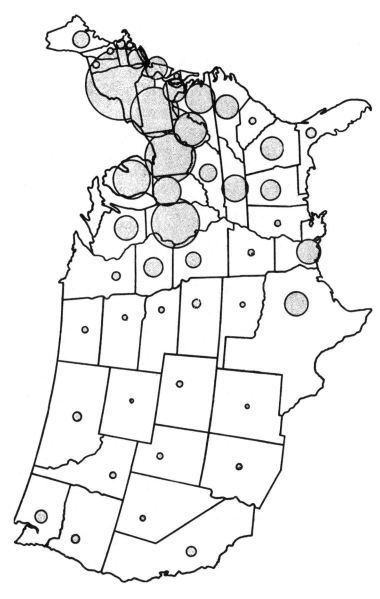

**Figure 4.5** Industrial water use, 1950. Figures represented by symbols for selected states are in millions of gal/day: New York, 16,280; Michigan, 5000; Louisiana, 1940; Utah 75. U.S. total 77,216. Amounts are withdrawals and do not take recirculation into account. Some industrial use of municipal water supplies is not included.
*Source.* Geological Survey estimates; Burrill (1956) copyrighted by the American Association for the Advancement of Science.

expect that they will change locations in the future. Thus, continued high demand for water should be expected from industrial sources in the northeast.

The question that must be answered is, "What will happen to the picture of industrial demand as industry begins to move to new locations because of markets, less expensive labor, or location of raw materials?" While in 1900 the presence of sufficient water of adequate quality might have been a major consideration in industrial location, it would not appear to be so at present. Many industrial processes can be carried out with much smaller volumes of water than originally thought. Thus, if there are other compelling reasons for the location of an industry, water supply would not seem to be a limiting factor in most cases. Of course, if water is plentiful, so much the better; but if it is not, most industrial plants can be designed to use very much less fresh water than they do at present.

As industries move to water-short areas where competition for available supplies from agriculture and domestic demands already exists, there will be increased pressure to conserve or to reuse water, which will result in serious reappraisal of the current uses of water in an area in order to determine if the greatest value is being obtained from the volume used.

At least five techniques are available to control water use within a factory.

(a) *Conservation of water in industry.* Many opportunities exist for conservation of water in industry. Conserve water by eliminating leaks, turn off unused water lines, install water control valves, and move products from one place to another without water. Too often production lines involving water are left on whether any product is being processed or not. Flowing water is often used to transport vegetables or other products from one process to another. Unused lines can be turned off and conveyor belts or other techniques used to move products. Leaks in pipes can be repaired at a considerable savings in water.

(b) *Multiple recycling of water.* This technique calls for water to be used over and over again for a particular process (such as for cooling). Some deterioration in water quality may occur but, with careful control of water movement in the plant, it is often possible to achieve several reuses of water before new water must be brought into the cycle. Clearly, this technique can only be used for operations that do not require fresh water at each step. It may also be possible to combine recycling with the addition of a small amount of fresh water in each cycle to keep the water at an acceptable pollution level.

(c) *Stepwise reuse.* In this technique, fresh water is first used for the process that requires the cleanest water. The slightly polluted water is then used in a process that does not require the freshest of water. In some cases, it might be possible to achieve a third or even a fourth reuse for processes that can accept quite polluted water. Stepwise reuse requires careful planning within the factory and the layout of water pipes in such a way that the water is shifted from one process to the next in the correct sequence. While more expensive during construction, savings in water may ultimately pay for the initial extra construction costs. The technique may not be economically feasible in an older factory with water pipes already installed.

(d) *Substitution of sources.* A very useful technique that has already been mentioned in connection with the Sparrows Point plant of Bethlehem Steel involves the replacement of fresh water with treated or saline water. Many operations within most factories do not require absolutely pure water. It is quite possible to satisfy the demand for water with water of lower quality with no great loss in product quality or value. It does require, however, a double system of pipes to bring the fresh and non-fresh water into the factory operation. Table 1.6 shows that saline water withdrawals, used mainly in manufacturing and steam electric generation, already total nearly 60 bgd, more than one-seventh of our total water withdrawals in 1975.

(e) *Water treatment and reuse.* This technique involves the actual treatment of the waste water in the factory and its recycling as fresh water to the factory operation. This might be especially useful for those industries that need special water and who have already installed their own treatment facilities. Once they have the facilities to obtain their special type of water, they may find it only slightly more expensive to treat their own waste water for reuse. Again, many industries are being forced to install treatment plants before they can release their waste into city sewer systems or to streams. With this facility available, they may ultimately find it more worthwhile to reuse their own treated water than to pay for new amounts of fresh water.

Many variations in these five techniques are available to reduce industrial use of water. They are being adopted with such speed that estimates of future water need for industry have shown the largest declines of any category of water use. For example, Table 1.6 shows that in the next 25 years withdrawals for manufacturing are expected to decline from the 1975 rate of 51 bgd to less than 20 bgd in year 2000. True, saline water use (in industry and steam electric generation) is expected to increase from 60 bgd to 119 bgd so that overall industrial demand will increase by some 28 bgd but it will result in a considerable savings in fresh water. Combining fresh water withdrawals for manufacturing, steam electric generation, and mineral industries, one finds a decrease of nearly 37 bgd (from 147 bgd to 110 bgd). Saline water uses will increase by 59 bgd so that total withdrawals for industry and power will be up but overall fresh water withdrawals are estimated to be down significantly.

Thus, the future picture for industrial water requirements is anything but bleak. Not only are a number of techniques available to reduce the water requirements of most industries but industries now moving to new locations need little encouragement to try to save water. As the costs for water increase, industry will be strongly encouraged to conserve. At the same time, as disposal of waste water becomes more expensive, because of the need for treatment, industry will again be encouraged to reduce water volumes in the factory. It seems quite logical to suggest that there will indeed be a reduction in the fresh water need by industry over the next two decades, that there will continue to be a movement of industry into water short areas, and that, of all water users, industry will have the most success in their efforts to conserve and reuse water.

## IRRIGATION USES

Almost since the beginning of agriculture, the practice of supplying water to

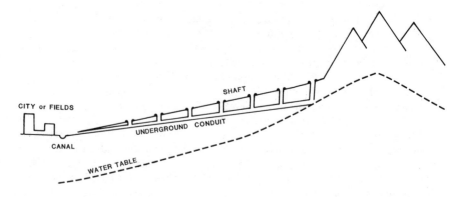

**Figure 4.6** Schematic diagram of simple qanat system. Series of vertical shafts connecting sloping conduit are used to remove material when conduit is constructed. Sloping conduit brings water by gravity from distant water table at reduced gradient until it reaches surface at point of use.

plants through some form of irrigation has been practiced. Hammurabi, the sixth king of the first dynasty of Babylon (about 2000 B.C.), developed laws designed to maintain and extend irrigation works upon which the people depended for their very lives. An Assyrian Queen, possibly living before 2000 B.C., is given credit for diverting Nile River water to irrigate the deserts of Egypt (Hansen et al., 1980). It is possible that canals built by order of this Queen are still operative today.

In China, the Hsia dynasty (about 2200 B.C.) was known for its work in water control and the Tu-Kiang Dam, which is still in use, was built about 200 B.C. Approximately half a million acres are irrigated by water from this reservoir. Irrigation in India was developed at about the same period.

The Spaniards found complex facilities for storing and transporting water in Mexico and Peru when they came to this hemisphere. The art of irrigation is quite old. Without the human application of water to the croplands of the Fertile Crescent, the great civilizations from Egypt to Iraq would not have been possible. In fact, the ancient water supply systems, whether to bring water to villages and cities or to bring water to parched fields, were often engineering feats of staggering size and complexity. One has only to consider the aqueducts of ancient Rome or the qanats of Persia to see truly magnificent examples of water engineering, accomplished with unsophisticated instruments but with great perseverance and skill (Fig. 4.6).

Although some have claimed that Spanish missionaries brought irrigation information to the southwestern part of the United States, there is good evidence that the Hohokam Indians who lived in what is now Arizona during the period from 300 BC to 1300 AD had a highly developed irrigation agriculture. They evidently constructed and maintained an irrigation canal of some 125 mi in length in the Salt River valley of Arizona and irrigated about 140,000 acres of land, enough to support at least 22 villages (Garstka, 1978). Modern development of an extensive irrigation agriculture in the Southwest began when the Mormons undertook cooperative irrigation enterprises in the Salt Lake Valley in 1847. They diverted small streams into low-lying areas bordering the water courses. In that year, some 5000 acres were brought under irrigation; the total rose to 16,000 acres by 1850. German farmers near Anaheim, California brought irrigation agriculture to the west coast in 1857. By the end of the Civil War, more than 1000 mi of canals were in service in the West supplying water to more than 150,000 acres of cropland (Armstrong, 1976).

Since agriculture without irrigation was not feasible in the dry western lands, water rights were prized even more highly than land itself as farmers, miners, and ranchers moved in to occupy the vast arid and semiarid regions of the country. Irrigation is not only a major emphasis in the water resources development of the western states but it has led to the significant overdevelopment of existing water supplies in many areas—the "mining" of water, the transportation of supplies from areas of surplus to areas of deficit, or the removal of so much fresh water from streams that the remaining water is too brackish for use without treatment. Development of irrigation remains vital to the western United States but that development cannot be allowed to exceed limits established by the available water supply. At the same time, a balance must be established between the various competing demands for water—irrigation, domestic, industrial, recreation—to achieve the most economic return on the limited resource.

Irrigation, of course, does more than just supply water to suffering crops. Hansen et al. (1980, p. 4) have listed the following purposes for irrigation:

(a)  to supply the needed moisture for plant growth;
(b)  to protect crops from short period droughts;
(c)  to cool the plant environment and to make it more favorable for growth;
(d)  to lower frost damage hazard;
(e)  to leach salts in the soil;
(f)  to soften or break down plow soles and soil clods;
(g)  to slow bud development by evaporative cooling.

Figures on irrigated lands by country (Table 4.9) indicate that the United States ranks third behind China and India with respect to total acreage irrigated. In terms of percentage of cultivated land irrigated, however, the United States ranks below the world average of 16%. Table 4.10 gives the acreage under irrigation by states for 1975 as well as an estimate for the year 2000. It is interesting that in 1975, California had more land under irrigation than all but the top 10 countries of the world (as shown in Table 4.9). The distribution of irrigation by states provides an insight into the problems and potentialities for irrigation in this country. For example, only one state east of the Mississippi River—Florida— appears in the top 18 states in terms of acreage under irrigation in 1975 and again in the year 2000. Interestingly, New Jersey, one of the most urbanized states in the United States despite its nickname as the "Garden State," appears as twenty-fourth on the list and third on the list of states east of the Mississippi River behind Florida and Mississippi. This is undoubtedly due to the great development of irrigated truck farms in southern New Jersey. The relatively low positions of

Table 4.9   Ranking of Countries Having More Than 1 Million Ha of Irrigated Land

| Country | Cultivated Land 1000 ha | Irrigated Land 1000 ha | Percent |
|---|---|---|---|
| China and Taiwan | 111,167 | 76,500 | 69 |
| India | 164,610 | 38,969 | 24 |
| USA | 192,318 | 21,489 | 11 |
| Pakistan | 21,700 | 12,400 | 57 |
| USSR | 232,609 | 11,500 | 5 |
| Indonesia | 18,000 | 6,800 | 38 |
| Iran | 16,727 | 5,251 | 31 |
| Mexico | 23,817 | 4,200 | 18 |
| Iraq | 10,163 | 4,000 | 39 |
| Italy | 14,409 | 3,500 | 29 |
| Thailand | 11,200 | 3,170 | 28 |
| Vietnam | 5,083 | 3,040 | 60 |
| Afghanistan | 7,980 | 2,900 | 31 |
| Egypt | 2,852 | 2,852 | 100 |
| Japan | 5,446 | 2,626 | 48 |
| Sudan | 7,000 | 2,520 | 25 |
| Spain | 20,626 | 2,435 | 12 |
| Turkey | 26,068 | 1,724 | 7 |
| Australia | 44,610 | 1,581 | 4 |
| Argentina | 26,028 | 1,555 | 6 |
| Chile | 4,632 | 1,500 | 32 |
| Peru | 2,979 | 1,116 | 37 |
| Philippines | 11,145 | 1,090 | 10 |
| Korea | 2,311 | 1,070 | 46 |
| Bulgaria | 4,516 | 1,001 | 24 |
| Venezuela | 5,214 | 1,000 | 19 |
| All others | 463,790 | 17,848 | 4 |
| Total | 1,457,000 | 233,637 | 16 |

*Source.* Hansen et al. (1980) with permission of John Wiley & Sons.

North Dakota and Iowa (numbers 29 and 35, respectively) are difficult to explain, occurring as they do below both New York and New Jersey.

Consideration of the change in extent of irrigated acreage anticipated between 1975 and 2000 suggests something of the patterns of irrigation that may develop as a result of changing availability and cost of water, changing agricultural demands, and changing population pressures. For example, among the top 18 states in 1975, 4 (Texas, Colorado, Montana, and Wyoming) are predicted to experience a reduction in irrigated acreage by the year 2000. Largest gains in irrigated acreage are expected to occur in Nebraska (600,000 acres), Kansas (700,000 acres), and Washington (420,000 acres). These are states with some available supplies of water already and where the development of additional supplies is still possible. The four states with expected decreases in irrigation are considerably drier climatically and may not have as large a developable supply of water within their own borders.

East of the Mississippi River, gains in acreage under irrigation are predicted to be less than in most of the western states but, of course, percentage gains in irrigated acreage may be much higher because of the small amount of land under irrigation in 1975. For example, it is predicted that Georgia will more than double its acreage under irrigation from 89,000 to nearly 205,000 acres in the last quarter of the century. Wisconsin, North Carolina, Michigan, Virginia, Illinois, Minnesota, Indiana, Delaware, Maryland, and Kentucky are expected to see more than a doubling of irrigated acreage between 1975 and 2000, as will North and South Dakota, Missouri, and Iowa west of the Mississippi River. Thus, the figures suggest that the very driest states will not experience growth of irrigated agriculture in the final part of this century and, in fact, some may even experience a decrease in irrigated acreage. However, states in the humid, subhumid, and possibly even semiarid areas may still experience a growth of irrigation agriculture, with the greatest growth coming in the largely agricultural states bordering the Mississippi River. Significant percentage gains in irrigation agriculture will occur in the more humid regions of the eastern United States although actual acreage increases will be relatively small. Table 4.11 sums up the totals of irrigated acreage for the United States, lumping the 17 western or Bureau of Reclamation states together and separating out Arkansas, Louisiana, and Florida from the remaining eastern states. It shows clearly that the greatest increase in irrigated acreage will still occur in the 17 western states although there will continue to be an increasing amount of new irrigation in the states east of the Mississippi River.

### Irrigation Development in the 17 Reclamation States

Table 4.12 provides information on the source of funds for the development of irrigation water supplies in the 17 western or Bureau of Reclamation states. It shows that non-Federal funds have been used to develop approximately 80% of the presently irrigated acreage in those states. This is to be expected on the basis of the type of water supplies developed. If at all possible, the farmer will develop his own water supplies first, whether from a nearby stream or pond or from the water table. Such sources are relatively inexpensive to use and the farmer has more direct control over supply and use of water. Only if such supplies are not available will most farmers consider water from more distant sources or water supplied by an irrigation district or from large reservoirs which have been built with Federal funds. Clearly, the individual farmer, or even a group of farmers, cannot afford the cost of a large water supply reservoir. Such projects are only

**Table 4.10**  Total Agricultural Acreage Under Irrigation in 1975 and 2000 (Estimated) by State

|                | 1975      | 2000      |
|----------------|-----------|-----------|
| California     | 8,731,115 | 9,002,069 |
| Texas          | 6,949,213 | 5,701,733 |
| Nebraska       | 3,330,884 | 3,933,899 |
| Colorado       | 2,907,727 | 2,754,360 |
| Idaho          | 2,896,328 | 3,251,037 |
| Montana        | 1,895,567 | 1,871,228 |
| Florida        | 1,599,741 | 1,923,376 |
| Oregon         | 1,595,474 | 1,597,601 |
| Kansas         | 1,588,760 | 2,280,612 |
| Wyoming        | 1,553,270 | 1,535,902 |
| Arizona        | 1,249,935 | 1,377,458 |
| Washington     | 1,219,089 | 1,639,071 |
| Arkansas       | 1,098,432 | 1,375,314 |
| Utah           | 1,065,053 | 1,124,464 |
| New Mexico     | 872,215   | 1,103,588 |
| Nevada         | 825,327   | 878,357   |
| Louisiana      | 654,354   | 683,350   |
| Oklahoma       | 524,909   | 780,878   |
| Mississippi    | 178,974   | 222,285   |
| South Dakota   | 150,421   | 347,396   |
| Missouri       | 146,199   | 338,714   |
| Hawaii         | 118,147   | 109,193   |
| Wisconsin      | 103,974   | 232,436   |
| New Jersey     | 98,410    | 168,531   |
| North Carolina | 96,651    | 199,080   |
| Puerto Rico    | 90,000    | 55,000    |
| Georgia        | 89,193    | 204,843   |
| New York       | 77,510    | 128,772   |
| North Dakota   | 77,386    | 317,029   |
| Michigan       | 77,028    | 158,725   |
| Virginia       | 52,888    | 106,440   |
| Illinois       | 43,326    | 101,460   |
| Minnesota      | 38,938    | 79,238    |
| Indiana        | 32,306    | 66,118    |
| Iowa           | 27,701    | 59,855    |
| Delaware       | 24,969    | 53,600    |
| Pennsylvania   | 23,654    | 39,043    |
| Maryland       | 23,550    | 50,233    |
| Ohio           | 23,350    | 42,663    |
| Kentucky       | 21,190    | 42,634    |

**Table 4.10**  (contd.)

|  | 1975 | 2000 |
|---|---|---|
| Massachusetts | 21,107 | 22,310 |
| South Carolina | 20,818 | 28,448 |
| Alabama | 15,432 | 24,480 |
| Tennessee | 13,927 | 19,679 |
| Maine | 13,374 | 21,354 |
| Connecticut | 10,780 | 13,112 |
| Alaska | 4,000 | 4,000 |
| West Virginia | 3,434 | 6,279 |
| New Hampshire | 2,290 | 4,229 |
| Rhode Island | 1,539 | 2,102 |
| Vermont | 755 | 917 |

*Source.* Data supplied through the courtesy of Marlyn Hanson and Rajinder Bajwa, National Resources, Economics Division, U.S. Dept. of Agriculture.

feasible with the commitment of massive amounts of Federal dollars. The Federal government spends large amounts of money for the total volume of water delivered because of the need to build expensive storage and transmission facilities.

For example, of the some 7 million acres that are irrigated with water developed from Federal funds, more than 60% receive water from storage facilities (reservoirs, lakes, ponds). About 30% of those acres are irrigated from water supplies from streams and rivers and the remaining few percent are irrigated by water from

**Table 4.11**  Summary of Irrigated Acreage in the United States, 1939–2000 (in million acres)

| Year | 17 Bureau of Reclamation States[a] | Arkansas, Louisiana, Florida | All Others | Total U.S. |
|---|---|---|---|---|
| 1939 | 17.2 | 0.7 | .04 | 17.9 |
| 1954 | 27.0 | 2.0 | .59 | 29.6 |
| 1975 | 37.4 | 3.4 | 1.40 | 42.2 |
| 2000 | 39.6 | 4.0 | 2.55 | 46.2 |

[a]All states west of the Mississippi River except for Louisiana, Arkansas, Missouri, Iowa.

**Table 4.12**  Sources for Funds for Irrigation Water Supplies in 17 Reclamation States (in million acres)

| Year | Non-Federal | Federal | Total | Cumulative |
|------|-------------|---------|-------|------------|
| Before 1900 | 6.7 | 0.4 | 7.1 | 7.1 |
| 1900–1909 | 3.6 | 0.9 | 4.5 | 11.6 |
| 1910–1919 | 2.8 | 1.0 | 3.8 | 15.4 |
| 1920–1929 | 1.0 | 0.5 | 1.5 | 16.9 |
| 1930–1939 | 0.9 | 0.6 | 1.5 | 18.4 |
| 1940–1949 | 3.0 | 1.4 | 4.4 | 22.8 |
| 1950–1959 | 6.0 | 2.0 | 8.0 | 30.8 |
| 1960–1969 | 1.7 | 2.7 | 4.4 | 35.2 |

groundwater supplies. The greatest amount of Federal money has gone into the building of large surface storage facilities. In contrast, of the more than 25 million acres of land irrigated by non-Federal sources of funds, over 40% receive water from surface streams and, of the remaining, less than 20% receive water from storage facilities developed from non-Federal funds. The emphasis has, and will continue to be, for non-Federal funds to be channeled into the more inexpensive sources of water (groundwater and surface streams) and Federal funds to be spent on the more expensive water supplies (storage facilities).

### Type of Irrigation and Crops Irrigated

It is difficult to generalize on the type of crops being irrigated today since almost all crops are irrigated in one place or another. The four most important categories, however, are hay (including forage and pasture), cereal grains, cotton, and vegetables. These four categories include some 87% of all irrigated land with hay (forage and pasture), by far the most important irrigated crop in terms of acreage. Approximately 43% of all irrigated land in the United States is devoted to hay, forage, sorghum, soybeans, and pasture, with about 27% devoted to cereal crops. Table 4.13

**Table 4.13**  Percent of All Irrigated Land in Different Crops, United States, 1975

| | |
|---|---|
| Hay and pasture (including forage, sorghum, and soybeans) | 43.2 |
| Cereal grains (including wheat, rice, corn, oats, barley) | 26.9 |
| Cotton | 8.6 |
| Vegetables (including potatoes, peas, beans) | 8.4 |
| Fruits and vineyard | 4.3 |
| Sugarcane and sugar beets | 3.1 |
| Others | 5.5 |
| | 100.0 |

gives the actual figures based on a survey by the U.S. Department of Agriculture completed in 1978. Clearly, these figures only apply nationwide since quite a different mixture of irrigated crops is found in each state and region of the country.

Wiesner (1970) has provided a complete summary of the different methods of irrigation including flood irrigation, furrow irrigation, sprinkler irrigation, and subirrigation. His summary permits comparisons of costs, efficiency, amount of water required, types of crops on which each technique may be used most successfully, and many other aspects of each major type of irrigation. For example, Wiesner estimates that controlled flood and furrow irrigation both cost the least to install ($60-70 per acre) while sprinkler irrigation might cost around $80-90 per acre. Subirrigation is, by far, the most expensive with installation costs running between $300 and $350 per acre. Annual costs (including interest on capital and operating costs) are somewhat reversed with subirrigation costing $4-6 per acre-foot, controlled flood irrigation $7-8 per acre-foot, furrow irrigation $8-10 per acre-foot, and sprinkler irrigation $12-15 per acre-foot.

While flood, furrow, and sprinkler irrigation may be adapted to most soils, subirrigation can really only be used on soils capable of lifting moisture into the root zone. Sprinkler irrigation can be undertaken on sloping topography although the other three types of irrigation require fairly uniform and generally quite gradual slopes. Wiesner (1970) suggests that subirrigation may have the highest irrigation efficiency (possibly up to 90-95%) although in practice it usually varies between 70 and 80%. Sprinkler irrigation also has very good efficiency (65 to 80%), followed by flood irrigation (50 to 90% efficiency) and then by furrow irrigation (50 to 70% efficiency). Flood and furrow irrigation both require some form of drainage facilities (often built into the original irrigation design) while subirrigation and sprinkler irrigation usually do not require any provision for drainage since the application of water may be more carefully controlled. Again, flood and furrow irrigation require fairly large supplies of water, while subirrigation can utilize a flow as small as 20 gal per minute. Sprinkler irrigation can also use fairly limited water supplies depending on the size and number of sprinkler nozzles. Soil or crop damage is quite unlikely with subirrigation while both types of damage are possible with sprinkler irrigation, especially if very large volume sprinkler heads are used. Furrow irrigation may often lead to furrow erosion while flood irrigation may produce general overwatering, puddling, or salt build-up in the soil (Wiesner, 1970).

## Types of Agricultural Drought

Drought occurs whenever the supply of moisture from precipitation or storage in the soil is insufficient to fulfill the water need of plants.* At first, failure of the climate and soil to provide sufficient water to the vegetation cover for evaporation and transpiration (evapotranspiration) is relatively unimportant, and this

---

*An important and comprehensive investigation of the problems of meteorological drought was undertaken by Palmer (1965).

early drought is usually overlooked. Before long, however, continued failure of the water supply to satisfy the water need results in some wilting of plant leaves and a reduction in growth and yield. Oftentimes, even these periods of drought go almost unnoticed by farmers. Plants have not been killed and yields are only reduced from some potential value which is seldom achieved at any time. It is only when the water supply fails to satisfy the water need for a fairly long and continuous period that many people begin to recognize that drought exists and there is a potential danger of crop failure. To many, this is now the time to bring out the irrigation equipment in the hopes of preventing a total crop failure.

Thornthwaite (1947; Anonymous, 1949) described four types of drought which result from a failure of water supply to satisfy plant need. He has called these permanent, seasonal, contingent, and invisible drought. *Permanent* drought is found in the desert areas where, in no season, does precipitation equal plant water need. Plants are adapted to the dry conditions and can survive only by elaborate means of controlling transpiration losses or by rapid growth and reproduction following the limited rains that occur. Agriculture is impossible without irrigation throughout the whole crop season. *Seasonal* drought occurs in regions with well-defined rainy and dry seasons. These droughts result from large-scale seasonal circulation changes and can be expected each year. Agriculture is possible during the rainy season or with the use of irrigation in the dry season. *Contingent* droughts result from irregular and variable rainfall. These droughts are characteristic of subhumid and humid areas as a result of the occurrence of significant periods without rain. They can occur at any season but are usually most severe during periods of greatest water need. They are serious because of their unpredictability and because periods of crop failure can result without supplemental irrigation.

While these three types of drought are fairly evident by the wilting of crops or the lack of much vegetative growth or appreciable yields, the fourth type of drought—*invisible* drought—is less easily recognized. Invisible drought can occur at any time, even during periods with precipitation, when the daily supply of moisture fails to equal the daily need of the plants. As a result, there is a slow drying of the soil and plants fail to grow at their optimum rate. To the casual observer, there is no appearance of drought. Plants continue to grow, little or no wilting is observed, and yields are "normal." Frequent rains lull the farmer into a false sense of security that growing conditions are satisfactory. He does not realize that plants cannot obtain all the water they need and so yields are less than those potentially possible. Invisible drought is almost entirely confined to humid regions, where frequent rains would seem to argue against the need for an irrigation agriculture. Invisible drought is the most insidious in reducing potential farm yields and can be most successfully combated with supplemental irrigation. At the same time, the economic need for supplemental irrigation to combat invisible drought is the most difficult to establish.

### Economic Feasibility of Irrigation in Humid Areas

Present and future demand for food and fiber require that we maintain a high and efficient agricultural production. To do this, it will undoubtedly be necessary to increase our irrigation agriculture. Serious thought must be given to the further development of irrigation agriculture in semiarid and arid areas. But water supplies are limited and expensive there. The costs of land management for irrigation are high.

Where then should we turn in our search for increased agricultural productivity? Certainly improved, more viable seeds, better farming practices, and increased use of fertilizers have contributed significantly to present rising farm production. Possibly a more overlooked aspect that deserves serious consideration in any assessment of our national agricultural potential is the increased use of irrigation in humid and subhumid areas.

To evaluate the possibility of increasing agricultural production in humid areas, it is first necessary to understand the real significance of drought in such areas and to determine whether the water deficiency that exists from time to time in humid and subhumid areas affects yields appreciably. We must then be assured that supplemental irrigation in those humid areas is economically feasible. Finally, we must clearly recognize that the problems and practice of irrigation in humid areas differ significantly from irrigation in arid or semiarid areas, chiefly in the need to be able to schedule the timing and quantity of supplemental irrigation to maximize profitability.

While supplemental irrigation will prevent the hidden and contingent droughts of humid and subhumid areas, it is feasible only if the expense of irrigation can be justified by the increase in value of the crop being irrigated. It is difficult to generalize on economic aspects since the fixed and operating costs of irrigation equipment vary so widely from farm to farm and year to year, depending on the type of equipment being used, the cost of the water supply, the frequency of use of the equipment, labor costs, and the cost of other farming practices which must accompany supplemental irrigation.

The need for a complete program of good farming practices to accompany irrigation has been clearly shown many times. For example, Table 4.14 summarizes

**Table 4.14**   Yield of Irish Potatoes Under Different Farming Programs in Alabama

| Treatment | Yield (bu/acre) |
|---|---|
| No treatment | 28 |
| 500 lbs fertilizer/acre (6-10-4) | 80 |
| 1000 lbs fertilizer/acre | 115 |
| 1000 lbs fertilizer/acre, organic material | 162 |
| 1000 lbs fertilizer/acre, organic material, irrigation | 229 |

*Source.* Ware and Johnson (1950).

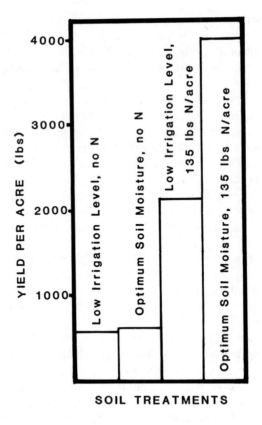

SOIL TREATMENTS

**Figure 4.7**  Wheat yields in Mexico, with different
soil treatments.
*Source.* Adapted from data reported in Richardson
(1960).

the farming program and the yields that resulted during a 4-year study of Irish potatoes in Alabama.

Cooke (1972) showed the effect on wheat yields of nitrogen fertilizer applied with and without irrigation in Great Britain. Without irrigation, maximum yields were obtained with 100 kg N/ha. While irrigation increased yields at each rate of application of nitrogen, he found that maximum yields occurred with a heavier nitrogen application. Thus, yields increased nearly 1 ton/ha by use of irrigation (a 20% increase) although to obtain this maximum increase nitrogen fertilizer application also had to be increased from 100 kg/ha to 150 kg/ha.

Similar results were also found in a study of wheat yields in Mexico in a combined fertilizer-irrigation experiment (Fig. 4.7). Optimum soil moisture conditions, without fertilizer, increased wheat yields only 5% over yields found under dry soil moisture conditions. Addition of 135 lbs of nitrogen per acre to the dry soil plots

**Table 4.15**  Effect of Nitrogen and Moisture on the Yield of Sweet
Corn

| Soil Moisture Condition[a] | Kg Nitrogen per Ha | | | |
|---|---|---|---|---|
| | 0 | 56 | 112 | 224 |
| | Yield in Tons | | | |
| Wet | 7.44 | 10.11 | 12.89 | 18.38 |
| Medium | 6.84 | 10.16 | 14.69 | 12.78 |
| Low | 6.88 | 11.43 | 13.14 | 12.69 |

[a]Wet plots irrigated when soil moisture in the root zone reached a tension of 0.3 bar. Medium plots irrigated when available soil moisture in root zone reached approximately 50% of field capacity. Low moisture plots irrigated only when soil moisture in the root zone approached the wilting point.
*Source.* Peterson and Ballard (1953).


increased yields by 3.7 times. Combining the increased nitrogen application with optimum soil moisture conditions increased yields almost seven times over the original yields. Such yield increases more than paid for the cost of the fertilizer and water applied.

The yield of sweet corn has been found to increase with nitrogen application and water just as wheat yield did (Table 4.15). Here, however, while the "wet" soil moisture condition gave the greatest yields with 0 nitrogen application and again with the heaviest application tested (224 kg N/ha), corn yields were higher on the medium and low moisture plots for intermediate values of nitrogen application. The greatest yield differences were found with the heaviest nitrogen application. Just as in the earlier wheat study, yield of corn decreased on the low moisture plots as nitrogen application was increased from 112 to 224 kg/ha.

Shipley and Regier (1975) have carried out an interesting irrigation experiment on grain sorghum on the High Plains of Texas. Over a 2-year test period, they applied one, two, three, and four irrigations per season in various combinations. The timing of the irrigation application was determined by the stage of development of the crop. All of the various plots were given an application of water prior to planting. The results are shown in Table 4.16.

As might be expected, highest yields were achieved with four applications of irrigation, one at each stage of development. In this program, the sorghum never suffered greatly from lack of water. With only one irrigation, the greatest yields were found when the irrigation occurred during the mid-to-late-boot period although nearly the same yields were found if the single irrigation had been applied between heading and flowering. Clearly, in these two stages of development, the crop is most water sensitive. Thus, if two irrigations are possible, highest yields

**Table 4.16**  Summary of Yields and Irrigation Treatment of Grain Sorghum Response to Irrigation Applications at Different Stages of Growth

| Irrigation Application at Stages of Growth | | | | Grain Sorghum Yield, kg/ha | | |
|---|---|---|---|---|---|---|
| 6–8 Leaf | Mid to Late Boot | Heading and Flowering | Milk to Soft Dough | 1969 | 1972 | 2-year Average |
| Preplant only | | | | 1615 | 3123 | 2369 |
| Preplant plus one 10 cm irrigation | | | | | | |
| X | – | – | – | 896 | 3186 | 2041 |
| – | X | – | – | 4505 | 4763 | 4634 |
| – | – | X | – | 3550 | 5502 | 4526 |
| – | – | – | X | 1279 | 3663 | 2471 |
| Preplant plus two 10 cm irrigations | | | | | | |
| X | X | – | – | 4102 | 4380 | 4241 |
| X | – | X | – | 4687 | 6401 | 5544 |
| X | – | – | X | 1412 | 4709 | 3060 |
| – | X | X | – | 5871 | 6257 | 6064 |
| Preplant plus three 10 cm irrigations | | | | | | |
| X | X | X | – | 7170 | 6715 | 6942 |
| X | X | – | X | 4166 | 6247 | 5206 |
| – | X | X | X | 6677 | 6681 | 6679 |
| Preplant plus four 10 cm irrigations | | | | | | |
| X | X | X | X | 7623 | 7603 | 7613 |

*Source.* Shipley and Regier (1975).

will be found with irrigations at the mid-to-late-boot period and the heading-to-flowering period. With the use of a third irrigation, yields are increased more if it occurs in the 6-to-8 leaf stage rather than later in the milk-to-soft-dough stage. The experiment suggests that there are definite times within the life of the plant when yields are more related to water applications than at other times.

Many crops have critical water stress periods (Table 4.17). If water is in limited supply and cannot be applied over the whole crop growing season, it would be wise to schedule the irrigation during these critical periods for maximum yield response. Cotton, while included in Table 4.17, has long been considered a dry land crop or at least one that can do well with a limited amount of water. For example, Christidis and Harrison (1955) showed that cotton growing at Shafter, California and irrigated 12 times in the year produced a larger yield than cotton irrigated only 7 times, while that irrigation program resulted in more seed cotton

**Table 4.17** Sensitive Growth Periods for Water Deficit

| | |
|---|---|
| Alfalfa | Just after cutting (and for seed production at flowering). |
| Banana | Throughout but particularly during first part of vegetative period, flowering and yield formation. |
| Bean | Flowering and pod filling; vegetative period not sensitive when followed by ample water supply. |
| Cabbage | During head enlargement and ripening. |
| Citrus | |
|   Grapefruit | Flowering and fruit set > fruit enlargement. |
|   Lemon | Flowering and fruit set > fruit enlargement; heavy flowering may be induced by withholding irrigation just before flowering. |
|   Orange | Flowering and fruit set > fruit enlargement. |
| Cotton | Flowering and boll formation. |
| Grape | Vegetative period, particularly during shoot elongation and flowering > fruit filling. |
| Groundnut | Flowering and yield formation, particularly during pod setting. |
| Maize | Flowering > grain filling; flowering very sensitive if no prior water deficit. |
| Olive | Just prior flowering and yield formation, particularly during the period of stone hardening. |
| Onion | Bulb enlargement, particularly during rapid bulb growth > vegetative period (and for seed production at flowering). |
| Pea | Flowering and yield formation > vegetative, ripening for dry peas. |
| Pepper | Throughout but particularly just prior and at start of flowering. |
| Pineapple | During period of vegetative growth. |
| Potato | Period of stolonization and tuber initiation, yield formation > early vegetative period and ripening. |
| Rice | During period of head development and flowering > vegetative period and ripening. |
| Safflower | Seed filling and flowering > vegetative. |
| Sorghum | Flowering yield formation > vegetative; vegetative period less sensitive when followed by ample water supply. |
| Soybean | Yield formation and flowering; particularly during pod development. |
| Sugar beet | Particularly first month after emergence. |
| Sugarcane | Vegetative period, particularly during period of tillering and stem elongation > yield formation. |
| Sunflower | Flowering > yield formation > late vegetative, particularly period of bud development. |
| Tobacco | Period of rapid growth > yield formation and ripening. |
| Tomato | Flowering > yield formation > vegetative period, particularly during and just after transplanting. |
| Watermelon | Flowering, fruit filling > vegetative period, particularly during vine development. |
| Wheat | Flowering > yield formation > vegetative period; winter wheat less sensitive than spring wheat. |

*Source.* Doorenbos and Kassam (1979) with permission of Food and Agriculture Organization of the United Nations.

**Table 4.18**  Influence of Irrigation Treatment on Fruiting and Yield of Seed
Cotton, Watkinsville, Georgia, 1956 to 1957

| Irrigation Treatment | Number of Fruit per 10 ft of Row | | | | Yield lb/acre | |
| | 7/23 | 8/27 | 7/30 | 9/3 | 1956 | 1957 |
| | 1956 | | 1957 | | | |
| --- | --- | --- | --- | --- | --- | --- |
| Irrigate at 60% AWC[a] | 456 | 250 | 643 | 280 | 3257 | 2759 |
| Irrigate at 30% AWC | 410 | 250 | 640 | 261 | 3463 | 3282 |
| Irrigate at observed wilting | 448 | 218 | 635 | 275 | 2911 | 2982 |
| Irrigate 5 days after observed wilting | 492 | 270 | 532 | 230 | 3306 | 3006 |
| Irrigate at 0% AWC | 416 | 191 | 633 | 251 | 3621 | 3071 |
| No irrigation | 467 | 181 | 400 | 140 | 1952 | 1844 |

[a] Available water capacity.
*Source.* Carreker and Cobb (1963).

than plants irrigated 21 times. Finally, Carreker and Cobb (1963) found the highest
yields in cotton that was irrigated when the soil was at 30% or less of available
water capacity (Table 4.18).

These experiments suggest strongly the need for careful scheduling of irrigation applications and the applying of just the right amount of water at the time
when the plant is most responsive. Considerable savings in water might be possible
with a more scientifically based irrigation program for all irrigated crops.

### Problems Associated with Saline Conditions

Salt build-up in irrigated soils, especially in arid and semiarid areas, has resulted
in the loss of productivity over vast areas. The salt problem usually results from
the practice of only moistening the root zone of soils. Salts brought in with the
irrigation water thus remain behind and build up in the upper soil layers. Even
heavier irrigation applications to try to leach the accumulated salt from the soil
may not help if the concentration of salt in the applied water exceeds that in
the leachate. In soils with water tables near the surface, saline groundwater may
flow upward by capillary action and result in an increase in the accumulation
of salt.

Hansen et al. (1980, p. 96) have listed the following techniques for temporary
control of salts on irrigated tracts: (a) deep plowing of salt crusts to mix soil;
(b) removal of surface salt accumulations; (c) neutralization of salts by addition
of other salts or acids.

More permanent rehabilitation of soils with salt accumulation involves: (a) lowering of water table to permit leaching and prevent salt accumulation from below;

**Table 4.19**  Increase in Salt Content of the Water in Typical Rivers Due to Seepage and Return Flow of Water from Irrigated Lands

| River | State in U.S.A. | Salt Content (mg/l) Upper | Lower | Increase (mg/l) | Distance (km) | Increase Per km (mg/l) |
|---|---|---|---|---|---|---|
| Colorado | Colorado | 110 | 1178 | 1068 | 32 | 33 |
| Jordan | Utah | 890 | 1970 | 1080 | 23 | 47 |
| Sevier | Utah | 205 | 831 | 626 | 97 | 6 |
| Sevier | Urah | 205 | 1316 | 1111 | 241 | 5 |
| Pecos | New Mexico | 760 | 2020 | 1260 | 48 | 26 |
| Pecos | New Mexico | 760 | 5000 | 4240 | 290 | 15 |
| Arkansas | Colorado | trace | 2200 | 2200 | 193 | 11 |

*Source.* Hansen et al. (1980) with permission of John Wiley & Sons.

(b) increasing water infiltration into the soil; (c) increasing rates of leaching of salts; (d) continued good management practices designed to achieve the long-term goal even at the expense of short-term productivity and profits.

Salinity in irrigation water from wells results from the prior leaching of salts to the water table. Vast areas of the United States are underlain with brackish or slightly saline water and continued use of this water in irrigation can result in serious salinity problems in the upper soil layers. Water obtained from surface streams can also be brackish or saline because of seepage and return flow of water to those streams. Also, phreatophytic vegetation along the river course removes fresh water from the stream resulting in increased salinity downstream. Table 4.19 illustrates the increase in salinity in some of the major dry area streams.

Crops differ appreciably in their tolerance to salt, as can be seen in Table 4.20. Use of the proper type of vegetation for the existing salinity conditions may allow lands already suffering from salt accumulation to be used profitably while more permanent rehabilitation of the land is progressing.

**Problems of Irrigation in Humid Areas**

Irrigation is absolutely necessary for large-scale agriculture in arid regions but its need in humid regions depends more on the distribution of precipitation in the given season and the economics of cost-benefit analysis. Hansen et al. (1980) suggest the need for careful consideration of many factors in planning humid-area irrigation that are quite unlike those found in more arid areas.

(a)  With irrigation and precipitation, provisions for drainage may become critical.
(b)  Precipitation shortly after irrigation may result in leaching of fertilizers.
(c)  Irrigation may lengthen the harvest season and result in greater wet weather hazard during harvest.

**Table 4.20**   Tolerance of Three Types of Crops for Salinity (in each group, most tolerant first, least tolerant last)

| Type of Crop | Salt Tolerance | | | |
|---|---|---|---|---|
| | Tolerant (Group I) | Moderately Tolerant to Moderately Sensitive (Group II) | | Sensitive (Group III) |
| Fruit | Date palm | Pomegranate Fig Grape Olive | | Lemon Grapefruit Pear Almond Apricot Peach Plum Apple Orange |
| Field and truck | Sugar beet Garden beet Milo Rape Kale Lettuce Alfalfa | Alfalfa Flax Tomato Asparagus Foxtail millet Sorghum (grain) Barley (grain) Rye (grain) Oats (grain) Rice | Cantaloupe Sweet potato Sunflower Carrot Spinach Squash Sugarcane Corn Onion Pepper Wheat (grain) Cotton Potato | Vetch Peas Celery Cabbage Artichoke Eggplant Sweet potato Potato Green beans Black walnut Pecan |
| Forage | Alkali sacaton Salt grasses Nuttail alkali Bermuda Rhodes Rescue Canada wild rye Beardless wild rye Western wheat grass Barley (hay) | White sweet clover Yellow sweet clover Perennial rye grass Mountain brome Barley (hay) Birdsfoot trefoil Strawberry clover Dallis grass Sudan grass Hubam clover Alfalfa (California common) Tall fescue Rye (hay) | Wheat (hay) Oats (hay) Orchard grass Blue grama Meadow fescue Reed canary Big trefoil Smooth brome Tall meadow oat grass Cicer milk vetch Sour clover Sickle milk vetch | White Dutch clover Meadow foxtail Alsike clover Red clover Ladino clover Burnet |

*Source.* Hansen et al. (1980) with permission of John Wiley & Sons.

(d) Since the margin for profit is smaller, good farm management is important and this includes the quality and amount of available labor and water.
(e) Farm machinery requirements are greater because fields are workable for a shorter time period.
(f) Irrigation supplements precipitation rather than providing all needed water so that cost per unit of benefit is greater.
(g) The increased wetness will result in greater opportunity for erosion and runoff.

Thus, in the humid area, irrigation may only add to a profit figure already existing because of natural rains. Irrigation must be more carefully scheduled in a humid area so as to put on the right amount at the right time since the margin for profit is much smaller in a humid area than in an arid area. Inefficient irrigation habits, while costly anywhere, might so reduce profit margins in a humid area that irrigation would not be economically feasible. This is much less true in an arid or semiarid area for the reason that no profit can exist without irrigation.

To utilize an irrigation system most efficiently and to maximize profitability, irrigators must have answers to the basic questions of when and how much to irrigate. Inasmuch as the value of irrigation is to prevent any reduction in yield through lack of adequate supplies of moisture, the most satisfactory program of irrigation would be one which calls for small, almost daily irrigations to supply the crops with those daily needs that are not supplied by precipitation. This irrigation program would eliminate all opportunity for invisible drought and the reduced yields associated with it. In practice, irrigation in such a manner is not feasible because the cost in labor and equipment would outweigh the profits from increased yields. Rather, a compromise irrigation program can allow a certain amount of drying of the soil to occur before irrigation water is applied. In climates where rainfall is possible in all seasons, just enough water should be applied to bring the moisture in the root zone of the soil to some value slightly below field capacity. Then an unpredicted rain immediately after irrigation will not lead to excessive leaching of fertilizers.

The irrigation program finally selected must balance the cost of a program of fairly frequent light irrigations to prevent invisible drought against the returns through increased yields as a result of the irrigation. While environmental conditions in different areas will suggest different values of permissible soil moisture deficit before irrigation becomes economically feasible, the need to have a simple, reliable system by which irrigation need can be determined on a day-to-day basis is still of greatest importance to the efficient use of irrigation equipment.

## Relation of Crop Yields to Water Deficit

Crop yields are related to a combination of climatic factors that express directly how satisfactorily the plant's requirements for moisture and energy have been met. The climatic water deficit gives, in quantitative terms, a measure of how well the plant's water need is satisfied. It is defined as the difference between the plant's climatic water requirements (potential evapotranspiration) and what the plant actually is able to obtain from the environment (actual evapotranspiration).

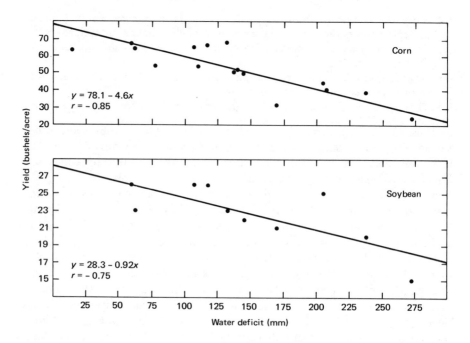

**Figure 4.8**  Relation between corn and soybean yield and water deficit, Seabrook, New Jersey, 1947–1961.
*Source.* Mather (1968), with permission of the Association of Pacific Coast Geographers.

If plants grow better and produce higher yields when they do not suffer a lack of water, a significant correlation should exist between yield and water deficit. Rouse (1962) found a correlation coefficient of -.71 between sugarcane yield in Barbados and the climatic water deficit, computed on a monthly basis, indicating a decrease in sugar yields as total moisture deficit for the crop-growing season increases.

The values of average crop yields by years for Cumberland County, in southern New Jersey, have been compared with the water deficit from a single station, Seabrook, in Cumberland County. Figure 4.8 shows the relation between crop yields and moisture deficit and the calculated regression line. The coefficients of correlation are -.85 for corn yield and -.75 for soybean yield.

In each case, crop yield is reduced as total water deficit during the growing season increases. In considering these results, several factors need to be stressed. First, over the period studied, there have been remarkable advances in the production of new, more viable and more productive seeds, and new fertilizers have become available. Cultivation techniques, methods of soil conservation, and techniques of rotation of crops to increase yields also have improved. The farmer has not been slow to make use of all these advances. Thus, superimposed on this relationship between crop yield and water deficit are the combined influences of improved seeds, fertilizers, and farming techniques which have tended to raise yields and mask climatic influences.

A way to adjust crop yield for the influence of changing weather and to reveal the influence of improved technology has been suggested by Shaw and Durost (1962). Between 1947 and 1954, corn yield in Iowa, adjusted to remove the influence of weather, increased from 45.5 bu/acre to 53.7 bu/acre. Between 1950 and 1959, weather-adjusted yield increased 9.8 bu/acre, from 54.2 to 64.0 bu/acre. These values of increased yields, holding weather factors constant, are quite similar to those found in New Jersey. A water deficit of 8.3 in. in 1947 is associated with a corn yield of 40 bu/acre, while a nearly similar deficit in 1954 resulted in a corn yield of 44 bu/acre, an increase of 4 bu/acre compared to an increase of 8.2 bu/acre in the corresponding period of Iowa. Similarly, 1950 and 1959 had essentially the same water deficits, yet the yield of corn increased by 12 bu/acre from 53 to 65 bu/acre. This compares favorably with the increase of 9.8 bu/acre found in Iowa when weather factors were removed.

Although the influence of improved technology has not been removed from the yield figures in Figure 4.8, there is a good correlation between yield and climatic water deficit. The scatter of the points around the regression lines is probably due, in large measure, to the influence of changing technology and the actual timing of the short periods of deficit through the year. Even without refining the present analysis to eliminate the influence of improved technology or the timing of the periods of deficit, the relatively small water deficits that occur in humid areas clearly do decrease crop yields. Supplemental irrigation, even in fairly humid areas, can increase these yields by decreasing the amount of water deficit.

## Future Irrigation Development

One can clearly anticipate a relocation of new irrigation enterprises from the most arid areas to the semiarid or subhumid regions of the country. Where possible, more irrigation agriculture will develop in the so-called humid regions of the country. These developments will, of course, call for more careful control over irrigation applications and a higher degree of irrigation efficiency.

In the more humid areas of the country, generally non-Federal funds will be used to influence irrigation developments since the cost of development will be low. In the arid and semiarid portions of the country, however, there should be a general decrease in the use of non-Federal funds for irrigation developments. Rather, there should be a pronounced shift, with nearly all future irrigation water supplies being financed by Federal funds in these areas. There will be an increase in schemes for large water-storage and water-transfer projects. One might question both the ethical and economic values of such large-scale water projects in dry areas. They will continue to require vast amounts of money to operate and, of course, once built they will establish a water-based economy in the area. Individuals attracted to the area because of the assurance of water cannot easily be removed if the water becomes increasingly expensive to supply. A moral obligation to continue to supply the area with water develops even though it might be more economically feasible to achieve the same agricultural production by using better agricultural techniques elsewhere. Great care must be exercised in establishing new irrigation enterprises to insure that they can be maintained with reasonable certainty for long periods without undue economic hardship on the rest of the country.

## 5

## MULTI-PURPOSE RESERVOIRS AND
## WATER NEEDS FOR POWER,
## NAVIGATION, AND RECREATION

## WATER USES IN MULTI-PURPOSE RESERVOIRS

### History

Early water projects in the West were developed almost entirely for the purpose of irrigation. As in the case of the Mormons in Utah in the late 1840s, such projects consisted largely of diversions of water from relatively small streams. Since water flows in such streams were inadequate to supply needs later in the dry summer season, early season crops had to be grown to take advantage of the higher spring and early summer streamflows. The acreage that could be so irrigated would still be somewhat limited by the late season flow of the stream. With the existence of periods of excess flow in the spring, capture, retention, and later use of some of the higher spring flow permitted both expansion of the area of cropped land and a longer growing season.

Where feasible, capture and retention started as off-stream storage in low-lying areas. This type of activity usually involved no more than a short canal and a few gates to control the flow of water and did not involve dams or locks on the main channel which could add significantly to the cost of the project and might disrupt transportation. Off-channel storage was feasible with the type of construction materials locally available although the size of the storage facilities was very limited.

As the importance of irrigation grew, and available off-channel sites became fully developed, small on-channel storage facilities were undertaken. Usually these involved deepening or enlarging pools in the main stream or the construction of small dams to retain larger volumes of water for later use. Always the activity was limited to the resources of a single farmer or a small group of farmers who banded together for the purpose. Competition for water was severe and many local fights developed over ownership of water rights or how to modify existing stream channels.

A great expansion in the demand for water in the West occurred in the last two decades of the 19th century. Some unproductive land was brought under cultivation and certain irrigation projects were poorly planned and built. Control reservoirs were not sufficiently large to retain spring floodwaters. More capital was required along with a better understanding of the hydrologic cycle.

By 1902, local development had gone about as far as it could in developing water supplies in some of the western areas. The passage of the National Reclamation Act of 1902 brought in Federal capital. A large number of new water resources projects were begun, essentially opening a new frontier where opportunity for economic development awaited the enterprising farmer. Many of the present economic centers in the West owe their very existence to the irrigation developments started with Federal assistance in the years following the passage of the Reclamation Act. One might include in such a list the Central Valley of California, the Yuma, Imperial and Coachella valleys of Arizona and California, the Yakima Valley and Columbia River basin of Washington, along with the Phoenix area of Arizona and the Boise Valley area of Idaho.

The 1902 Reclamation Act provided funds for irrigation only although it was not long before it was recognized that water power might be developed as an important by-product of any reservoir storage activity. Thus, 4 years after the initial act, the legislation was broadened to permit water power and municipal water developments to be included in projects funded under the Reclamation Act. Since water storage for irrigation and for water power are not entirely compatible, irrigation was still the prime control of reservoir activities and the development of water power was relegated to a secondary role.

With the influx of Federal capital, it became possible to develop middle-sized streams such as the Salt River in Arizona, the North Platte in Colorado, the Snake in Idaho, and the Yakima in Washington. Irrigation agriculture increased significantly in the first several decades of the 20th century, doubling the total acreage that had existed under irrigation before the turn of the century.

Water power was not to be denied, however. While waterwheels to drive grist and sawmills were a characteristic of early east coast settlements, it was only with the development of electricity in the 1880s that hydroelectric developments began. The first hydroelectric plant in the United States was built in Appleton, Wisconsin in 1882, producing 12.5 kW from a direct current generator, a far cry from today's 600,000 kW units at Grand Coulee and other large installations.

## Changing Ideas About Dam Construction

Dams to create the reservoirs for storing irrigation water and the head for power generation evolved slowly over time. In the 19th century, dams were generally limited to less than 100 ft in height. Development of engineering knowledge about the strength of materials, better understanding of loads and stresses within structures, construction of foundations, and dissipation of the energy of flowing water made it possible to increase dam heights and to enlarge reservoir capacities. The Roosevelt Dam in Arizona, in 1911, went to a height of 280 ft. Knowledge gained in that construction was applied to the building of the 350-ft high Arrowrock Dam in Idaho in 1915, which remained the world's tallest for 17 years. The 417-ft Owyha Dam in Oregon, in 1932, was followed quickly by the 726-ft high Hoover Dam in 1936, which was the world's highest for more than 20 years. It is still the highest concrete dam in the United States. Actually, there are six higher dams in the world now, the record holder being Dixence Dam in Switzerland built in 1962 to a height of 935 ft.

Even up to the 1930s, it was accepted that earth dams could not exceed 100 ft in height. However, significant advances in soil mechanics along with improved control of soil compaction techniques for earth embankments made it possible to exceed this long-standing limit. In 1940, the Fort Peck earth dam in Montana was built to a height of 250 ft, followed in 3 years by the Green Mountain earth dam in Colorado built to a height of 305 ft. The Anderson Ranch Dam in Idaho, completed in the late 1940s, rose to a height of 456 ft, the highest earth dam in the world for more than 20 years. Even these large earth dams have been over-shadowed by recent developments—the Oroville Dam in California built to a height of 770 ft in 1968, the Mica Dam on the Columbia River in Canada built to a height of 794 ft, and two Russian earth dams now under construction to heights of 1040 and 1066 ft. Scientific engineering principles have been used in analyzing the construction methods and materials and give assurance of their continued safety in spite of several failures of small dams in recent years.

### Problems in the Operation of Reservoirs

Hydroelectric power became an important consideration for reservoir construction by 1930. By this time, dams were sufficiently high to develop the head needed to generate significant amounts of electric power. With the larger dams, it was possible to develop larger streams which had heretofore been relatively undeveloped.

But the rise in hydropower developments created real problems in terms of the operation of reservoirs. For irrigation, water had to be stored in the reservoir and only released in the summer months as needed. For power, however, a steady release of water was needed year round. This steady release would draw down the reservoir too far before summer irrigation demand rose to the annual peak and so limit the value of the reservoir for irrigation uses. Conflicts developed and the problems of the reservoir manager became much more complex where water was to be used for both irrigation and power.

Floods were an increasingly significant problem in the 1920s, culminating in the massive Mississippi River flood of 1927 (see Chapter 11). Finally, in 1936, Congress passed the Flood Control Act in an effort to reduce future flood losses. Although the Act was a response to a real national need, it has, unfortunately, over the years resulted in an exacerbation of the flood problem rather than its gradual elimination. Action under the Act has indeed decreased the frequency of floods and greatly reduced the loss of life from flooding (better communications and transportation facilities to move individuals out of the paths of floods have probably been responsible for the real reduction in the loss of lives). But the existence of the Act and the work of the Corps of Engineers to try to control floods have greatly encouraged building on the floodplains. While 9 out of 10 floods on a river may have been eliminated, the flood that ultimately will come will do a significant amount of damage because flood control activities have encouraged high density occupancy of the floodplains.

One of the actions of the Corps of Engineers in implementing the Flood Control Act of 1936 was to construct dams (or to enlarge existing dams) for the purpose of temporary storage of floodwaters. Control of flood crests by a series of reservoirs would make it possible to spread out the crest and to reduce the possibility for downstream damage.

But flood control activities compete with both irrigation and power activities within a reservoir. For flood control, it is desirable to have the reservoir as empty as possible at the start of the spring flood season. For irrigation, the reservoir should be as full as possible at the start of the summer irrigation season. If the spring floods fill the reservoir, all is fine; but if the spring floods are not sufficient to fill the reservoir, there will be insufficient water for all of the summer irrigation need. And, of course, power requirements can best be met with a full reservoir all through the year and a steady release of water.

So the concept of multi-purpose reservoirs grew over the years but satisfactory solutions to the problems that arose when a reservoir was to be used for irrigation, flood control, and power all at the same time were not forthcoming. Clearly, operating a multi-purpose reservoir had an advantage because costs could be spread over a number of important activities, but there were also problems to be faced for none of the users could be satisfied fully.

Since the Second World War, one other use has developed for reservoirs—namely, recreation. With the increase in income and leisure time that has come to the American worker since the mid-1940s, campers and boats have become a basic part of many households across the country. Water-based recreation is a significant aspect of American life and, in many parts of the country, the reservoirs and lakes created for irrigation, power, and flood control have been utilized by the vacationer. Unfortunately, the demands of the boater, swimmer, and fisherman do not fully accord with the need for irrigation and flood control. Operation of a reservoir for power and for recreation would probably not cause any problems for both require a fairly constant reservoir level. For flood control, a nearly empty reservoir is needed although this is usually before the peak recreational use so that it may not conflict with recreational use. Irrigation demand, however, will draw down the reservoir significantly during the recreational season, leaving extensive mud slopes and docks and marinas far from the edge of the water surface. Achieving a workable compromise among all four of the major demands for reservoir storage is a difficult and demanding task.

## Seasonal and Cyclic Reservoirs

Reservoirs for irrigation are often divided into two types—those providing storage for an individual season and those providing storage for several seasons. The seasonal storage reservoirs, which usually have a fairly small capacity in relation to stream-flow, seek to equalize the flow over a given year. The so-called cyclic storage reservoirs, which seek to equalize the flow of the stream over a series of years, usually have a large storage capacity in relation to the streamflow. Thus, there can be significant variation in the streamflow from year to year but there will be sufficient storage in the reservoir to even out these variations and to supply a steady amount of water for irrigation purposes. The trend today is toward the building of cyclic storage reservoirs. In part, this is because (1) the trend is toward basin development of water resources rather than just the development of a single reach of the stream, and (2) the cost of storing water varies inversely with the capacity of the reservoir. The larger the reservoir capacity, generally the lower will be the cost of any unit volume of water stored within the reservoir.

## Losses in Water from Reservoirs

Of course, while reservoirs are useful because they provide storage capacity for water for irrigation, power, flood control, and recreation, they also result in some waste of water because of the various losses that can occur from reservoirs.

Water stored in a reservoir is subject to a number of losses. To achieve the maximum amount of water for use, the goal of the project operator is to keep these losses to a minimum. Several types of losses occur in the reservoir itself. For example, there is the continual loss of storage capacity in the reservoir as it fills with sediment. All streams carry sediment. A large portion of that sediment will settle out in the quiet water of the reservoir, decreasing reservoir capacity and reducing the ability of the reservoir to regulate the flow of water in the stream. Reservoirs are designed to have a certain amount of sediment storage—usually 50 years or more—based on the sediment load in the contributing streams. Thus, all reservoirs are slowly filling and will have a limited life. Of course, the sediment load of a stream can be partially controlled by land management practices on the watershed itself and sediment can be cleared from the reservoir by flushing or even by dredging. It is helpful to pass as much sediment through the reservoir as possible, not only to clear the reservoir but also to maintain a load of sediment in the stream below the reservoir. Without such a sediment load, the stream below the dam will begin to scour its channel and pick up a new sediment load. Thus, to keep downstream channel cutting to a minimum, it is desirable to pass through the reservoir as much sediment as possible even though the nature of the reservoir is to remove sediment.

Other losses in the reservoir include water losses by evaporation from the water surface itself and evapotranspiration from the vegetated areas around the edges of the reservoir. Large, shallow reservoirs will lose more water by evaporation than will narrow, deep reservoirs of the same capacity. Large, shallow reservoirs will also generally be bordered by gently sloping shorelines, providing an opportunity for the development of a dense cover of water-loving vegetation feeding on the nearby water table. Seepage from the bottom of the reservoir can also occur although under most conditions this is a fairly small portion of the overall losses from the reservoir itself. Although it is hard to generalize, it is often assumed that evaporation from the reservoir surface is about 2 to 3% of the inflow to the reservoir while transpiration from surrounding vegetation varies from 2 to 20% of the inflow but averages about 5% of the inflow. Seepage loss is about 1% of the inflow to the reservoir. Thus, losses of water from the reservoir may reach 25% of the inflow although the average value of losses is about 8%.

A second set of losses occurs in the distribution of water from the reservoir to the irrigation system itself in the canals and laterals. If the canals are unlined, there will be both seepage losses from the canal and losses by evapotranspiration from vegetation growing along the banks. These losses can be quite significant, depending on the length of the canal and the area through which it flows. Such losses can be greatly reduced by lining the canal. This is often done when the value of the water saved exceeds the cost of lining. When significant amounts of vegetation develop along canal banks, clearing may also be necessary just to maintain the flow of water in the canal.

In 1975, transmission losses (including operational losses, evaporation from the water surface, evapotranspiration, and seepage) from all water releases from Bureau of Reclamation projects amounted to 21% of the amount of water released. The largest loss from canals is by seepage. Seepage losses from unlined canals may average 40% of the water flowing in the canal. Worstell (1976), on the basis of over 700 seepage tests, found that losses in canals ranged from 0.1 to 2 $ft^3$ per square foot per day.

Canal linings may be divided into four general categories: earth, exposed, buried membrane, and miscellaneous (Bureau of Reclamation, 1963). Earth materials include the natural soil as well as the use of clays such as bentonite to mix with the natural soil to reduce seepage. Exposed linings include various types of asphalt, cement, soil-cement mixtures, or blocks of different materials. Buried membranes include hot-applied asphalt, rubber or plastic materials, or bentonite all covered by a layer of soil or gravel material. Buried membranes, while more expensive to install, have the advantage of added protection from perforation by growing plants or puncture by grazing animals.

As of September 1977, the Bureau of Reclamation had constructed some 314 irrigation canals with a total length of just under 6000 mi. Counting canals they had rehabilitated and had under construction, the total came to 376 canals with a total length of 7010 mi. Since the Bureau supplied water to only about 18% of the irrigated acreage in the United States, one can begin to appreciate the extent of the system used for the conveyance of irrigation water (Bureau of Reclamation, 1977).

One final loss in the system, called operational waste, results from the inability of the operator of the distribution system to forecast exactly the needs of all of the irrigators on the canal. Clearly, enough water must flow in the canal so that the last irrigator will have a sufficient flow to satisfy his needs. This often means that some water may be left over after he has taken his share—this is operational waste. If irrigators further up the canal do not take their full share because they do not feel the need for it, this additional water will also appear as operational waste, Water, once in the canal or distribution lateral, cannot generally be recovered and returned to the reservoir but must be allowed to flow unused to some lower water body. Operational waste can vary appreciably depending on the skill of the manager and the varying needs of the irrigators but may be anywhere from 5 to 50% of the water inflow to the canal.

Reservoirs have considerable utility in evening out the flow of water in a stream and in making available more water during low flow periods than would otherwise be possible. They do have certain losses associated with them so that a time is reached when additional reservoirs on a stream will actually not make any more water available for other uses. The losses associated with the reservoir may exceed the water made available by the storage reservoir. Along certain reaches of the Colorado River, we are at a stage now where additional reservoirs, if installed, can no longer provide new supplies of water for use by irrigators or industry. Under such conditions, it is time to stop reservoir development on the stream.

## WATER FOR POWER

Rivers have been used to turn wheels to grind grain and to lift water for many centuries. The method is attractive because water is not destroyed in the process of producing power. Once used and returned to the river, it can always be used again.

A water power project developed on a river having no reservoir or storage facilities, so that the power is obtained from the natural flow of water in the stream, is called a run-of-river project. The amount of power that can be developed under such conditions varies appreciably from season to season and from year to year. The minimum flow of water in the stream will determine the minimum amount of power that can be developed during the year—this is called the prime power. Any power that is generated from flows above the minimum flow is called secondary power. Of course, most run-of-river projects have the capacity to produce some secondary power. Since it is expensive to have the equipment necessary to produce secondary power installed but standing idle when river flows are not high enough, the amount of secondary power actually generated is usually not great.

Storage of water by means of a reservoir evens out the river flow during the year and thus can increase significantly the amount of prime power that can be produced by the streamflow. Storage, in other words, makes a run-of-river project more efficient by increasing minimum flows and, thus, increasing the amount of prime power that can be developed.

Power is created by a stream because of the flow of water between two different elevations (or the differences in "head"). The energy of the water turns a waterwheel which, in turn, drives a generator which produces electricity. A thousand cubic feet of water falling 42.6 ft in a controlled descent and without friction will generate 1 kw-hr of electrical energy. The reverse is also true for it will take 1 kw-hr of energy for a perfectly efficient electric pump to lift 1000 ft$^3$ of water 42.6 ft. The higher the dam and thus the greater the controlled fall of water, the greater will be the amount of energy produced by a given volume of water (Deming, 1975). The capacity of a hydropower system is measured in terms of kilowatts while the energy actually produced is given in kilowatt-hours. The ratio of the average demand for electricity over a particular time period to the peakload that might occur in that time period is known as the "load factor." This is an important aspect, for equipment must not only be available to supply power for the peakload period but actually somewhat above that in order to allow for outages or maintenance of equipment.

### Theoretical, Technical, Economic Potential Power

Since the development of water power depends on the volume of water and the amount of fall (head), it is possible to determine what might be called a "theoretical hydroelectric potential" for a region or country from information on streamflow volume and total vertical drop of all streams. The expression for the theoretical potential is $\Sigma 9.81$ qh (in kilowatts) where q is the arithmetic mean flow in the stream (cubic meters per second) at the midpoint of the watercourse averaged over a number of years and h is the head or difference in elevation between the origin and mouth of the stream (Schurr and Netschert, 1960).

The theoretical hydroelectric potential can never be realized since it assumes equalization of seasonal flows, complete utilization of flow volume and head, and 100% efficiency in the conversion of water power into electricity. If this provides a theoretical limit, a lower value—the "technical potential"—can be defined as the total water power that can be developed within a given technological state-of-the-art. Technical potential starts with the theoretical potential and then tries to account for losses due to alternative water uses, leakage, poor flow regulation, turbine and generator inefficiencies, and lack of continuous operation through the year. In addition, of course, the technical potential for a stream may be somewhat lower than the theoretical potential merely because it is not possible to locate a power site everywhere on a stream because the head may be too small or the geology of the site may not be favorable for use (Schurr and Netschert, 1960).

Another and more practical term, the "economic potential," may be defined as the water power that can be developed under the given state-of-the-art in technology and under given relative conditions of costs. This latter phrase can be assumed to mean the hydroelectric capacity that would have a competitive margin over other available energy sources. The economic potential may change over time as economic conditions and competitive energy prices change.

Each of the three foregoing potentials has a useful significance. The theoretical potential marks the limit imposed by the environment while the technical potential is the limit of water power that can be developed with a given technical state-of-the-art but without consideration of economic factors. The economic potential considers both technical and economic factors and suggests the limit of water power development under the existing practical limits of money and technical achievement. Of the three, the theoretical potential is most easily determined,

**Table 5.1**   World Potential and Developed Hydroelectric Power Capacity

| Region | Potential Hydroelectric Power ($10^3$ megawatts) | Percent of Total | Developed Hydroelectric Power Capacity 1967 ($10^3$ megawatts) | Percent Developed |
|---|---|---|---|---|
| North America | 313 | 11 | 76 | 23.0 |
| South America | 577 | 20 | 10 | 1.7 |
| Western Europe | 158 | 6 | 90 | 57.0 |
| Africa | 780 | 27 | 5 | 0.6 |
| Middle East | 21 | 1 | 1 | 4.8 |
| Southeast Asia | 455 | 16 | 6 | 1.3 |
| Far East | 42 | 1 | 20 | 48.0 |
| Australia | 45 | 2 | 5 | 11.0 |
| USSR, China and satellites | 466 | 16 | 30 | 6.4 |
| World | 2857 | 100 | 243 | 8.5 |

*Source.* Hubbert (1974) with permission of the American Water Resources Association.

although less realistic. Measurement problems become more serious when trying to evaluate the technical and economic potentials. Installed (or installable) capacity bears little relation to mean flow in the stream since it is more influenced by such things as range of flow conditions and how a particular site can be integrated into a whole power system.

### Developed Hydroelectric Capacity

Table 5.1 provides information on the world potential and developed hydroelectric power capacity while Table 5.2 summarizes total hydroelectric and thermal power generating capacity and production around the world. Table 5.2 shows that world-wide about 26% of installed power capacity is hydroelectric and 74% is thermal. North America is not particularly well endowed in terms of potential hydroelectric power, having only about 11% of the total world potential capacity. Of this, some 23% of the potential is already developed. As opposed to North America, Western Europe, with only 6% of the world potential hydroelectric power capacity, has developed some 57% of that capacity, an actual value in megawatts in excess of the developed capacity in North America. Africa has the greatest potential hydropower capacity (some 27% of world capacity) although it has developed just 0.6% of that capacity, the lowest percentage of any of the regions shown in Table 5.1.

**Table 5.2** World Hydroelectric and Thermal Power Generating Capacity and Production (Data as of Year End 1968) by Continent

| Geographical Division | Installed Capacity (MW)[a] | | | Energy Production (Gwh.)[b] | | | Population (1000) | Kwh. Per Capita |
|---|---|---|---|---|---|---|---|---|
| | Hydro | Thermal | Total | Hydro | Thermal | Total | | |
| North America | 79,772 | 272,230 | 352,502 | 373,468 | 1,261,731 | 1,635,199 | 269,261 | 6,073 |
| Central America | 483 | 577 | 1,060 | 2,104 | 1,875 | 3,979 | 15,563 | 256 |
| West Indies | 187 | 3,643 | 3,830 | 487 | 14,507 | 14,994 | 23,825 | 629 |
| South America | 11,396 | 12,623 | 24,019 | 47,914 | 41,226 | 89,140 | 180,017 | 495 |
| Europe[c] | 120,920 | 330,682 | 451,602 | 436,084 | 1,337,665 | 1,773,749 | 692,902 | 2,560 |
| Africa | 4,942 | 13,697 | 18,639 | 18,702 | 51,100 | 69,802 | 329,216 | 212 |
| Asia | 35,424 | 70,645 | 106,069 | 136,562 | 299,738 | 436,300 | 1,939,831 | 225 |
| Oceania | 5,683 | 10,653 | 16,336 | 18,133 | 39,468 | 57,601 | 18,041 | 3,193 |
| World | 258,807 | 715,250 | 974,057 | 1,033,454 | 3,047,310 | 4,080,764 | 3,468,656 | 1,176 |

[a] Megawatts = thousand kilowatts.
[b] Gigawatt-hours = million kilowatt-hours.
[c] Includes all of USSR.
*Source.* Federal Power Commission (1968).

**Table 5.3**  Hydroelectricity Production (Million kwh) and Installed Capacity (in 100 kw) of Leading Producers

| Country | Production 1965 | 1966 | Hydro % of Total Electricity Output (1965) | Installed Hydro Capacity |
|---|---|---|---|---|
| United States | 197,001 | | 17 | 44,492 |
| Canada | 117,063 | | 81 | 21,711 |
| USSR | 81,431 | | 16 | 22,244 |
| Japan | 76,739 | | 40 | 16,279 |
| Norway | 48,858 | | 99.8 | 9,783 |
| France | 46,429 | 50,736 | 49[a] | 12,683 |
| Sweden | 46,423 | | 95 | 9,278 |
| Italy | 42,367 | 44,000 | 51[a] | 13,955[b] |
| Brazil | 25,515 | | 85 | 5,391 |
| Switzerland | 24,015 | | 98 | 8,120 |
| Spain | 19,550 | | 62 | 8,141 |
| Austria | 16,083 | 17,327 | 73[a] | 4,054 |
| W. Germany | 15,365 | 16,800 | 9[a] | 4,072 |
| India | 14,807 | | 41 | 3,331[b] |
| Finland | 9,488 | 10,516 | 67[a] | 1,857[b] |
| Yugoslavia | 8,985 | | 58 | 2,265 |
| Mexico | 8,609 | | 50 | 2,327 |
| New Zealand | 8,588 | | 81 | 1,910 |
| Australia | 8,367 | | 23 | 2,092 |
| United Kingdom | 4,625 | | 2 | 1,760 |
| Czechoslovakia | 4,456 | | 13 | 1,540 |
| Portugal | 3,983 | | 86 | 1,311[b] |
| Chile | 3,954 | | 64.5 | 710 |
| S. Rhodesia | 3,864 | | 94 | 705 |
| Colombia | 3,218 | | 63 | 793[b] |
| Peru | 2,625 | | 68 | 680 |
| Turkey | 2,167 | | 44 | 510 |
| Bulgaria | 2,001 | | 20 | 768 |

[a] For year 1966.
[b] For year 1964.
*Source.* Beckinsale (1969) with permission of Methuen and Co., Ltd.

Table 5.3 provides comparative figures for hydroelectric production and installed capacity by country as of the mid-1960s. The breakdown by country reveals that in North America, Canada obtains some 80% of its electricity from hydropower plants as opposed to just 17% for the United States and 50% for

**Table 5.4**  Water Used for Hydroelectric Power in the United States, 1970

| State | Mgd | 1000 Acre-feet Per Year | State | Mgd | 1000 Acre-feet Per Year | State | Mgd | 1000 Acre-feet Per Year |
|---|---|---|---|---|---|---|---|---|
| Alabama | 130,000 | 150,000 | Maine | 80,000 | 90,000 | Oregon | 350,000 | 390,000 |
| Alaska | 780 | 870 | Maryland | 23,000 | 25,000 | Pennsylvania | 30,000 | 34,000 |
| Arizona | 20,000 | 23,000 | Massachusetts | 17,000 | 19,000 | Rhode Island | 58 | 65 |
| Arkansas | 25,000 | 28,000 | Michigan | 59,000 | 66,000 | South Carolina | 41,000 | 46,000 |
| California | 84,000 | 94,000 | Minnesota | 18,000 | 20,000 | South Dakota | 59,000 | 66,000 |
| Colorado | 4,000 | 4,500 | Mississippi | 0 | 0 | Tennessee | 130,000 | 150,000 |
| Connecticut | 6,600 | 7,400 | Missouri | 9,000 | 10,000 | Texas | 9,300 | 10,000 |
| Delaware | 0 | 0 | Montana | 72,000 | 80,000 | Utah | 3,000 | 3,300 |
| Florida | 11,000 | 13,000 | Nebraska | 29,000 | 32,000 | Vermont | 15,000 | 16,000 |
| Georgia | 45,000 | 50,000 | Nevada | 4,200 | 4,800 | Virginia | 16,000 | 18,000 |
| Hawaii | 330 | 370 | New Hampshire | 22,000 | 24,000 | Washington | 720,000 | 810,000 |
| Idaho | 84,000 | 94,000 | New Jersey | 120 | 130 | West Virginia | 25,000 | 28,000 |
| Illinois | 10,000 | 12,000 | New Mexico | 430 | 480 | Wisconsin | 66,000 | 74,000 |
| Indiana | 18,000 | 21,000 | New York | 270,000 | 300,000 | Wyoming | 6,500 | 7,200 |
| Iowa | 35,000 | 39,000 | North Carolina | 88,000 | 98,000 | District of Columbia | 6.0 | 6.7 |
| Kansas | 520 | 590 | North Dakota | 14,000 | 15,000 | Puerto Rico | 610 | 680 |
| Kentucky | 79,000 | 88,000 | Ohio | 370 | 410 | | | |
| Louisiana | 0 | 0 | Oklahoma | 20,000 | 22,000 | United States[a] | 2,800,000 | 3,100,000 |

[a]Including Puerto Rico.
*Source.* Murray and Reeves (1972).

Mexico. Installed hydro capacity in the United States is considerably greater than in other countries—essentially twice that of the installed capacity of the next two leading countries, Canada and the USSR.

There is, of course, no easy way to determine the amount of hydropower still left to be developed in an area. Economic possibilities for development vary with differences in power demand from place to place. The capital and operating costs of generating power at different sites also vary appreciably. Whether storage sites are developed on a river or not will greatly influence the amount of prime power that potentially can be developed on the river. A 1960 Federal survey by the Senate Select Committee on National Water Resources described two different governmental estimates of U.S. hydropower potential. One had been made by the U.S. Geological Survey while the other resulted from estimates by the Federal Power Commission. The Geological Survey estimated technical potential hydropower without considering economic constraints and came up with 86,174,000 kw of hydropower potential based on mean river flow. They also suggested a value of 48,479,000 kw based on capacity available 50% of the time. The Federal Power Commission used estimates based on rated capacities of generators that normally would be installed assuming reasonable regulation of flow by means of storage. They arrived at 27,854,000 kw already developed and installed and 90,242,000 kw of undeveloped capacity. Other figures for hydropower potential may be found in the literature. The variation in the estimates result from both inexact calculations and the use of different definitions of potential power.

Various attempts have been made to reconcile the Geological Survey and the Federal Power Commission figures not only with each other, but also with the other estimates (Select Committee on National Water Resources, 1960). The results suggest that a feasible potential hydropower development in this country, excluding Alaska and Hawaii, would be between 50 and 60 million kw.

While hydroelectric power generation does not consume or even pollute any appreciable amount of water, clearly large volumes of water are required to produce the power. Table 5.4 provides information on the total amount of water needed, by state, in the production of hydroelectric power in that state. The total for the country comes to some 2,800,000 million gal/day. This water is, of course, available for other uses once it has been returned to the stream channel.

Thermal power plants also require large volumes of water for cooling purposes. Table 5.5 provides estimated values of water requirements (in 1970) for thermoelectric power generation by state. The data are divided into the volumes needed for condenser cooling, for all other uses, and the water actually consumed in power generation—both fresh and saline. The data reveal that some 98% (170,000 mgd out of a total water requirement of 173,300 mgd) of the water needed for thermoelectric power production is for condenser cooling; only about 0.6% of total water requirements result in consumed water (1040 mgd out of a total of 173,300 mgd). Saline water constitutes a growing proportion of the total amount of water used in generating thermal power (46,016 mgd out of a total of 173,300 mgd, or 27%) while groundwater supplies are used for only about 0.8% of all requirements. The water used in the generation of thermal power is only about 6% of the volume

**Table 5.5** Water Used for Thermoelectric Power Generation in the United States, 1970

[In Million Gallons Per Day; Partial Figures May Not Add to Totals Because of Independent Rounding]

| | Condenser Cooling | | | | | Other Uses | | | | | Water Consumed | |
|---|---|---|---|---|---|---|---|---|---|---|---|---|
| | Self-supplied | | | | | Self-supplied | | | | | | |
| | Fresh Ground-Water | Surface Water | | Public Supplies | Self-supplied and Public Supplies | Fresh Ground-Water | Surface Water | | Public Supplies | Self-supplied and Public Supplies | | |
| State | | Fresh | Saline | | | | Fresh | Saline | | | Fresh | Saline |
| Alabama | 0 | 4,600 | 190 | 0 | 4,800 | 2.6 | 250 | 4.0 | 0 | 260 | 25 | 0.4 |
| Alaska | .9 | 2.5 | 1.0 | 0 | 4.4 | .5 | 66 | 0 | 0 | 66 | 0 | 0 |
| Arizona | 40 | 2.0 | 0 | 0 | 42 | 0 | 0 | 0 | 0 | 0 | 32 | 0 |
| Arkansas | 3.0 | 950 | 0 | 0 | 950 | 1.0 | 1.0 | 0 | 0 | 2.0 | 3.0 | 0 |
| California | 300 | 1,200 | 8,200 | 130 | 9,900 | 0 | 1.0 | 0 | 10 | 11 | 24 | 66 |
| Colorado | 30 | 83 | 0 | 0 | 110 | 0 | .1 | 0 | .1 | .2 | 8.0 | 0 |
| Connecticut | 0 | 1,000 | 1,800 | 0 | 2,900 | 0 | 5.0 | 0 | 3.0 | 8.0 | 2.0 | 0 |
| Delaware | .6 | 1.4 | 720 | 2.0 | 730 | .1 | 0 | 0 | .3 | .4 | 0 | 0 |
| Florida | 12 | 1,700 | 9,300 | 2.6 | 11,000 | 1.0 | 0 | 0 | 2.5 | 3.5 | 20 | 86 |
| Georgia | 4.4 | 3,900 | 140 | 0 | 4,000 | 2.3 | 28 | 1.1 | 9.5 | 40 | 39 | 1.1 |
| Hawaii | 82 | 46 | 860 | 0 | 980 | 0 | 0 | 0 | 0 | 0 | 0 | 0 |
| Idaho | 0 | 0 | 0 | 0 | 0 | 0 | 0 | 0 | 0 | 0 | 0 | 0 |
| Illinois | 0 | 11,000 | 0 | 1.0 | 11,000 | 7.0 | 320 | 0 | 3.0 | 320 | 5.0 | 0 |
| Indiana | 1.0 | 4,700 | 0 | 26 | 4,700 | 0 | 110 | 0 | 1.0 | 110 | 5.0 | 0 |
| Iowa | 0 | 1,300 | 0 | 7.7 | 1,400 | 0 | 42 | 0 | .2 | 42 | 20 | 0 |
| Kansas | 38 | 220 | 0 | 0 | 260 | 0 | 0 | 0 | 0 | 0 | 32 | 0 |
| Kentucky | 1.0 | 3,600 | 0 | 0 | 3,600 | .9 | 230 | 0 | .7 | 240 | 21 | 0 |
| Louisiana | 34 | 2,600 | 140 | 0 | 2,700 | 1.8 | 140 | 7.2 | 0 | 140 | 170 | 25 |
| Maine | 0 | 20 | 180 | 0 | 200 | 1.0 | 0 | 0 | 1.0 | 2.0 | 0 | 0 |
| Maryland | 0 | 500 | 2,700 | 0 | 3,200 | 0 | 12 | 3.0 | 1.0 | 16 | 0 | 0 |
| Massachusetts | 0 | 550 | 2,100 | 0 | 2,700 | 0 | 29 | 0 | 2.5 | 32 | 1.0 | 1.0 |
| Michigan | 0 | 9,700 | 0 | 0 | 9,700 | 0 | 49 | 0 | 0 | 49 | 0 | 0 |
| Minnesota | 280 | 1,400 | 0 | 0 | 1,700 | 0 | .2 | 0 | 0 | .2 | .2 | 0 |
| Mississippi | 34 | 400 | 490 | 0 | 920 | 1.0 | 12 | 0 | 0 | 13 | 35 | 0 |
| Missouri | 4.0 | 2,400 | 0 | 2.0 | 2,500 | 2.0 | 5.0 | 0 | 3.0 | 10 | 13 | 0 |

| | | | | | | | | | | | | |
|---|---|---|---|---|---|---|---|---|---|---|---|---|
| Montana | 0 | 60 | 0 | 0 | 60 | 0 | 0 | 0 | 0 | 0 | 0 | 0 |
| Nebraska | 270 | 620 | 0 | 84 | 970 | 0 | 0 | 0 | 0 | 0 | 8.4 | 0 |
| Nevada | 7.1 | 49 | 0 | 0 | 56 | 0 | .1 | 0 | 0 | .1 | 7.5 | 0 |
| New Hampshire | 0 | 250 | 160 | 0 | 410 | 0 | 0 | 0 | 0 | 0 | 26 | 0 |
| New Jersey | 0 | 1,000 | 3,200 | 0 | 4,200 | 8.0 | 0 | 0 | 18 | 26 | 26 | 0 |
| New Mexico | 8.6 | 19 | 0 | .1 | 28 | 0 | .9 | 0 | 0 | .9 | 27 | 0 |
| New York | 0 | 4,400 | 8,500 | 22 | 13,000 | 130 | 260 | 0 | 3.9 | 390 | 9.6 | 17 |
| North Carolina | 0 | 4,300 | 170 | 0 | 4,500 | 1.0 | 41 | 0 | 0 | 42 | 30 | 0 |
| North Dakota | 1.0 | 350 | .9 | 0 | 350 | 3.0 | 0 | 0 | 0 | 0 | 1.2 | .8 |
| Ohio | 46 | 13,000 | 0 | 300 | 13,000 | 0 | 590 | 0 | 16 | 610 | 14 | 0 |
| Oklahoma | 2.8 | 180 | 0 | 7.8 | 190 | 1.0 | .1 | 0 | .5 | 1.6 | 28 | 0 |
| Oregon | 0 | 22 | 0 | 0 | 22 | 0 | 0 | 0 | 0 | 0 | .1 | 0 |
| Pennsylvania | 0 | 12,000 | 0 | 1.9 | 12,000 | 0 | 270 | 0 | 7.0 | 280 | 8.9 | 0 |
| Rhode Island | 0 | 0 | 310 | 0 | 310 | .7 | 0 | 0 | 0 | 0 | 0 | 0 |
| South Carolina | 0 | 2,500 | 68 | 0 | 2,600 | 0 | 25 | 0 | 0 | 26 | 14 | 0 |
| South Dakota | .7 | 2.7 | 0 | .2 | 3.6 | 0 | 0 | .3 | 0 | 0 | .6 | 0 |
| Tennessee | 0 | 4,500 | 0 | 0 | 4,500 | 0 | 340 | 0 | 0 | 340 | 62 | 0 |
| Texas | 58 | 5,800 | 3,800 | 3.7 | 9,600 | 1.9 | 2.1 | 0 | .1 | 4.4 | 120 | 24 |
| Utah | 0 | 87 | 0 | .5 | 87 | 0 | .2 | 0 | .4 | .6 | 3.8 | 0 |
| Vermont | 0 | 0 | 0 | 0 | 0 | 1.0 | 4.0 | 0 | 0 | 5.0 | 0 | 0 |
| Virginia | 0 | 3,100 | 770 | 0 | 3,900 | 0 | 0 | 0 | 0 | 0 | .8 | 0 |
| Washington | 0 | 4.0 | 0 | 0 | 4.0 | 0 | .3 | 0 | 0 | .3 | 0 | 0 |
| West Virginia | 0 | 4,800 | 0 | 0 | 4,800 | 0 | 120 | 0 | 0 | 120 | 1.1 | 0 |
| Wisconsin | 0 | 5,300 | 0 | 0 | 5,300 | .6 | 0 | 0 | 0 | 0 | 0 | 0 |
| Wyoming | .4 | 200 | 0 | 0 | 200 | .2 | 1.0 | 0 | 0 | 1.6 | 5.3 | 0 |
| District of Columbia | 0 | 1,000 | 0 | 0 | 1,000 | 0 | 60 | 0 | 0 | 60 | 0 | 0 |
| Puerto Rico | .2 | 0 | 2,100 | .5 | 2,100 | .2 | 0 | 0 | 1.3 | 1.5 | .4 | 0 |
| United States[a] | 1,300 | 120,000 | 46,000 | 590 | 170,000 | 170 | 3,000 | 16 | 85 | 3,300 | 820 | 220 |

[a]Including Puerto Rico.

*Source.* Murray and Reeves (1972).

of water used to generate hydropower even though a much greater proportion of our power comes from thermal plants.

Except for some small amounts of evaporation of water in cooling towers or in reservoirs or ponds, most of the water requirements for power production are nonconsumptive. However, in arid and semiarid areas of the country, if sufficient water is not available, it makes little difference whether the use is consumptive or not. Whether the water volume is actually present for use when needed is more significant.

### Pumped Storage

The idea of pumped storage—the use of off-peak power which is available in a given system to pump water uphill to a storage reservoir from which it can be released later to generate hydropower during peakload periods—is fairly old although it has only been used for the past few decades in this country. In use for more than 60 years in Europe, there are over 100 such installations there. Prior to 1950, there was only one such installation in the United States. Even now, there are only a handful of installations.

In a pumped-storage power plant, the generator is designed to work at night as a motor and the turbine as a pump. During the day or when peak power is needed, the machinery serves as a turbine-generator that supplies synchronized power to the system.

The Ludington pumped-storage project of the Consumer Power Company located near Lake Michigan has a generating capacity of some 1009 megawatts. It is the largest privately owned pumped-storage project at present in the United States. The project with the highest pumped head is located on Cabin Creek near Georgetown, Colorado. Some 455 mgd of water are involved in that pumped-storage project, and, when released, this water adds some 324,000 kw of additional capacity to the system. Garstka (1978) has estimated that about one-third of the energy is lost in a pumping cycle in an average pumped-storage system.

Environmentalists are fighting the use of pumped-storage systems because of the great level changes that occur daily in the high-level storage reservoir. The operation of the system calls for the reservoir to be filled at night during the off-peak period. Excessive demand during the day would lead to the rapid lowering of the level for hydropower purposes. This would limit the use of the storage reservoir for recreational or other purposes.

The pumped-storage idea has many positive aspects. The technique would not be feasible if the electricity to pump the water to the upper storage reservoir had to be paid for, although it may be practical if the power is already available as unused capacity. The availability of cheap hydropower to augment thermal power during peakload periods provides a most economically attractive way to add flexibility to the whole system, to obtain many of the advantages of hydropower operation in a space often unsuitable for full-scale hydropower development, and to match system capacity to load.

A question may really be asked whether pumped-storage constitutes hydro-power. The term usually implies energy created by the natural downward flow of excess precipitation. Pumped storage is created energy due to the elevation of water by some means other than by nature. But it does produce electricity by means of falling water, so it can be thought of as a particular form of hydro-power. There would seem to be compelling reasons for the increased use of pumped storage if at all possible to expand system capacities to cover peakload periods at minimum cost. Loane (1957) has estimated that many systems could use pump-ed storage to increase system capacity by 10 to 15% at fairly low cost.

### The La Grande Hydroelectric Complex

One of the world's largest hydroelectric complexes is now being developed some 900 mi north of Montreal, Quebec, Canada by the Société d'energie de la Baie James (SEBJ). To place this project into proper perspective, consider that the Société employed some 18,000 workers at the time of peak construction activity. More than 500,000 tons of cement have been used in the construction along with 68,180 tons of explosives, and some 262 million $m^3$ of excavated material and fill have been moved (enough to build the Great Pyramid of Egypt some 82 times).

The overall project has been divided into four phases. Phase I involves the construction of three powerhouses on the La Grande Rivière and the diversion of water from the Eastmain, Opinaca, and Petite Opinaca rivers from the south and Caniapiscau river from the east into the La Grande to augment its flow and to permit the generation of more than 10,000 MW of power (Fig. 5.1). The drainage basins have little relief so that major reservoirs must be built along with more than 200 dikes and dams to control water flow.

The La Grande hydroelectric project was begun in 1971 and Phase I is expected to be completed by 1985 even though the difficult terrain, Precambrian rocks of the Canadian Shield, average annual temperatures of 26°F (-4°C), and heavy snowfall make delays inevitable and limit the length of the construction season. Careful planning and coordination of building activities by SEBJ have resulted in construction being on schedule or even ahead of schedule generally throughout all phases of the project.

As of mid-1982, the four projects in Phase I—construction of the La Grande 2 powerhouse, the diversion of the Eastmain and Opinaca rivers, construction of La Grande 3 powerhouse, and the diversion of the Caniapiscau river to power the La Grande 4 generating station—were more than 80% complete. Already La Grande 2 was able to operate at its full capacity (5328 MW). The first generating unit at La Grande 3 was also commissioned in mid-1982; the reservoir behind the dam was being filled and should reach its maximum level by late summer 1982. When fully operational, La Grande 3 will have an installed capacity of 2304 MW.

La Grande 4 is about 60% completed. The powerhouse has been installed, water structures are in place, and the impounding of the water of the Caniapiscau

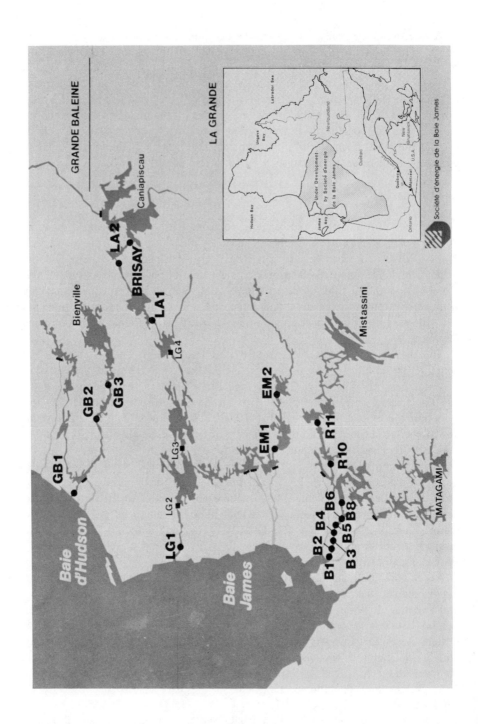

GRANDE BALEINE

LA GRANDE

Société d'énergie de la Baie James

is scheduled for fall 1982. Already the natural flow of the river to Ungava Bay has been halted and a new lake of immense dimensions has been created. The Caniapiscau reservoir will take 2 years to fill.

Phases II, III, and IV of the La Grande complex are all under way although completion dates in some cases are several decades away. Phase II includes additional projects in the basin of the La Grande Rivière not already included in Phase I. These include building five new generating units (La Grande 1, Eastmain 1, Laforge 1 and 2, and Brisay). These units together will have an installed capacity of some 3623 MW, although completion dates range betwen 1989 and 1993 (Fig. 5.1).

Phase III involves the Grande-Baleine complex located well to the north of the La Grande basin. This complex will have three powerhouses on the Grande Rivière de la Baleine whose flow will be increased by diversion of water from the Petite Rivière de la Baleine. Three reservoirs, of which Lake Bienville is the head, will be created. Preliminary studies, campsite and roadway layouts, and airports are presently completed. Final installed capacity will total 2891 MW when the units are completed in about 2000.

Phase IV involves the construction of some 11 powerhouses (8740 MW capacity) as part of the Nottaway-Broadback-Rupert complex to the south of the La Grande basin. This complex will develop the basins of those three rivers that cover some 8% of the area of Quebec. More than 18 dams and 127 dikes are planned to retain the river waters in five reservoirs.

Because of wet and freezing weather in the area covered by the various projects, earth fill embankment construction work is limited to about 75 to 100 days per year. This short construction season must be fully utilized. The existence of wet conditions much of the time has forced one contractor to install a large rotary kiln dryer to reduce the natural water content of the till by 2% at a rate of not less than 500 tons per hour (Bergevin and Seemel, 1976).

The cost of the La Grande (Phase I) hydroelectric complex is estimated at $14.6 billion, a figure that includes the costs of the needed infrastructure (airports, telecommunications, housing, roads) as well as the hydroelectric installations, power stations, dikes, and the 735 kV transmission network (Hamel and Nixon, 1978). About one-quarter of the total cost of the project is for the construction of 16 switching and transformer stations and 5 power lines having a total length of more than 3000 mi.

The concept of the La Grande hydroelectric complex is almost beyond imagination. The progress that has already been made is a tribute to the organizational efforts of SEBJ and the work efforts of all of the construction workers under very trying climatic conditions in a remote area without much in the way of preexisting facilities.

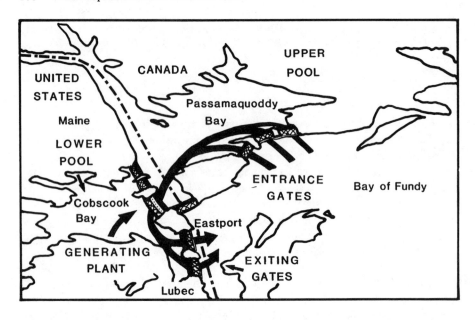

**Figure 5.2** Sketch map of proposed dams, generating plant, and storage pools in the Passama-quoddy Bay hydroelectric project.


## Passamaquoddy—An Idea That Will Not Die

In the days of the New Deal and Franklin Roosevelt, a grandiose scheme was developed to utilize the very high tides that roll every day into the Bay of Fundy and Passamaquoddy Bay located between Maine and New Brunswick, Canada. The project, estimated to cost $30 million in the early 1930s, was considered to be impractical in 1936. But, as most grandiose schemes, it has never really been allowed to die. Revised in the 1960s and now with a price tag estimated at over $1 billion, it would seem to be more useful or practical than it was 40 years earlier.

The project would involve constructing some 7 mi of ocean dams and two pools (bays separated from the ocean by dams) (Fig. 5.2). The high pool (Passa-maquoddy Bay) would be filled by the high tides arriving in the Bay of Fundy. Water from the high pool would then be released through the turbines in the hydroelectric plant near Eastport into the low pool of Cobscook Bay. Between Eastport and Lubec, emptying gates would allow the water to flow out of the low pool during low tides and so ready the system for operation again on the next high tide. The most recent Corps of Engineers study has estimated that the project could supply 1 million kw of electricity at peak times as well as some 250,000 kw of firm power at costs well below the existing prices for power in the region. One might question the various estimates since actual construction costs usually exceed initial estimates and the need for that much power in the Passamaquoddy area can certainly not be justified. Without long-distance transfer of the power developed, the project would not be able to operate at anywhere near capacity on the basis of local demand.

Another significant factor in the project which should cause some concern is the suggestion that the United States might finance the whole project although more than half of it would be physically located on Canadian soil. Canada would buy any power that it uses. The 7 mi of ocean dams would be considered "engineering marvels" and would be expected to bring thousands of tourists into the area. This argument hardly seems to justify a costly and questionable hydropower development.

### Future of Hydropower

Realistic figures for potential hydropower development in the United States of some 50 to 60 million kw are about four times the 1960 level of actual development. Inexpensive sites have already been developed so that future hydropower development will be more costly. However, it is clear that our sources for hydropower can be greatly expanded without running into serious resource limitations. The limitations that may ultimately slow hydro development will be (a) competing consumptive uses of water and (b) economic factors in new site development, rather than the lack of feasible sites.

Pumped storage should continue to be developed as a technique to add additional power and flexibility to thermal power systems. Through its use, it might be possible to increase by at least 20% the previous figure of 50 to 60 million kw of hydropower potential.

One cannot talk about the future of hydropower without considering the present situation of high oil prices, rising prices for coal, and increasing air pollution. With such economic and environmental problems increasing, hydropower becomes even more attractive than it was previously. However, the need to take land out of other uses and to submerge it beneath the waters of reservoirs has stirred many conservation groups to action. Higher fuel prices will increase pressure for hydropower development but increased awareness of the environmental losses resulting from reservoir construction will place serious limitations in the way of rapid development of many choice sites. As a result, slow future development of selected hydropower sites may be anticipated following long legal and political skirmishes between opponents and proponents of hydropower.

## WATER FOR NAVIGATION

Water transport played an important role in the early development of the United States. Rivers have always served as links between places, mountains as barriers. While the highway, the railroad, and now the airplane have contributed their own links, river transportation still plays a significant role in the industrial and commercial life of the nation.

**Table 5.6**  Commercially Navigable Waterways of the United States (by Depths), 1979[a]

| Waterway System | Under 6 ft | 6 to 9 ft | 9 ft and Over | Total | Percentage by System |
|---|---|---|---|---|---|
| | | | Miles | | Percent |
| Atlantic coast | 1,426 | 1,241 | 3,103 | 5,770 | 23 |
| Atlantic intracoastal (Norfolk, Virginia to Key West, Florida) | – | 65 | 1,169 | 1,234 | 5 |
| Gulf coast waterways | 2,055 | 647 | 1,590 | 4,292 | 17 |
| Gulf intracoastal (St. Marks, Florida to Mexican border) | – | – | 1,137 | 1,137 | 4 |
| Mississippi River system | 2,020 | 969 | 5,965 | 8,954 | 35 |
| Pacific coast waterways | 730 | 498 | 2,347 | 3,575 | 14 |
| Great Lakes | 45 | 89 | 356 | 490 | 2 |
| All other | 76 | 7 | 8 | 91 | b |
| All waterways | 6,352 | 3,516 | 15,675 | 25,543 | 100 |
| Percent of total system | 25 | 14 | 61 | 100 | |

[a]Mileage represents all commercially navigable channels of the United States.
[b]Less than 0.05%.
*Source.* American Waterways Operators, Inc. (1979) with permission.

In the early days of the United States, not only were rivers used extensively for travel and commerce but many miles of canals, often paralleling the rivers, were constructed and used for transportation. The Erie Canal built by the State of New York was so profitable that it led, in the early 1800s, to the construction of a vast network of privately owned canals. Steamboats or packets were used extensively on the larger rivers during the first half of the 19th century, delivering large tonnages of cargo. However, the Civil War disrupted Mississippi River steamboat traffic and provided an opportunity for the railroads to capture much of the river traffic. Mississippi River traffic did not recover its pre-Civil War volume until well into the 20th century. The need for movement of large amounts of materials during World War I and the previous development of more efficient engines and larger shallow-draft barges and towboats resulted in a resurgence of river traffic. Beginning in 1920, the Federal government undertook a program to deepen and widen river channels and to modernize the waterways of the United States to keep pace with developments being achieved in boats and engines.

### Commercial Waterways of the United States

Table 5.6 summarizes information on the commercial waterways of the United States, including the Great Lakes, as of 1979. This waterways system totaled more than 25,000 mi in length, of which 61% had a depth of 9 ft or greater. Of the over 25,000 mi of commercial waterways, about 20,000 mi had been improved by the Federal government.

Figure 5.3 illustrates the location and depths of the commercially used waterways. The major components of the present inland and intracoastal waterways network are as follows:

(a) The Mississippi River and tributaries consist of about 9000 mi of commercial waterways of which about two-thirds has a channel depth of at least 9 ft.
(b) The Gulf Intracoastal waterways and other gulf coastal waterways involve some 5400 mi of navigable channels of which about half have depths of at least 9 ft.
(c) The Atlantic Intracoastal, New York, and New England waterways system has a length of about 7000 mi with not quite two-thirds of it having depths of at least 9 ft.
(d) The Central California and Columbia–Snake River waterways total some 3600 mi in length with two-thirds of that total at least 9 ft in depth.
(e) The Great Lakes Waterways and St. Lawrence Seaway system connects Great Lakes ports to the Atlantic Ocean. It has about 1700 mi of waterways with depths sufficient for ocean-going vessels.

Gradual increases in domestic waterborne traffic are to be expected as waterways are improved and transportation costs for competing methods increase. For example, barge traffic has risen from 562 million tons in 1970 to 657 million tons in 1979 in the United States. The 1979 figures showed that about 12% of the total intercity freight in ton-miles was carried on inland waterways as opposed to 37% by rail, 26% by highway, and 25% by oil pipeline.

The USSR and the United States have the greatest length of navigable waterways in regular use, the USSR figure being some three times greater than the United States total. In terms of ton-miles of freight carried, however, the United States figure is several times greater than the total for the USSR. Of the other countries of Western Europe, West Germany, with less total length of navigable waterways than either France or the Netherlands (the two countries with the highest totals of navigable waterways), still has more ton-miles of freight carried than both of those countries combined. Table 5.7 shows the ton-miles of freight carried on the different inland waterways of the United States in 1978 and 1979.

The Corps of Engineers estimated that some $3.2 billion had been spent to improve the inland waterway system of the United States up to 1971. The rate of expenditures for operating and maintaining the waterways was running at about $80 million annually. Under present policies, the Federal government pays the full cost of improvements to waterways for commercial use although private or other public groups must provide lands, right-of-way spoil areas, as well as public terminal and transfer facilities.

### Barges and Towboats

In 1979, over 1800 transportation companies operated nearly 34,000 vessels on inland and coastal waterways. About 85% of these were unpowered barges.

CONTROLLING DEPTHS

9 FEET OR MORE
UNDER 9 FEET
AUTHORIZED EXTENSIONS

**Figure 5.3** Waterways of the United States showing authorized depths, 1979. *Source.* American Waterways Operators, Inc. (1979), with permission.

**Table 5.7**  Ton-mileage of Freight Carried on the Inland Waterways of the United States in 1978, 1979

| System | 1978 | 1979 |
|---|---|---|
| Atlantic coast waterways | 30,493,637 | 31,929,470 |
| Gulf coast waterways | 37,060,981 | 38,052,138 |
| Pacific coast waterways | 13,571,186 | 14,144,892 |
| Mississippi River system (including Ohio River and tributaries) | 209,270,717 | 218,776,279 |
| Great Lakes system | 118,919,613 | 121,666,371 |
| | 409,316,134 | 424,569,150 |

*Source.* U.S. Department of the Army, Corps of Engineers (1979).

Barges now in use have a capacity of from 1000 to 3000 tons, the upper figure being the load that could be carried by a fully loaded train of 55 average-sized cars or 30 of the newer big cars. Fifty years ago, large barges could carry only 800 tons of freight and many carried only 200 tons. Common barge sizes are 26 by 175 ft or 35 by 195 ft. These dimensions permit three of the larger or four of the smaller barges connected abreast to go through the 110-ft wide locks found on most major waterways.

Barges are usually joined into tows (although barges are more often pushed by towboats, the word tow is still used). Tows of 20 or more barges carrying 20,000 tons of freight are possible on the larger river systems. Such tows carry the equivalent of nearly 400 average-sized railroad cars, equivalent to several good-sized train loads. The development of many types of special purpose barges has greatly increased the variety of products that can be moved on the inland waterways.

Developments have not been confined to the barges themselves for significant changes in the power and size of the towboats have also occurred in the past half-century. The modern towboat of 4000 hp has about 20 times the power of the largest river packets of the 1850s. Such a towboat can handle 20,000 tons of freight with ease. The most powerful towboat on the Mississippi River, the United States, boasts an 8500 hp diesel engine. This boat is capable of moving 40 loaded barges. The development of diesel engines as well as changes in the way power is carried to the propellers, in the design and setting of the propellers, and even in boat designs have all led to more efficient operation of modern towboats. The development of the reversing-reduction gear which allows the towboat to reverse direction without reversing the engine itself, and the use of the "tunnel stern" which enables the top of the propeller to be placed above the top of the water level in the stream but to be completely submerged in operation, are developments of great significance. Use of three propellers instead of two and improved rudder systems have resulted in more efficient, safer operation of towboats

Table 5.8 Navigation Water Requirements for Major Inland Waterways of the United States

| Major Waterway | 1965 | | Required Flow with Authorized Improvements[b] (cfs) |
| | Freight Movement (thous. tons/day) | Required Flow[a] (cfs) | |
| --- | --- | --- | --- |
| New York State Barge Canal | 12 | 525 | 630 |
| Savannah River below Augusta | 72 | 6,100 | 6,100 |
| Cross Florida Barge Canal | | | |
| to Atlantic Ocean | — | — | 305 |
| to Gulf of Mexico | — | — | 427 |
| Okeechobee Waterway | | | |
| to Atlantic Ocean | 9 | 19 | 19 |
| to Gulf of Mexico | — | 25 | 25 |
| Black Warrior-Tombigbee Waterway | | | |
| Mobile to Tuscaloosa | 13 | 120 | 418 |
| Upper Mississippi River | | | |
| mile 858 to 218 | 38 | 1,600 | 1,900 |
| mile 218 to 203 | 88 | 1,600 | 1,900 |
| Illinois Waterway at the lower 5 locks | 84 | 350 | 800 |
| Missouri River | | | |
| at Kansas City | 10 | 35,000 | 32,500 |
| at St. Louis | 9 | 40,000 | 37,500 |
| Upper Mississippi River, mile 203 to 0 | 90 | 54,000 | 54,000 |
| Ohio River | | | |
| at mile 939 | 90 | 230 | 1,550 |
| at mile 607 | 75 | 450 | 2,500 |
| Tennessee River, mile 650 to 0 | 48 | 3 | 3 |
| Lower Mississippi River | | | |
| Cairo to White River | 164 | 150,000 | 150,000 |
| Baton Rouge to Gulf of Mexico | — | 150,000 | 150,000 |
| Columbia River | | | |
| above Bonneville | 7 | 110,000 | c |
| at Bonneville | 7 | 58,000 | 58,000 |
| Willamette River above Portland | 3 | 6,000 | 6,000 |
| Columbia River below Portland | 39 | 77,000 | 77,000 |

[a]Flow and lockage water necessary to maintain waterway at authorized depth under 1965 traffic conditions.
[b]Flow and lockage water necessary to maintain waterway at depth of authorized improvement under maximum traffic conditions.
[c]Slack water with nominal flow required immediately below dams.
*Source.* Todd (1970), after U.S. Water Resources Council (1968).

with larger cargoes. The developments in barges and towboats would not have occurred without changes in the depths and widths of river channels and locks. Thus, the recent increase in number of ton-miles of cargo carried on our waterways represents a close allegiance between the private waterways operators and the Federal government through the Corps of Engineers who are responsible for the maintenance of the stream channels and locks.

## Water Requirements

The U.S. Water Resources Council has published estimates of the rates of flow in the various inland waterways needed to insure their efficient use for navigation (Table 5.8). In general, it is found that the flow necessary for navigation on a river constitutes a relatively small percentage of the average flow in the stream. While the newer towboats may push greatly increased amounts of cargo, they do not necessarily require that much more water. Actions already completed by the Corps of Engineers in clearing snags and sandbars have made channels passable even at river flows well below average. Thus, flows needed to fulfill other demands for water, such as for power or for pollution abatement, generally exceed the flows required for navigation purposes. Usually, water does not have to be released from reservoirs just for the maintenance of navigation. This conclusion may not hold for the Missouri River where releases for navigation may limit power production, but it holds generally for most other streams. Water released for other purposes would, of course, be helpful for navigation in periods of low flow.

## Should Inland Waterways Be Developed?

While the Corps of Engineers has been entrusted with the task of maintaining our streams and improving them for the purposes of navigation, it must be pointed out that there is not universal agreement that they should be as active as they have been in the pursuit of their goals. Many arguments have been raised against developing our streams to the degree that they have been.

From a geological point of view, channelization of streams heavily ladened with sediments or located in geologically unstable areas is economically unfeasible. The channel can only be kept open by repeated dredging, while banks of the channel will continue to cave in and obstruct navigation.

Until the authorization of work on the Arkansas River, ultimately designed to bring deep water facilities to within 15 mi of Tulsa, Oklahoma, channelization of sediment-laden rivers had not been attempted. Kazmann (1972) reports that the sediment load of the Arkansas River at Little Rock, Arkansas is estimated to be 100 million tons annually (some 30,000 to 40,000 acre-feet per year of sediment accumulation). To make the channelization project more feasible, it was decided that a number of upstream reservoirs should be constructed for the main purpose of trapping and storing sediments. Without effective storage of sediment, the main channel would rapidly silt up and need redredging. Even with the extra storage capacity in the tributary and upstream reservoirs, there is no guarantee that human activities will not so increase the sediment load of the river as to rapidly reduce useful storage capacity. The upstream storage reservoirs, seven in number, have a total storage capacity of some 11,800,000 acre-feet. If the annual sediment load in the river is 40,000 acre-feet, it is clear that some years will pass before the storage capacity of these reservoirs is filled with sediment but it will not take many years before a significant portion of the reservoir capacity will be lost to sediment storage.

The cost of the entire Arkansas project was estimated to be some $1.2 billion— all borne by the Federal government. Kazmann (1972) suggests that if even 20% of the cost of the project had had to be borne by local interests, the project would probably never have been started. He feels that the project will be short-lived because of the sedimentation problem and that both economically and geologically the project is doomed to failure. He also stresses the potential entrapment of industries and individuals as a result of such a project. Large amounts of Federal funds expended in the initial construction of large reservoirs or waterways activities suggests permanence. Businesses and individuals move into such an area to take advantage of the opportunities created. A moral obligation on the part of the government to keep the project operating is also implied. In some way, the Federal government must continue to supply these same opportunities even if, at a later time, it is decided to end the initial water project. Such entrapment can lead to the wasteful expenditure of vast sums of money to keep a project going long after it is clear that there is no reasonable economic basis for its continuation.

Kazmann (1972) goes even further in his discussion of navigation by saying that there may be no economically justifiable reason for the maintenance of navigation on rivers. He believes that there are other inexpensive ways to move materials such as by pipelines, low draft vessels, seasonal transport when water levels are high, air cushion vehicles, as well as by the more conventional truck and railroad.

## Future Needs and Developments

The final report of the National Water Commission (1973) discussed the present Federal inland waterways program in some detail and found several weaknesses. First, it pointed out that there were deficiencies in the procedures used to determine whether a particular waterway program would provide a worthwhile addition to the overall national transportation system. Second, beneficiaries of the waterway program are not required to share in the cost of constructing or operating Federal waterway projects. Third, there is not overall planning to show how the inland waterway system can be made an integral part of the national transportation system.

To correct these deficiencies, the Commission recommended that through a combination of fuel taxes and user fees, the government should try to obtain sufficient money from waterways already constructed to cover annual Federal expenditures for operation and maintenance. For new waterway projects, the Commission recommended that before construction begins, groups other than the Federal government must agree to repay over time the entire construction costs with interest.

The Corps of Engineers had earlier suggested that those waterways currently operating at below average efficiencies because they were not deep or wide enough to utilize modern equipment should be modernized to increase their operating efficiencies. They argued that more money should be spent on better maintenance of existing waterways while those waterways not contributing as viable components of the national transportation system should be removed from the Federal waterways system. Terminals should be improved in efficiency in order to reduce costs of handling goods carried by water. The Corps listed the following waterways projects that had attracted more than ordinary public interest:

(a)  Connection of the Ohio River with Lake Erie.
(b)  A canalized waterway following the Trinity River to Dallas, Texas.
(c)  Extension of the Arkansas River project beyond Tulsa to the vicinity of Oklahoma City.
(d)  Extension of the Columbia River waterway.
(e)  Channelization of the Big Sandy River to provide for barge traffic into the coal regions of eastern Kentucky and western West Virginia as well as channelization of the Kaskaskia and Big Muddy rivers for coal traffic in southern Illinois.

The National Water Commission recommended that any future waterway project must include an estimate of the true economic cost and benefit to the nation including a comparison of costs through use of waterways vs. costs through use of the least-cost alternate method. Cost schedules must be revised to include payment by users or other beneficiaries of the project. The Department of Transportation must increase its efforts to improve and extend its national transportation policy and integrate the waterways program into it in such a way that overall costs of transportation to the nation are minimized. We cannot afford the luxury of piecemeal development as in the past, controlled more by politics than by rational planning.

## WATER FOR FISH, WILDLIFE, AND RECREATION

Fish and wildlife are natural resources of far more than economic value to the nation requiring adequate water supplies for their continued development. The great increase in demand for water for industrial and agricultural purposes poses a serious threat to the maintenance of even minimal water supplies for fish and wildlife since they command less economic power than do other competing demands. It is difficult to evaluate the economic benefits of a thriving fish, wildlife, and recreation facility to the nation's well being.

A few statistics may serve to indicate the importance of fish and wildlife in the nation's life. It is now estimated that commercial fishing represents a $7 billion industry worldwide with nearly 500,000 man-years of employment. More than two-thirds of the total commercial and recreational fish caught and consumed in

**Table 5.9** Comparison of Publicly Administered Water-Oriented Recreation Areas to Total Publicly Administered Recreation Areas in the United States (Data as of 1965)

| Administering Agency | Total Public Recreation Areas | | Water-Oriented Recreation Areas[a] | | | |
|---|---|---|---|---|---|---|
| | | | Number of Areas | | Acreage | |
| | Units | Thou. acres[b] | Units | Percent | Thou. acres[b] | Percent |
| Federal | 2,127 | 447,558 | 1,512 | 71 | 362,520 | 81 |
| State | 18,614 | 39,702 | 7,529 | 40 | 32,272 | 81 |
| County | 4,048 | 2,977 | 2,142 | 53 | 2,688 | 90 |
| Municipal[c] | 27,104 | 1,238 | 23,149 | 85 | 1,007 | 81 |
| Total | 51,893 | 491,475 | 34,332 | 66 | 398,487 | 81 |

[a]Having a body of water within the area or accessible adjacent water outside the area.
[b]Based on a sample of selected municipalities.
[c]Includes both land and water area within an area but not the accessible adjacent water outside the area.
*Source.* Todd (1970).

the United States depend, in part or entirely, on fresh water or estuaries where freshwater sources meet the sea. In 1954, some 1 billion lbs of fish were caught annually by sport fishermen, three-fourths of which were freshwater species. By 1970, it was estimated that some 1.5 billion lbs of fish were caught by marine sport fishermen alone. The number of marine sport fishermen had grown in the United States from 6.3 million in 1960 to 9.5 million in 1970.

In 1959, it was estimated that some $40 billion (more than 8% of the gross national product) was spent by Americans on leisure-type activities (Coughlan, 1959). The recreational boat business took in some $2.5 billion in 1959, four times what it did in 1951. Based on estimated future growth in travel, personal income, and leisure time, along with estimates of future population growth, there is the suggestion that the number of visitors to national parks will increase from 63 million in 1959 to 240 million by 1980 and to over 400 million by the year 2000. Total person-day visits to all national parks and forests, as well as state, county, and municipal parks, may exceed 8 billion per year in 1980 and 13 billion by 2000. It is clear that a majority of these visits will be associated in some way with water-based activities. Around 66% of our public recreation units, units which constitute some 81% of all the acreage devoted to public recreation, have a body of water within the area or accessible immediately adjacent to the area (Table 5.9).

The demand for adequate opportunities to utilize and enjoy fish, wildlife, and recreational resources is increasing at a rate considerably faster than the increase in population itself, as disposable income and amount of leisure time rise.

This would not necessarily pose a problem if it were not for the fact that other competing demands for water are severely limiting or actually destroying the water and land habitat so necessary for fish and wildlife development.

The basic solution to this dilemma is not simple. The first step, of course, is for an increased recognition of the need to assign to fish, wildlife, and recreation uses a fair and adequate share of our water resources. Thus, water resources development programs should not be undertaken solely to satisfy irrigation, flood control, navigation, power, or industrial purposes without consideration of fish, wildlife, and recreation aspects. It may even be necessary to modify development plans or to make them serve these other needs less efficiently if, at the same time, they are to provide adequate recognition of fish and wildlife needs. It must also be recognized that fish and wildlife needs cannot necessarily be satisfied cheaply.

A significant step toward a solution to the dilemma posed by the increasing need for recreational type activities and the decreasing land and water resources to support such activities has already been achieved by the passage of a Wildlife Coordination Act and several subsequent amendments. Originally passed in 1934, the Act authorized general studies as well as assistance by cognizant Federal agencies in the fish and wildlife field. Water impounded by the Federal government could be utilized for fish hatcheries and migratory bird refuges under provisions of the Act. Amended in 1946 and again in 1958, the Act now authorizes both Federal and state fish and game agencies to study all Federal construction projects in order to determine the effect of the project on fish and wildlife resources and to incorporate in the construction and operation plans aspects which will eliminate or reduce the adverse effects of the plan on fish and wildlife as well as to develop those aspects which might encourage the production of fish and wildlife.

### Impact of Water Projects on Fish and Wildlife Resources

The law now requires that the impact of water-use projects on fish and wildlife resources must be evaluated before approval for these projects can be granted. The building of dams and reservoirs can have many positive as well as negative effects on fish and wildlife. Dam building may block the passage of anadromous fish species up the river to spawning grounds unless adequate fish ladders are available. Reservoirs may cover spawning grounds, change the flow patterns or the amount of silt in rivers and so affect feeding or spawning grounds, and change temperature and oxygen conditions in the river. Fish eggs may be stranded by reservoir drawdown at critical times. Reservoirs may also inundate wildlife habitat, in areas which are often the only suitable wildlife habitat for some distance, or they may force wildlife to migrate to other nearby areas where they must compete with other animals.

While the law can require construction and maintenance of facilities to allow the transport, protection, and spawning of fish (such as the salmon), history argues against the efficacy of such activities. The original dams were built on the Columbia River in the late 19th century to provide power and irrigation water. The first dams cut off significant spawning areas so laws were passed to require fishways or fish ladders to allow the salmon to move upriver to spawning sites. Young salmon migrating seaward were often killed either when diverted into irrigation canals or in passing through turbines of hydroelectric plants. Screening was required to eliminate the problem with irrigation canals but screening could not be used to keep the salmon out of the turbines. A loss of about 10% of the young salmon through the turbines of a low dam might be acceptable but such a loss becomes serious when there is a series of dams to traverse. Spawning grounds (with flowing water and proper temperature conditions) were also lost in the creation of forebays behind reservoirs.

Even with the creation of all the facilities for transport and protection that human ingenuity could devise, salmon production in the Columbia River slipped significantly. More than 40 hatcheries were built along the lower part of the Columbia River to help compensate for the loss of spawning and nursery facilities in the river. As of the early 1960s, there was no real evidence that these artificial facilities were able to keep up with the decline in natural productivity in the river.

The total cost to try to protect the salmon industry on the Columbia River has exceeded $200 million in facilities and there is a continuing expenditure of almost $30 million a year for operations, debt services, and new facilities. Yet, in spite of this expenditure, the annual commercial salmon catch from the Columbia River is estimated to be less than one-fourth the average found during the period 1900–1930. Our laws and financial commitments have slowed the decline of the salmon fishing industry on one river but they have not prevented it from occurring (Royce, 1967).

Benefits from reservoirs include the fact that the fishing opportunities can expand manyfold (although stream-type fishing will have to give way to lake-type). Controlled release of water from the reservoir may result in improvement of the fishing opportunities downstream. Tailwater fishing at the foot of the dam is often excellent. Waterfowl and aquatic fur animals are often increased in number although waterfowl may cause considerable damage to agricultural crops in the vicinity of the reservoir.

### Channelization

Channelization—the deepening and straightening of streams to help control the velocity of runoff, improve navigation, and reduce flooding—often destroys the habitat for fish by dredging, increased siltation, or reduction in the amount of overflow on nearby lands which may be used as fish spawning or nursery areas. Channelization can also be harmful to animals by draining or filling marshlands and destroying tree or shrub cover along stream banks. Most results from channelization are harmful to fish and wildlife although the creation of cutoff meanders or oxbow lakes may result in some improved fish and wildlife habitat.

Arthur D. Little, Inc. (1973) has estimated that in the past century and a half more than 200,000 mi of waterways have been modified to reduce danger from flooding, to improve navigational use of the stream, and to increase drainage of nearby lands for agricultural purposes. The Little report estimated that some 130 million acres of wetlands have been drained as a result, often with little regard to sound conservation measures. In the three decades from 1940 to 1970, some 34,000 mi of waterways have been channelized in the United States by the Corps of Engineers and the Soil Conservation Service. Nearly 80% of this work was undertaken in eight southern and five midwestern states.

In recent years, there has been growing controversy concerning the real value of channelization in view of the demonstrated negative environmental impacts that result from stream modification. While flood reduction and land drainage may result from channelization, their worth must be balanced against losses in fishery and wildlife resources, and in increased sediment flow in the stream channels due to bank erosion.

Channelization increases the capacity of waterways to carry off high streamflows before floods develop. If the channel is in cohesive soils, little erosion may result from the channelization work; but if the work is performed in noncohesive soils, there is a great likelihood that a significant amount of erosion will occur before the banks can be stabilized. The Little (1973) report estimated that the $15 billion invested in channelization has reduced flood damage by about $1 billion annually.

The effect of channelization on fish life has been documented in a number of studies. Leaves from trees supply considerable food materials needed by fish for biological activities. Trees are removed during channelization and so an important food source is lost. Straightening and removing obstructions in streams also increases stream velocity which reduces the ability of food sources, such as algae, to develop in the stream. Other important links in the food chain can be eliminated by removing obstructions, rotting logs, and bankside vegetation. For example, in a study of the Missouri River, Groen and Schmulbach (1978) found fish harvests in unchannelized portions of the river were 2 to 2.5 times greater per mile than in channelized portions.

Sampling a number of coastal plains streams in North Carolina, Tarplee, Louder and Weber (1971) found the average weight of fish per surface area in natural streams was 155 lbs as compared to 49 lbs for the average weight of fish from channelized streams. Game fish were affected more by channelization than were nongame fish. They estimated that, in this humid region, it would take about 15 years to restore fish populations to former levels if no further channel maintenance were undertaken.

The Blackwater River of Missouri has been channelized, in part, for more than 60 years. Jahn and Trefethen (1973) reported a yield of 565 lbs of fish per acre in the unchannelized portions of the river, 449 lbs/acre in portions that were lightly channelized, and 131 lbs/acre in a heavily channelized portion of the river. These differences were even more pronounced when fish shorter than 12 in. were removed from the sample.

Schoof (1980) suggested that the controversy over the effects of channelization has led to a general moratorium on such work in the past decade. Other activities, including diking to prevent floods, should be considered as viable alternatives to channelization although some stream modification work under carefully controlled conditions might still be necessary in certain situations. Channelization work seems to have less permanent impact on fish populations in moist climates than it does in dry climates.

Dredging and filling are particularly harmful to shellfish and fish spawning and feeding habitats. Silt resulting from dredging operations will move great distances downstream covering shellfish beds and food sources while the dredge spoils, deposited along the stream banks or in low-lying marshy areas, may eliminate fertile source areas of food and valuable spawning grounds.

Every year more fish and wildlife habitats are made unproductive by pollution and siltation than are created by the various public agencies through programs to restore fish and wildlife lands. Water quality is greatly impaired in many streams and lakes resulting in the elimination of many desirable fish species or in the introduction of less desirable species. Pollution may be the single most negative influence in estuaries. Runoff from agricultural lands, effluents from cities and industries, wastes from ships—all have contributed to the pollution of wetland areas bordering estuaries and to a marked destruction of fish and wildlife habitat in the estuary itself.

Increasing demand for water for irrigation, industry, power, and human use have placed additional pressure on fish and wildlife needs. Not only is the supply of water often not sufficient to meet all needs, but the construction of dams and reservoirs and the building of irrigation canals and diversion facilities have contributed to reducing the amount of valuable fish and wildlife habitat available. Yet consideration of the needs of fish and wildlife in the planning of these other water-use projects can easily turn a negative impact into a strongly positive one for fish and wildlife. For example, an irrigation enterprise introduced into a formerly arid or semiarid area can provide many benefits for fish and wildlife. Crop materials left from the production of irrigated land can serve as a food source for wildlife. Vegetation that develops around the edges of irrigated tracts naturally (or purposefully within the tract) can serve as the home for many new wildlife species. Ponding areas along irrigation canals, fencing of canal and drainage areas to protect wildlife, building crossovers for wildlife to use to cross canal areas to reach food sources, control of toxic herbicides and insecticides to protect wildlife and fish can go a long way to increase the attractiveness of irrigation enterprises for fish and wildlife resources.

Drainage of wetlands is removing some of the more productive habitat for the use of fish and wildlife. At one time there were as many as 127 million acres of wetlands in the nation. This figure has now been reduced to only 74 million acres of which just 22 million acres can be considered as good or better. It is estimated that some 300,000 acres of wetlands are being lost in the United States each year. The reduction in acreage is due to filling of wetlands for industrial, municipal, or highway uses, drainage to control mosquitoes, or to improve the appearance for aesthetic purposes. Clearly, programs to protect and preserve the wetlands of the nation are vital if the fish and wildlife resources are to be brought back to the level needed to serve the growing needs of the nation.

**Table 5.10**  Estimated Water Requirements for Wildlife and Fisheries in 1954, 1980, and 2000 (in $10^9$ gallons)

| Activity | 1954 | 1980 | 2000 |
|---|---|---|---|
| Wildlife | 31,264 | 54,672 | 62,500 |
| Fisheries | | | |
| Sport | 29,104 | 63,146 | 89,161 |
| Inland commercial | 10,083 | 16,232 | 24,091 |
| Estuaries | 329,260 | 329,260 | 329,260 |
| Total—freshwater need for wildlife and fisheries (excluding flow to estuaries) | 70,361 | 134,050 | 175,752 |
| Total—all freshwater needs for wildlife and fisheries | 399,621 | 463,310 | 505,012 |

*Source.* Select Committee on National Water Resources (1960a).

Small watershed projects designed to reduce flooding, improve drainage, and reduce sediment damage oftentimes can work against fish and wildlife interests. Measures to reduce flooding can include drainage, straightening channels, and levee construction as well as soil conservation measures. While the latter will probably improve wildlife habitat, the former three measures could work against fish life by reducing the amount of feeding and spawning grounds. It is, of course, relatively straightforward to incorporate measures to improve the habitat for fish life into small watershed projects but since these activities might cost money, a conscious effort to protect and encourage fish life has to be made.

Highway construction has done a tremendous amount of damage to the nation's fish and wildlife resources. For example, road culverts can serve as barriers to fish movement if not properly installed, highways are often built across productive wetlands areas since the land is inexpensive and easily accessible, rerouting of streams through dredged cuts may limit fish production or development, and modifying stream channels in the process of building bridges can eliminate or harm feeding and spawning grounds. Proper highway construction methods can preserve and even enhance fish and wildlife resources.

**Water Requirements for Fish and Wildlife**

Total freshwater requirements for fish and wildlife are difficult to estimate. Water needs for fish farms, hatcheries, and animal raising areas can be determined with some precision but the other part of the water need—that required in streams and lakes to support present or anticipated fish and wildlife populations—is more difficult to estimate.

Table 5.10, prepared for the Select Committee on National Water Resources

(1960a), provides estimates of the freshwater requirements for sport and commercial inland fishing, for wildlife, and for freshwater inflow to estuaries for 1954, 1980, and 2000. The figures are based, in part, on the average daily catch, the number of fishermen, and estimates of the productivity of fishing waters in terms of pounds of fish per surface-acre (later converted into acre-feet of water). The wildlife figures were based on the amount of water needed to maintain wetland vegetation of significance to wildlife populations. Water requirements are expected to grow significantly from 1954 to 2000, with those for sport fishing increasing essentially by a factor of 3 (the number of fishermen was also predicted to increase by a factor of 3) and the wildlife need for water increasing by a factor of 2 over the 46-year study period.

Freshwater flow to estuaries was not considered to change over the study period but was kept at $329 \times 10^{12}$ gal, a figure nearly five times the 1954 requirements for other fisheries and wildlife activities although less than double the figure for these needs in the year 2000. It represents an estimate of the average annual freshwater flow into all habitable estuaries and would not be expected to change over the near future.

The figures in Table 5.10 are far from exact, based as they are on a number of estimates—of population, number of fishermen, pounds of catch, and especially of the amount of productivity per surface area. They are only included here to give some idea of the order of magnitude of the need for fresh water to support our fish and wildlife resources.

### Water Resources for Outdoor Recreation

Use of river resources in outdoor recreation has been studied from a number of viewpoints by planners, managers, sociologists, and engineers. Most commonly, the interview technique is employed despite its obvious shortcomings of sampling a biased cross-section and asking directed questions. Still, the interview technique provides some information on what is looked for in a recreation area, what should be preserved, and some idea of the value of rivers and water bodies as recreational resources.

For example, in a survey of 80 small streams in Wisconsin, some affected by the discharge from wastewater treatment plants and others not so affected, Kalnicky (1976) found that streams in which treated waste water discharge occurred experienced only 25 to 50% as much recreational use as streams in which no discharge occurred. The discharge of treated wastewater either degraded the stream waters (or they were perceived to have degraded them) in such a way as to make those streams less desirable for recreational use.

Interviews with other than the traditional river water users—for instance, with those who just wanted to sit or stroll along the river bank—showed Kaplan (1977) that there were many individuals who appreciated the presence of the river because it represented a sense of orderliness, of involvement, of mystery, or of tranquility. Kaplan pointed out that since these "passive" users experience benefits similar to the more active users of the rivers, their requirements must be considered

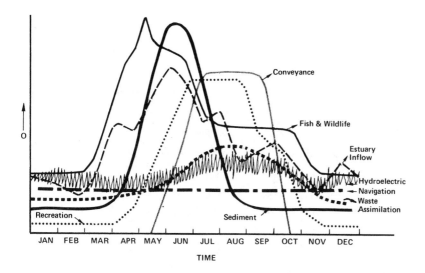

**Figure 5.4** Seasonal course of instream-flow uses for a hypothetical United States river.
*Source.* U.S. Water Resources Council (1978).

in planning and management decisions. The benefits that accrue to these users must also be factored into any benefit-cost analysis of the recreational value of rivers.

A study by the Maine Department of Conservation (1974) of the use of the Allagash Wilderness Waterway suggested that some problems of congestion and user conflicts existed. To alleviate these problems, planning efforts should be directed toward trying to redistribute river use over both time and space and to separate small group and large group users. It was also felt that sites should be developed for vehicle camping and picnicking separate from those used by river floaters.

Obtaining firm information on economic benefits to recreational users of water is most difficult. From interviews with users, Walsh (1977) estimated that decreasing the pollution loading of U.S. streams would increase recreational benefits to users by about $7.3 billion. Over $4 billion of that total would come from savings in travel and time costs to reach recreational sites. The study left open the question as to how much improvement in the quality of the river would be necessary to increase recreational use of the river. In other words, since improving river quality would be expensive, it would be desirable to optimize the benefit-cost relation in such a way that expensive upgrading activities are continued only as long as user benefits continue to accrue at a more rapid rate.

The term "instream use" or "instream-flow need" may be defined as the amount of water flowing in a stream channel that is needed to sustain instream values at acceptable levels. Instream values in this case refer to all uses made of the water

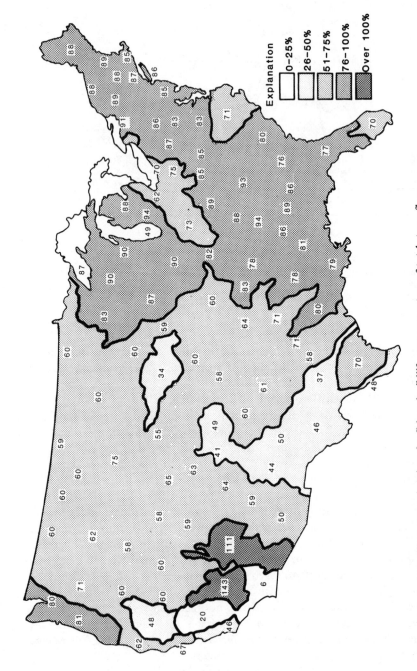

**Figure 5.5** Instream flow approximations for fish and wildlife, as a percentage of total streamflow. *Source.* U.S. Water Resources Council (1978).

as it flows in the channel. These might include fish and wildlife, recreation, navigation, power, waste assimilation, and ecosystem maintenance. Evaluation of the minimum amount of water needed in the stream channel to meet these various needs establishes the acceptable level of instream-flow need. Of course, at a particular place on a river, not all of those possible needs for water would exist at the same time so that the actual value of instream-flow needs is either based on the dominant needs or on some mix of needs within the particular stream sector.

The Second National Water Assessment issued by the U.S. Water Resources Council attempted to look at this problem of instream use both qualitatively and quantitatively. Figure 5.4 provides an estimate of the seasonal pattern of instream-flow needs for all of the possible uses to which water in the stream channel can be put. Fish and wildlife use for streamflow was the dominant use in every region. The recreation and fish and wildlife needs combined would probably be sufficient, in most cases, to establish the minimum instream-flow level.

Quantitative estimates of the instream-flow need for fish and wildlife (as a percentage of total streamflow) were attempted for each water resource subregion. The estimates (Fig. 5.5) are for the flow at the outflow point of each subregion. Two basins, in the southwestern part of the country, clearly do not now have sufficient instream water for fish and wildlife needs. The remaining subregions vary from needing 80–95% (generally in the most moist eastern part of the country) to 50–75% of their total streamflow needs for fish and wildlife (generally in the Great Plains and Rocky Mountain areas). Several basins needing less than 50% of their total streamflow for fish and wildlife are also found in the drier southwestern part of the country (Texas, New Mexico, and California). Great care should be exercised in interpreting these distributions inasmuch as this represents only a first attempt and the results contain a number of approximations and assumptions.

Outdoor recreation is usually divided into two forms—water-dependent (swimming, boating, fishing, water skiing) and water-enhanced (picnicking and camping). Table 5.11 provides estimates of the number of activity occasions (defined as participations of persons 12 years or older in the activity without regard for the length of the activity) in 1975, 1985, and 2000. The 2 billion activity occasions in 1975 should increase by about 34% to some 2.7 billion by the year 2000.

**Table 5.11** Outdoor Recreation Activity Occasions—1975, 1985, 2000 (million activity-occasions)

|                  | 1975 | 1985 | 2000 |
|------------------|------|------|------|
| Water-dependent  | 1379 | 1567 | 1852 |
| Water-enhanced   | 663  | 752  | 889  |
| Total            | 2042 | 2319 | 2741 |

*Source.* U.S. Water Resources Council (1978).

The Second National Water Assessment estimated that, at present, only about one-fourth of approximately 85 million acres of surface-water area available in the conterminous United States are used for recreation, the remaining 65 million acres being either inaccessible, polluted, or otherwise restricted from recreational uses. Since a number of recreational problems existed in all of the water resources subregions, the Second Assessment estimated that 8 million more acres of water-surface area were needed in 1975. This number would be 10 million acres by 1985 and 13 million acres by 2000. These needs could, of course, be met by reducing existing pollution or improving access to more water bodies.

The recreational water problems which were found to exist in the water resources subregions included: (a) preserving free-flowing stream values; (b) retaining floodplains, coastal beaches, and wetlands; (c) improving water quality; (d) increasing recreation opportunities at reservoirs; (e) providing public access to water (U.S. Water Resources Council, 1978).

**Future Needs**

Clearly, the most significant future need from the standpoint of fish and wildlife is not for water per se but for recognition of the problem of decreasing habitat for fish and wildlife. The Federal Acts mentioned earlier speak to this problem but probably do not go far enough. It must be recognized that the fish and wildlife resource is important to us as a nation and that it will take time and effort on the part of all citizens, as well as money, to provide the proper environment for a developing fish and wildlife resource. Ways will have to be devised to evaluate the benefits from fish and wildlife in the same way as we evaluate the benefits of power, irrigation water, or agriculture. Ways must be found to put an economic value on fishing and hunting. This is difficult to do but since our society reacts most strongly to benefit-cost studies, it is clear that ways will have to be devised to bring fish and wildlife into such a framework.

Very little has been said about the recreational need of water outside of the realm of fish and wildlife. Certainly it is clear that water skiing and boating are important also. These needs are not incompatible with the fish and wildlife needs—in fact, the same water can be used for both purposes. As visitor use of parks increases in the future, water use for all purposes will increase significantly.

The Select Committee on National Water Resources (1960c) came up with a series of recommendations to help provide for the future demand for water for recreation. These included:

(a) At least 15% (excluding inlets and islands) of our ocean and major inland shorelines should be acquired by some level of government and made available for public recreational purposes. Only about 6.5% of the Atlantic and Gulf shoreline is now under Federal or state ownership.
(b) Certain selected streams should be preserved in their natural free-flowing condition because of their scenic, scientific, aesthetic, or recreational value. These include streams such as the Allagash River in Maine, the Rogue River in Oregon, and the Current and Eleven rivers in Missouri.

(c) Every effort should be made to prevent pollution of streams, rivers, lakes, and shoreline areas.

(d) Congress should adopt a policy stating that the recreational potentials of all Federal multi-purpose reservoirs should be developed for public use. The Secretary of the Interior should prepare plans for the recreational development of all Federal reservoirs. The Secretary should be authorized to procure the land needed for access roads and facilities for safety, health, and protection of visitors.

(e) The Secretary of the Interior should be authorized to construct water-control projects having public recreation as their primary purpose. Reservoirs associated with such projects should not exceed 2500 acres in surface area.

(f) As far as possible, recreation should be recognized as an appropriate use of domestic water supply reservoirs and watersheds. Such use will have to be strictly controlled in order to protect the water supplies from possible contamination.

Enactment of such recommendations along with a growing public awareness will go a long way to insure the maintenance of adequate water and land habitats for our fish and wildlife resources both now and in the future.

### The Wild and Scenic Rivers Act

Passed in 1968 by Congress, the Wild and Scenic Rivers Act stated that it was the policy of the United States to preserve certain rivers possessing scenic, recreational, fish, wildlife, historic, or cultural values for the benefit of present and future generations. Thus, the Act created a Federal wild and scenic river system. The system, if it can be called that as yet, has grown slowly and painfully, for several reasons, until now it includes portions or all of 27 rivers totaling 2300 mi in length (Goldfarb, 1980).

Rivers can become designated as "wild and scenic" in one of two ways. The legislature of the state through which the river flows can request the governor to petition the Secretary of the Interior to include the river in the Federal system. The river to be included must be administered for all times by the state without further expense to the United States (except when federally-owned lands exist in the area so designated). Clearly, states have little to gain by having a river included while they lose all rights to other developmental options on the river. As a result, state legislatures have been slow to adopt this method to include rivers in the wild and scenic system.

The second method for inclusion in the system requires that Congress designate a particular river as a "study river." Following a protracted study by the Secretary of the Interior (utilizing the Heritage, Conservation, and Recreation Service), a report is made to the President after both Federal and state water resources groups are allowed adequate time for their evaluations and testimony. If the President recommends inclusion of the river in the system, Congress must pass an Act actually admitting the river. The study process can be quite lengthy (even as long as 10 years) so that only rather noncontroversial rivers have been admitted so far. Goldfarb (1980) has suggested that the study process is somewhat self-defeating since the river is not protected from private activities during the study period. Both land and recreational developments resulting from the increased publicity often contribute to the destruction of the very resources that are to be protected and enhanced by inclusion in the system.

A further shortcoming of the Act is the fact that the area of the river system to be protected by the Act is limited to just 320 acres per mile—essentially ¼ mi on each side of the river. The Act is unable to protect the river or basin from private development outside this narrow corridor although it is quite clear that such developments, if uncontrolled, could result in considerable degradation of the river.

The designated "wild and scenic" river does receive considerable protection against further Federal development activities although it is not well protected from private activities. The Act states that Federal dams, conduits, reservoirs, powerhouses, transmission lines, or other projects or works which might affect the river directly cannot be authorized and that no department or agency of the government may assist in any way in the construction of water resources projects that might adversely affect the river. Even here, however, it should be recognized that these limitations apply only to the portion of the river designated as "wild and scenic." Such developments either up or downstream from the designated portion of the river are permitted.

About half of the states have their own programs of river preservation. They differ markedly in how they are administered and in what is being protected but since the admission process is more direct and under local control, it is quite possible that a strong state program may actually provide more protection, more rapidly, than the Federal program. Inclusion in a state program does not, of course, protect against later Federal developments on the river.

# PART C:
## OBTAINING ADDITIONAL
## SUPPLIES OF WATER

A review of the various demands for water for industrial, agricultural, recreational, and municipal purposes, and an understanding of the increase in those demands over the past few decades, ultimately must lead to the possibility of obtaining more water—either by tapping new sources of fresh water, by improving conservation, or by introducing other water savings techniques so that the current supplies will go further. Many different techniques exist for making additional supplies of water available in a particular local situation. Since water is not short globally or, for that matter, in a particular continent but rather only unequally distributed, one obvious technique to make more usable water available would be to move it from one place to another by means of interbasin transfers. No new water is created overall but the distribution problem is resolved and the available supply of water is put to better use.

Techniques involving reduction in the waste of water also must come to mind. Some 70% of our precipitation is lost through evapotranspiration. Certainly some of this occurs from standing water bodies, reservoirs, lakes, and ponds (and so serves no useful purpose). If it could be eliminated or reduced, more water would be available. Another large amount of transpiration occurs from noneconomic vegetation. Removal or control of this vegetation would reduce these useless transpiration losses and thus make more water available for higher-valued purposes.

Over 97% of all our water is stored in ocean basins, essentially unusable because of its salt content. Techniques to remove the salt from a portion of this water would provide great new supplies of fresh water. While they might only be economically used in coastal locations because of the cost of transportation, still such techniques would be of great value in many arid and semiarid coastal areas of the world.

Precipitation augmentation—rainmaking—is a technique that has given promise of providing new supplies of water for the past several decades. As more is learned of the physics of the air, that dream is becoming closer to a reality. The reality is not so much "push-button weather," as some would have us believe, where we

183

can dial whatever kind of weather conditions we want, but it does seem now that under certain conditions a 10 to 15% increase in precipitation might be possible by careful use of a cloud-seeding technique.

Great quantities of water are made unfit for further use as a result of pollution. This problem is compounded when the polluted water is mixed with relatively fresh water in order to dilute it because then larger volumes of water are contaminated, even though to a lesser degree. Renovation of waste water, thus, looms large as a useful technique to make available additional supplies of water. We are continually reusing water since essentially no new water is being created on the earth. The planned renovation and reuse of water, however, involves the conscious effort to purify and reuse water from sewage treatment plants. Not only does it involve economic considerations but also health and aesthetic aspects. Despite these problems, it would seem to be a technique whose time is about to come.

Other techniques involving vegetation control, forest management, and reduction of seepage of water from canals and reservoirs can be suggested as possible ways to provide additional sources of water for use. All of these techniques will be discussed and compared in the following chapters in an effort to understand their costs, as well as their real potential for adding to our limited water supplies.

## 6

WEATHER MODIFICATION AND
SALINE WATER CONVERSION

## WEATHER MODIFICATION

The term "weather modification" covers a wide variety of activities although its primary meaning to most people is rainmaking. Scientific rainmaking has really been practiced only since the mid-1940s yet various activities to produce additional supplies of rain have been practiced since almost the dawn of civilization. Prayers to various gods for special weather were common occurrences in ancient Greece while human and animal sacrifices to various weather gods were often practiced in other civilizations. Indian rain dances, now more relegated to the role of tourist attractions, were fairly common throughout the arid and semiarid southwestern part of our country. The Zunis of New Mexico named their chief gods the Vivanami, or the rainmakers (Oliver, 1977). Some cultures have believed that by imitating conditions associated with precipitation they might induce nature to provide the real thing. For example, Oliver (1977) reports the delightful practice in Estonia where three men climb a selected tree during dry periods. One knocks two burning sticks together, the second beats a drum, the third scatters water all around. The actions—which were human imitations of lightning, thunder, and rain—were supposed to encourage nature to produce a real thunderstorm.

For many centuries, people believed that loud noises would induce precipitation. The ancient Chinese were said to have used firecrackers or other noise makers to encourage rainfall. Statistical studies have been carried out to offer proof that rainfall often followed battles—supposedly related to the noise of the cannonades.

Congress even considered the noise-rainfall association and, in 1891, authorized some $9000 to a professed rainmaker to set off a dynamite explosion in Washington, D.C. as well as in Texas in order to make rain. The Washington experiment gave some suggestion of success. Even though light showers occurred before the explosion and the period was hot and muggy, the showers also continued well after the experiment was over. Congress, however, was not too impressed with the effort. In Europe, noise has also been used to suppress unwanted meteorologic events such as hail or severe thunderstorms. Some believed that positive relations existed between cannonades or the ringing of church bells and reduced crop damage by hail but there is little rationale for such a relation. The technique did have some negative aspects, also, since some individuals were killed by misfiring cannons or by lightning strikes on bell towers. As a result, it was outlawed in some places. In more recent years, rockets have been fired into thunderstorms with the idea that the sound waves from the explosions within the thunderstorm will somehow crush or crack the hailstones as they are forming and so result in smaller or less damaging hail from the storm.

The continued desire on the part of farmers and others for rain during dry periods or in regions of low and variable precipitation has produced a legion of self-styled rainmakers with wild and ingenious devices to coax moisture out of the atmosphere. To be a rainmaker was often an easy way to make money; even if your device or technique had no influence on the weather whatsoever, rain might still fall and credit as well as payment could be accepted. Success in one place, whether real or imagined, would do wonders for building one's reputation and increasing the size of one's fees.

There was probably no single rainmaker as successful as Charles Hatfield whose activities covered mainly western United States and Canada in the early days of the 20th century (Spence, 1980). Although little was known about the device that Hatfield used to make rain, it consisted generally of galvanized iron tanks mounted on a 25-ft tower. Copper wires connected the tanks with the ground and a chemical solution filled the tanks. Hatfield's success was truly magnificent. In a drought situation in Los Angeles in 1904, his efforts were followed by more than 11½ in. of rain over a 2-day period. Later, in San Diego, use of his apparatus was followed by more than 20 in. of rainfall, enough to wash out the city reservoir and to drown 17 inhabitants. The irate citizenry forced Hatfield to leave town without his fee in that case.

Rainmaking is on a more scientific basis now than it was during Hatfield's day. But modern, scientific rainmaking cannot guarantee success in a drought situation. Let us consider the reasons for this by studying the physical processes involved in the formation of rain and how human actions attempt to compensate for a deficiency of nature.

### The Physics of Rainmaking

As a parcel of moist air rises in the atmosphere, pressure decreases and the parcel cools by expansion. If sufficient cooling occurs, the air will be brought to saturation (100% relative humidity) because the amount of moisture that can remain in the air in vapor form decreases as the air temperature decreases. Thus, sufficient cooling of air will always result in the air becoming saturated. Any further cooling of the air will cause the air to be supersaturated—having more water in vapor form than can exist at that given temperature—and moisture will begin to condense

**Table 6.1**  Values of Saturated Vapor Pressure (in mm of Hg) with Respect to Ice and Water at Different Temperatures

| Temp °C | Saturated Vapor Pressure (mm of Hg) | |
|---|---|---|
| | Over Ice | Over Water |
| 0 | 4.579 | 4.579 |
| −2 | 3.880 | 3.956 |
| −4 | 3.280 | 3.410 |
| −6 | 2.765 | 2.931 |
| −8 | 2.326 | 2.514 |
| −10 | 1.950 | 2.149 |
| −12 | 1.632 | 1.834 |
| −14 | 1.361 | 1.436 |

out on small hygroscopic particles (salt, dust, and other water-attracting particles that are always found floating in the air). A cloud will form, made up of these tiny condensation nuclei surrounded by tiny water droplets. But cloud droplets are quite small and will continue to float in the air until they become large enough to fall. It will take as many as 1,000,000 small cloud droplets to make one average-sized raindrop.

The condensation process results from cooling of moist air in the presence of water-attracting particles in the air. This is a slow process; for cloud droplets to grow to the size of ordinary raindrops in this manner might take anywhere from 12 to 24 hrs. Yet we know that heavy rainstorms can develop from cumulonimbus clouds that did not even exist 2 hrs before the rain started. How, therefore, is the precipitation process different from the condensation process?

As the air temperature in a cloud drops below freezing, some of the water particles are converted into ice crystals but even at temperatures well below freezing not all of the cloud droplets are frozen. Between −10 and −20°C, there may be as many water droplets as ice droplets in the cloud, the water in a supercooled form. This is the key to the process suggested by the Scandinavian meteorologist, Bergeron.

At any given temperature, the moisture (or vapor) in the air at saturation will exert a definite and physically reproducible pressure. This is known as the saturation vapor pressure of the air at that temperature. The saturation vapor pressure decreases with decreasing temperature but it still has a particular value even at temperatures below freezing. It has been found, however, that at a given temperature below freezing there are two different saturation vapor pressures—one with respect to water (supercooled) and the other with respect to ice—at that temperature. Table 6.1 provides a few values of the saturation vapor pressures over ice and water at different temperatures below freezing.

In a cloud with both water and ice droplets at the same temperature below freezing—say $-12°C$—an interesting thing will happen. The air in the cloud can be considered to have an average vapor pressure between the saturation vapor pressure with respect to ice and that with respect to water at that temperature ($-12°C$ in our example). From Table 6.1, this is seen to be 1.632 mm of mercury with respect to the ice droplets in the cloud and 1.834 mm of mercury with respect to the water droplets in the cloud. The average vapor pressure in the cloud is 1.733 mm of mercury. The air in the cloud is supersaturated with respect to the ice particles and appears to be unsaturated with respect to the water droplets in the cloud. Under those unique conditions, the water droplets tend to evaporate into the unsaturated air while vapor in the supersaturated air around the ice particles will rapidly condense onto the ice. The ice droplets will grow rapidly in size at the expense of the water droplets in the cloud. Upon reaching a size large enough, the ice droplets will start to fall out of the cloud. If they pass through warmer air layers on the way down, they may melt and fall to earth as rain drops—or, if they are small in size, they may even evaporate on the way down and never reach the ground—but the cloud may have disappeared since moisture has been removed from the cloud. In this example, the reason for the precipitation is the presence of the ice crystals which provides the physical mechanism—the difference in saturated vapor pressure—to cause the ice crystals to become larger rapidly. One could, therefore, argue that the lack of rain from a cloud containing an adequate amount of water in cloud droplet form is due to the lack of sufficient ice crystals to initiate the so-called ice-crystal precipitation process. There may not be enough ice crystals in the clouds, of course, because of the lack of freezing nuclei in the air. There are two types of particles, the so-called condensation nuclei around which vapor will condense as the humidity approaches 100% and freezing nuclei around which vapor will also condense but will result in ice crystals if the temperature is somewhat below freezing. Without freezing nuclei, clouds may form with the rising and cooling of air but no precipitation may result because the ice-crystal precipitation process will not operate.

Thus, to obtain precipitation from a cloud, one might specify the following requirements: (a) an adequate and continuing supply of moist air rising in the atmosphere which is being cooled sufficiently to bring the air to saturation; (b) the presence of freezing nuclei in the cloud. One without the other will usually not result in significant amounts of precipitation. Introducing freezing nuclei in thin, stratiform clouds not well endowed with a supply of moisture may only result in some drizzle or very light rain. The occurrence of growing cumuloform clouds, indicating a large supply of moist air rising vertically upward and being cooled, might not result in precipitation if freezing nuclei are absent as the cloud droplets would have to grow to raindrop size by condensation and coalescence alone.

It should be mentioned here that the ice-crystal process is not the only process capable of initiating precipitation in the atmosphere. It does seem to be one of the more efficient precipitation mechanisms in mid-latitude areas, but it is also clear that heavy precipitation can result from cumuloform clouds in tropical and subtropical areas when the cloud may be entirely at temperatures above freezing. The ice-crystal process does not operate in those cases, yet precipitation can develop fairly rapidly. Clearly, some process other than slow condensation is at work. It is known that different-sized condensation nuclei can exist in the air, some many times larger than others. Vapor condensing around the large condensation nuclei will create large cloud droplets. These are much more efficient in collecting additional moisture by coalescence and capture as they move around in the cloud coming into contact with the many smaller cloud droplets. When small droplets hit together, they often just bounce off one another without merging together. Thus, it is felt that rain can be produced, in some cases, from clouds that have quite different-sized condensation nuclei just by the process of coalescence and capture. Size variation in nuclei seems to be more a factor in tropical and subtropical clouds than in mid-latitude clouds.

Another possible precipitation factor that we have known about since the days of Lord Rayleigh, who did some initial experiments on the subject, is the role of electrical charges or electrical fields on the coalescence of cloud droplets. Rayleigh showed that cloud droplets that were oppositely charged would coalesce if they came into contact with one another while uncharged droplets not in an electrical field would merely bounce apart. More intensive work in the past two decades (Sartor, 1969) has provided much more information about the effect of electrical charges. By shooting two streams of small droplets so that they will intersect with one another, Sartor has conclusively shown that charged droplets will merge with greater frequency while uncharged droplets will merely bounce apart. Moreover, he has shown that uncharged droplets in an electrical field will also have a tendency to merge if they come into contact. Since electrical fields are often present in growing cumuloform clouds or the cloud droplets may have electrical charges, it is quite possible that electrical forces ultimately play a role in the rapid formation of large cloud droplets and rainfall. Some modern theories of rainmaking involve the establishment of electrical potentials in the atmosphere although field tests of the technique have failed to give real evidence that the idea will be practicable.

Of the three possible ways shown to be effective in the growth of cloud droplets—the ice-crystal process, size variation in condensation nuclei, and electrical forces—the ice-crystal process has been repeatedly demonstrated to work in the laboratory. Vincent Schaefer's original work in 1946, in the General Electric Laboratories in New York, is well known to all meteorologists. Dropping small pieces of dry ice into the supercooled air of a freezing cabinet, he saw ice crystals form and fall to the bottom of the cabinet. The dry ice caused the supercooled water in the air of the cabinet to migrate to the ice particles and to fall out. Repeating the experiment in a supercooled, stratiform cloud caused moisture to fall out of the bottom of the cloud and a large hole to develop in the cloud. Later, seeding of cumuloform clouds with an adequate supply of vapor continually feeding into the cloud seemed to produce significant amounts of precipitation. Vonnegut, also working at General Electric, later showed that another substance, silver iodide, could be added to a cloud with the same result.

How do we know whether cloud seeding has been successful? Just because rain falls out of a cloud that has been seeded does not mean success. Do we know that the cloud would not have given us precipitation in approximately the same quantity if we had not seeded? Since cloud seeders admit that they cannot be successful seeding all clouds and since they confine their activities to only certain clouds, considerable bias is added to the whole problem of verification of success. It is quite possible that cloud seeding has only speeded up the formation and fall of precipitation by a few minutes but has not changed the amount at all. Statistically relevant answers to these and other questions about cloud-seeding operations still must be achieved.

Many field tests using randomized seeding procedures when the right type of clouds exist, of seeding clouds in one basin but not in nearby similar basins, or by seeding in certain time frequencies, have suggested that some degree of willful control of rainmaking in certain areas is possible. However, the more professional rainmakers will still not promise any given amount of rainfall at any time, or that drought conditions can be eliminated by the flip of a switch to turn on a silver iodide generator.

## Results of Rainmaking Activities

Because early publicity promised too much and many practitioners of cloud seeding did not fully understand the problems in cloud modification work, there have been both successes and failures over the years. At present, we know that while we have modified climates inadvertently in many different ways, we believe also that we have a modest degree of willful control over certain aspects of our weather.

A 1978 report of the Weather Modification Advisory Board to the Secretary of Commerce concludes that we have a usable technology for significantly increasing precipitation, although it is careful to point out that the technology will not be effective at all times and in all parts of the country. The report also concludes that we have good evidence that:

(a) Supercooled fog and stratus can be dissipated by seeding to increase visibility;
(b) Stratus clouds can be dissipated by seeding to increase solar radiation receipts at the ground;
(c) The depth of snowpack can be increased by seeding wintertime orographic clouds in some mountainous areas;
(d) Rainfall can be increased from certain types of summer cumulus by dynamic seeding of growing cumulus turrets;
(e) Rainfall can be increased in Israel (and possibly California) by seeding winter cumulus clouds;
(f) Hail suppression experiments, while attempted in many areas, have been given credit for reductions of up to 80% in the Soviet Union although little convincing evidence is available from the United States;
(g) There is no firm evidence that increases in precipitation in one area have resulted in decreases elsewhere.

In all, scientific cloud-seeding activities have been carried out in 74 countries. In the United States, $4 to $6 million a year is spent on such activities with about

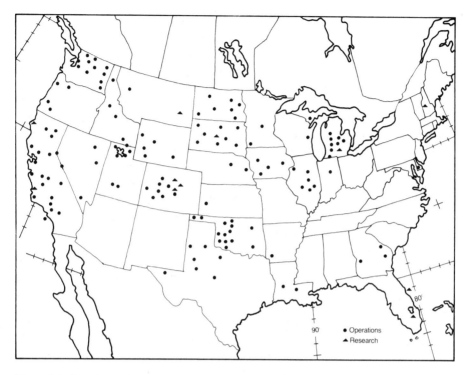

**Figure 6.1** Locations of research and operational weather modification projects in the United States, 1973–1977. Two operational projects were in south-central Alaska and no projects were in Hawaii.
*Source.* Weather Modification Advisory Board (1978), courtesy of NOAA, Weather Modification Reporting Program, Rockville, Maryland.

12 companies carrying out the actual seeding activities for a wide range of clients (Fig. 6.1). The Weather Modification Advisory Board reports that in 1977 there were 88 cloud-seeding projects in 23 states and these projects covered about 260,000 mi² (7% of the area of the United States). The two types of seeding activities with highest priority are increasing snowpack depth by seeding winter orographic clouds and the seeding of growing-season cumulus clouds in agricultural regions.

**Strategies for Seeding Cumulus Clouds**

In terms of rainmaking, we have already learned a great deal as a result of three decades of careful field and laboratory studies. First, we have learned that we may never be able to make significant amounts of rain from stratiform-type clouds. These clouds—stratus, stratocumulus, fog, altostratus, cirrostratus—are often formed from turbulent motion or stirring of the air or by the slow ascent of warm air flowing up the gentle incline of an advancing warm front. Usually the clouds are not particularly thick and so, if warm moist air is not continually flowing into them, any seeding which causes ice crystals to grow rapidly does nothing more than produce a limited amount of moisture to fall out of the cloud and the cloud to dissipate. Of course, we also know that we are not going to make rain if there is no supercooled water in the cloud. The ice-crystal process cannot operate without supercooled water.

Over the past three decades, there has been a change in our ideas about seeding cumulus clouds. At the beginning, a technique known as static seeding was employed. In this technique, it was assumed that precipitation was not occurring from the cloud because it was naturally deficient in ice-forming nuclei. The deficiency resulted in a great amount of supercooled water at temperatures below freezing and not enough ice crystals. Thus, it was reasoned, addition of ice-forming nuclei by seeding should result in a better balance and produce more precipitation. Use of ice nuclei or silver iodide generators either on the ground or in aircraft upwind of the clouds should cause the ice nuclei to enter the cloud and begin the process of release of precipitation.

The results of static-seeding experiments have ranged from negative through inconclusive to positive. In Project Whitetop in Missouri from 1960 to 1964, for example, there seemed to be a net decrease in rainfall, which was attributed at first to overseeding since there was a high ice particle density naturally in the clouds. Later studies of the Whitetop data suggested that there were increases in precipitation when cloud tops on radar were in the 20–40,000 ft range. With higher tops, precipitation decreased. Generalized statements about increases or decreases should not be made without careful analysis of all aspects of cloud height, temperature, liquid water content, and stability.

A study of 14 short-term operational projects using static seeding suggested to a National Academy of Sciences panel (1966) that precipitation increases of from 0 to 33% might be obtained. While certain of the results had some statistical significance, the study did not provide conclusive evidence of an increase in rainfall due to static seeding.

The technique of dynamic seeding was first suggested in 1948 although the major development and use of this technique was delayed until the late 1960s and early 1970s. The delay was partly due to the need for numerical cumulus cloud models able to simulate cloud processes and the need for improved pyrotechnic seeding devices. Dynamic seeding attempts to increase the cloud's potential for rain by making the cloud grow larger than it might have without seeding, or by making several nearby cumulus turrets grow together and hence produce increased intensity and/or duration of rainfall. Dynamic seeding involves massive seeding of cumulus-type clouds with silver iodide generated by pyrotechnics (Fig. 6.2). Two ambitious tests of the dynamic seeding concept were carried out in Florida in the late 1960s and again through most of the 1970s with seemingly positive results. For example, precipitation measured by radar from individual cumulus clouds showed increases of 2 to 3 times, significant at the 0.5% level. Later, in the Florida Area Cumulus Experiment (FACE), randomized seeding of growing cumulus clouds showed that average rainfall increases of 20 to 70% might have been achieved (Woodley, Simpson, Biondini, and Jordan, 1977).

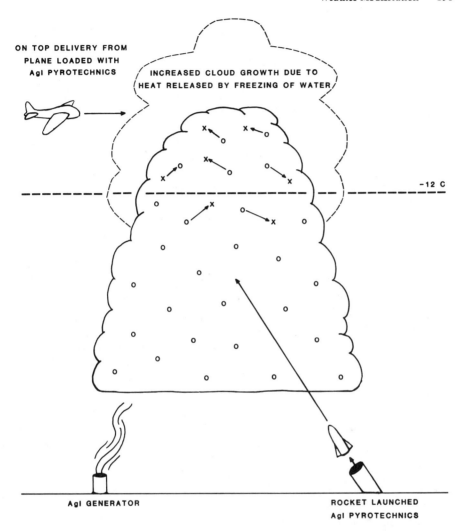

ON TOP DELIVERY FROM
PLANE LOADED WITH
AgI PYROTECHNICS

INCREASED CLOUD GROWTH DUE TO
HEAT RELEASED BY FREEZING OF WATER

-12 C

AgI GENERATOR

ROCKET LAUNCHED
AgI PYROTECHNICS

**Figure 6.2** Cumulus seeding with silver iodide. Silver iodide delivered to cloud in smoke from ground-based generator or by shooting Ag I pyrotechnics into cloud from ground or from on top. Rapid freezing of supercooled water releases heat which should cause substantial cloud growth and more precipitation.

In experiments with warm season cumulus in South Dakota, increases in rainfall were attributed to dynamic seeding although seeding rates were considerably below those used in FACE. Thus, there appears to be no clearly defined line between static and dynamic seeding in terms of rates of seeding although the basic idea of each type of seeding is different.

## Rainmaking by Initiating Convective Activity

Islands in the oceans have long been recognized as places over which clouds form. Convection activity resulting from the somewhat higher temperatures (5 to 10°C) occurring over the island than over the surrounding sea often produces clouds over the island that may drift in lines downwind. This, of course, would suggest that ground coatings of asphalt or some other substance to absorb heat might similarly be able to increase convective activity and cause clouds to develop and so increase rain. Suggested sizes for the asphalt covered areas are of the order of 40 mi$^2$. It is estimated that such coated areas would have a temperature excess of some 11°C over surrounding grass areas. The warmer temperatures would exist for 20 of the 24 hours in the day.

Use of the technique in desert areas where it might be feasible to cover vast areas with asphalt raises certain questions. First, the desert areas are often regions of air subsidence and few natural clouds so that convective activity would be more difficult to initiate. Also, of course, the surroundings are apt to be sand or vegetation-free so that temperature differences between the asphalt cover and the surroundings might not be as great as in the case of an island in the midst of a sea. Still, extensive modeling studies have been carried out to learn more about the technique. Some researchers cite the urban increases of precipitation as proof that the technique works, but the urban example is different in many respects from a large asphalt sheet in the midst of a dry barren area.

## Hail Suppression

Two hypotheses have been used to explain why cloud seeding might suppress hail. The first is based on the idea of glaciating the upper portion of cumulus clouds in order to reduce the amount of supercooled water present and available for hail growth. The second is based on the idea of adding more hail embryos so that the available supercooled water will be spread over a greater number of hailstones, each thus achieving a smaller size.

During the 20-year period from 1958 to 1977, some 71 hail suppression field trials were carried out in the United States, generally in the Great Plains. In spite of the large number of trials, and the reports of some reduction in hail occurrence and especially in hail-loss damage claims in some areas, the Statistical Task Force of the Weather Modification Advisory Board (1978) found inconclusive evidence for hail suppression. This conclusion is in strong conflict with findings in the Soviet Union where claims of 60 to 80% effectiveness in reducing hail have been made. While the great difference in the results of hail suppression work is difficult to explain, it must be pointed out that techniques of seeding differ in the two areas, with U.S. seeders generally using seeding from aircraft either above or below the thunderstorm clouds, sometimes supplemented with ground-based silver iodide generators, while the Soviet technique is to introduce the seeding agent into the cloud by means of ground-based rockets. It is generally felt that the difference in seeding techniques is not the cause of the difference in the results achieved, however.

**Figure 6.3** Schematic illustration of a vertical slice through a hurricane before cloud seeding (above) and after (below) if the hypothesis of outward displacement of the eyewall by seeding is correct.
*Source.* Weather Modification Advisory Board (1978), courtesy of R. H. Simpson, Simpson Weather Associates, Charlottesville, Virginia.

## Hurricane Modification

Hurricanes are superior rain producers. While they are accompanied by devastating winds (and also tornadoes) and storm surges that bring great destruction and death, they still often result in 8 or more inches of precipitation falling over wide areas to recharge water tables and fill reservoirs. Annual economic losses from hurricanes striking the United States are estimated to be about $800 million with a loss of several hundred lives annually (Weather Modification Advisory Board, 1978). Thus, while we can see great value in reducing the destructive power of hurricanes, we would hope that such activities would not significantly reduce the total precipitation as well.

Several models of hurricane modification by cloud seeding exist. The basic approach has been to try to cause the inner wall of clouds around the eye to move outward. This action, it has been felt, would lead to a reduction in the windspeed around the eye and so make the storm less dangerous. One model suggested that seeding in the ring of clouds around the eye would result in glaciation in those clouds. The heat released by glaciation would decrease the temperature difference

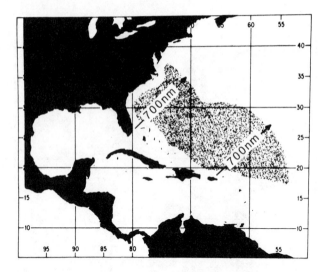

**Figure 6.4** Permitted experimental seeding area for Atlantic hurricanes. Even when a storm is in this area, it is not eligible for seeding if the forecast probability that it will strike land within 24 hours is 10% or higher.
*Source.* Weather Modification Advisory Board (1978), courtesy of the National Hurricane and Experimental Meteorology Laboratory, NOAA, Coral Gables, Florida.

across the eyewall and thus reduce the pressure gradient force and the windspeed around the eye. Later modeling studies suggested that dynamic seeding of clouds at some distance from the eyewall might cause them to grow much taller and essentially to develop as a new eyewall. Warm moist air feeding into the old eyewall would be cut off and it would weaken. The new larger diameter of the eye would result in reduced wind velocities (Fig. 6.3).

Three Atlantic hurricanes were seeded to test this latter hypothesis and in all cases a reduction of wind velocities occurred—10% in Esther in September 1961, 14% in Beulah in August 1963, and 30% and 15% on two different days of seeding in Debbie in 1969. There is always some possibility that the changes noted might have occurred by chance, but since the changes did occur and in the direction postulated by the model, it is highly suggestive that seeding hurricanes can ultimately lead to some modest control of their windspeed. Since there is only a limited area in the North Atlantic Ocean in which experimental seeding of Atlantic hurricanes is permitted, in order to insure safety for coastal inhabitants, opportunities for experimental seeding are often few and far between (Fig. 6.4). Five hurricanes per year occurring in the permitted seeding area is about the maximum that might be anticipated although the average is closer to two. Thus, it may take many years to accumulate enough experimental evidence to be able to state with any assurance that seeding of hurricanes has produced statistically significant results.

**Potential for Rainmaking**

As a result of early experimental work, the general conclusions of responsible scientists during the 1960s were:

(a) Seeding of winter-type storm clouds in the mountainous areas of western United States could result in a 10 to 15% increase in monthly precipitation.
(b) In nonmountainous areas, in any season of the year, there was no statistically valid evidence that any increase or decrease in precipitation accompanied cloud-seeding operations.
(c) So little had been learned about cloud seeding for hail modification that no conclusions could be drawn.

As a result of later cloud-seeding activities, especially in southern Florida, these early and very cautious conclusions have been modified somewhat. For example, we now believe that seeding of winter cumuloform clouds in Israel and in California may produce a 10 to 15% increase in rainfall. Also, seeding of summer cumulus clouds in FACE has seemed to result in some increases in rainfall. Seeding in small, inactive clouds will result in their dissipation. Injection of water drops or salt into active, warm, (nonsupercooled) clouds may also release some rain. Seeding of certain types of fog will be effective in causing the fog to dissipate for a period of time. Cloud seeding, therefore, acts to trigger the release of precipitation from existing clouds. Release of substantial amounts of precipitation requires the existence of a large moisture supply (moist air currents) as well as active cloud-forming processes.

The Weather Modification Advisory Board (1978) suggests that in the next two decades our weather modification capabilities will increase rapidly. They believe that, with careful seeding *under favorable conditions* (this caveat is always important for we will not be able to increase rainfall where no clouds or the wrong type of clouds exist), we might expect to increase crop yields 5 to 10% by improving rainfall conditions, to lower hurricane winds by 15 to 20%, and to cut hail losses by 30 to 60%. Changes of these magnitudes might well enable us to meet some of the anticipated weather—and climate—induced problems of the year 2000.

The Board points out that the economic value of 10% more growing season rainfall would far outweigh the benefits resulting from a 50% decrease in hail and wind losses. In fact, a 30% reduction in hail loss with no increase in rainfall at the same time would have little economic value, they concluded. It is estimated that a 10% increase in growing season precipitation (in the major crop growing areas) would increase farm income by over $200 million. Other individual economic benefits noted by the Weather Modification Advisory Board (1978) include:

(a) A 3 to 34% increase in farm income in southeastern South Dakota depending on the year and type of farming.
(b) A $1.3 million cost advantage (over the cost of the next best alternative) of an estimated 15% increase in precipitation over the Connecticut River Valley (over 9 months) due to increased water for domestic and industrial use.

(c) A 4% increase in snowpack runoff in rivers in California due to seeding would have values ranging from $340,000 up to $1.3 million annually depending on whether the extra water was used for irrigation or power generation.

(d) Benefits of $12.8 million from the estimated 2 million acre-feet of water that could be added to the Colorado River Basin by enhancement of the winter snowpack by seeding orographic clouds.

(e) Net benefits of from $1 to $3 million due to modification of precipitation over the Great Lakes due to improved shipping and hydroelectric power generation conditions over and above the damage to shore properties.

Table 6.2 provides some estimates of changes in farm income in northwestern Kansas that might be anticipated as a result of different percentage changes in rainfall and crop-hail damage. It shows that a 20% change in growing season rainfall (from -10 to +10) results in a greater change in net income per harvested acre (about $6.00) than an 80% change in crop-hail damage (about $4.50 per harvested acre). Combining both the increase in rainfall with the greatest possible reduction in crop-hail damage results in the greatest increase in net income (about $7.30 per harvested acre). Thus, hail reduction is of less importance to crop production than even a small increase in rainfall.

### Downwind Effects

The question may be asked whether cloud-seeding operations, if successful, would be taking precipitation away from other areas downwind that were destined to receive precipitation naturally. Conclusive answers are still not possible but a review of the information on the hydrologic cycle (Chapter 2) or any consideration of the vast amounts of water moving in the atmosphere would suggest that we have little to fear in this regard. Yet the layman is unconvinced and more than one cloud-seeding airplane has been shot at by distrustful and concerned farmers.

Actual evidence of downwind effects has shown that wintertime seeding of convective clouds in California also increased precipitation up to 155 mi downwind of the principal target area. Another winter project showed increases in precipitation up to 62 mi downwind although the project was not specifically designed to study downwind effects. In a Swiss hail-suppression study, positive precipitation effects were found several hundred kilometers downwind from the generator site while, in a project in the Necaxa Basin in Mexico, there seemed to be a positive precipitation effect downwind but a negative effect upwind. As mentioned previously, negative effects were found downwind of the Project Whitetop plume while negative effects were claimed more than 100 km downwind of an Arizona summer cumulus project. The results, therefore, are quite mixed and unconvincing at present.

### Project Skywater

Many of the earlier results and estimates have been obtained from work carried out under Project Skywater, a research program of the U.S. Bureau of Reclamation.

**Table 6.2**  Net Income Changes for Wheat Production in Northwestern Kansas Predicted from Various Degrees of Postulated Hail and Rainfall Modification

| Percent Reduction in Crop-Hail Damage | Percent Change in Growing Season Rainfall | Average Net Income Per Harvested Acre (dollars) |
|:---:|:---:|:---:|
| 0 | -10 | 21.56 |
| 0 | No change | 24.58 |
| 0 | +10 | 27.31 |
| 20 | -10 | 22.60 |
| 20 | No change | 25.74 |
| 20 | +10 | 28.47 |
| 50 | -10 | 24.34 |
| 50 | No change | 27.35 |
| 50 | +10 | 30.11 |
| 80 | -10 | 25.98 |
| 80 | No change | 29.11 |
| 80 | +10 | 31.88 |

*Source.* Changnon, Davis, Farhar, Hass, Ivens, Jones, Klein, Mann, Morgan, Sonka, Swanson, Taylor, and Van Blokland (1977).

The Bureau began its weather modification activities in 1961 with some funds for cloud-seeding research. The work has grown over the years to include hundreds of scientists and technicians from both academic and private groups as well as Federal and state governments under the general guidance of more than 30 specialists from the Bureau who direct Project Skywater.

Project Skywater employs both static and dynamic seeding approaches and is directed toward seeding both convective (cumulus) clouds and orographic clouds. Thus, the project attempts to increase summer rainfall needed for agricultural activities in the western states as well as to augment winter snowpack. Three major field projects have been undertaken under Skywater.

The Colorado River Basin Pilot Project (CRBPP) commenced seeding experiments in 1970 and the project was completed in 1975. The results of that experimental seeding of winter orographic clouds suggested that, in spite of the complex nature of such storm clouds, seeding under favorable conditions could increase the winter snowpack by about 10 to 15%. Cloud temperatures as well as stability of the air, the windspeed, and the location of the seeding generators all play important roles in the type of results achieved (Fig. 6.5).

The High Plains Cooperative Program (HIPLEX) began in 1975 with the goal of developing techniques to manage precipitation from summertime cumulus clouds. Planning went on for several years prior to the actual start of seeding of individual clouds in 1977 and was based on the results of many small research efforts that had been carried out over the previous decade in the High Plains region.

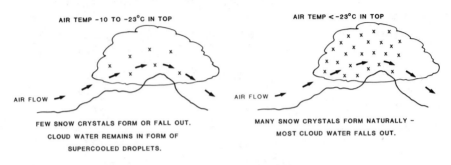

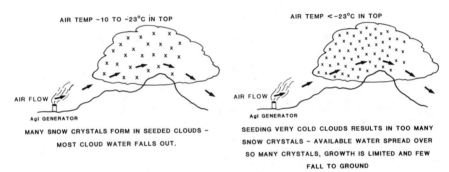

**Figure 6.5** Seeding of orographic clouds should lead to increase in number of snow crystals in clouds that are naturally deficient. If done carefully, and in clouds with correct temperature structure, all cloud water will be utilized to make crystals grow and to fall out of clouds.

HIPLEX involves joint efforts between the Bureau and the states of Montana, North Dakota, Nebraska, Colorado, Kansas, and Texas. The actual experiments are being conducted at three places—Miles City, Montana; Colby-Goodland, Kansas; and Big Springs, Texas.

The third project is the Sierra Cooperative Pilot Project (SCPP) which began in 1972 and involved winter cloud seeding in the Sierra Nevada Mountains. It is a joint program between the Bureau and the state of California designed to build upon the information gained under the CRBPP work.

One significant aspect of Project Skywater has been to evaluate the potential for increasing water supplies over river basins in the West by cloud seeding on an operational basis. Based on studies of 12 major western river basins, the potential annual increases in water supply were found to range from 32 to 57 million m$^3$ for the small San Luis Valley in the Rio Grande Basin to 1739 to 2292 million m$^3$ for the Sacramento River near Sacramento, California. The greatest benefits from increasing summer rain will accrue to agriculture while increasing winter precipitation will bring benefits to power generation, municipal needs, recreation, energy development, as well as irrigation.

## Ecological Effects of Successful Cloud Seeding

Cooper and Jolly (1969), working on the premise that weather modification will ultimately be feasible and will result in increases in precipitation over an area, undertook to study possible ecological effects that might result from an increase in precipitation. Specifically, they anticipated that:

(a) Organisms would respond to increased precipitation as individual species rather than as communities.
(b) Changes in vegetation composition would result from changes in relative abundance of the species already present. Significant invasions of outside plant species would not be expected.
(c) Major changes in insect population or plant diseases would not be anticipated.
(d) Increased precipitation would increase plant biomass; within 5 to 10 years, a new insect population and plant disease rate would become established.
(e) Weeds would increase roughly in line with the increase in precipitation.
(f) Both large and small mammals might decrease in number due to loss of food or habitat or to increase in winter snow cover. Big game population might be reduced by loss of winter range with increased snow cover.
(g) Plant and animal communities already moderately disturbed would be more responsive to weather modification than would either stable or continuously disrupted communities.
(h) In regions with less than 12 in. of precipitation annually, increased rainfall would increase sediment yield; in regions with more than 12 in. of precipitation annually, erosion should decrease. (This is possibly too simplified, for intensity and frequency of precipitation are as important as magnitude in determining sediment yield. The use of a 12-in. figure may be questioned since there is evidence of increased sediment yield at annual precipitation values say above 60 in. and factors of slope, soil characteristics, and vegetation cover need to be considered).

From their analysis, Cooper and Jolly did not expect catastrophic changes in either animal or plant communities but rather slow modifications in response to the altered climatic conditions. Combining increased precipitation with other environmental stresses (such as air pollution and pesticide applications) might make the total change greater than the change resulting from the sum of the individual contributions themselves. Thus, it becomes difficult to interpret the magnitude of an overall change by just summarizing small individual changes.

## Other Considerations

From a hydrologic viewpoint, quite significant changes might result from the initiation of a large-scale successful cloud-seeding program. First, the increased precipitation might generally result from shower-type rainfall. Since stream channels in the area would be in equilibrium with the former rainfall amounts, stream flooding would occur. It might be argued that cloud seeding would only be utilized in very dry times when stream levels were naturally low. Increased runoff should then be able to be contained within the stream banks. While this should certainly be the case much of the time, the increased precipitation, infiltration, and percolation of water should result in somewhat higher water tables in the area which, in turn, should result in more frequent flooding situations from heavy rainstorms occurring even at times when no seeding was being utilized. Thus, provision must be made for increased flood control as well as improved flood prediction and warning.

If seeding is directed toward increasing winter snowfall in mountainous watersheds, there will, of course, be the opportunity for increased avalanche activity. Great care must be exercised in avalanche-prone areas to insure that the increased snowfall does not increase hazards to inhabitants. The increased snowfall will also increase snow removal activities by highway crews and increase the costs to keep roadways open. Use of mountainous roadways and watershed areas by winter tourists, skiers, and campers will have to be more carefully regulated to prevent increased loss of life as a result of the augmented winter snowfall due to seeding.

The question of pollution as a result of seeding activities has largely been passed off as nonsignificant and it may well be. However, it has been estimated that the winter orographic seeding project in the Upper Colorado Basin would release almost 8000 lbs of silver iodide each winter. In the presence of sunlight, silver iodide will break down to form metallic silver and iodine. Insoluble silver might enter into food chains and could possibly build up to dangerous concentrations in higher life forms. At present levels of usage, silver iodide may not be a serious pollutant but widespread utilization in programs to increase precipitation over many different areas, both in the United States and abroad, might lead to the development of pollution conditions previously unsuspected.

Nothing has been said so far about possible conflicts of interest in the matter of cloud seeding but they have already become quite apparent even with the present low level of seeding attempts. If we could make rain successfully, there would be the need for a huge Federal bureaucracy to administer the various questions that would then arise. Do we want snow or rain? Do we want precipitation in the daytime or at night? On weekdays or on the weekends? As a few heavy showers or many light rainfalls? In certain areas but not in others? Can anyone imagine that these and other complex social and economic problems could ever be resolved? It would seem that the only real profit to be made from cloud seeding might occur to lawyers for the number of lawsuits would rise to staggering proportions since all interested parties could not be satisfied.

## SALINE WATER CONVERSION

The ancient mariner may have gazed seaward and seen nothing but undrinkable water but, to a growing number of water engineers, the vast supplies of ocean and brackish water available in many parts of the world would seem to offer a golden opportunity to increase our limited supplies of fresh water. Saline water conversion (desalination, as it is also called) has been actively practiced for many years and a number of different techniques are available to remove the salt content from ocean or brackish water. Many large desalination plants are in operation; in some cases, the work has progressed from theory through the pilot plant stage to practical, on-line processing plants. In fact, by the early 1970s, there were over 800 land-based desalination plants in operation or under construction throughout the world, each having a capacity to desalt more than 25,000 gal/day (Office of Saline Water, 1972, 1973). The total desalting capacity of these plants was some 350 million gal/day of which 93% utilized a variation of the distillation process, 6% used a membrane process, and the remaining plants utilized some type of freeze separation. While the United States had 338 desalination plants at that time, not quite half of the world total, their capacity was rated at only 64 million gal/day or about 18% of the world desalting capacity. The U.S. plants were basically smaller in capacity than the plants built abroad.

Because of wide variations in types of water to be desalinated, in energy costs, in availability of heating sources, in final uses for the treated water, in alternative sources of water, in costs of removal of the waste products, and many other factors, there has been no effort to standardize on just one or a few types of treatment plants. Each new idea or technique may have application in a particular situation and, of course, there may be possibilities of combining portions of several different treatment techniques into a single plant that will produce potable water more cheaply than will a plant that utilizes only a single treatment technique.

Because of the unlimited supply of saline water, the goal of the desalination program, unlike the goals of many other programs to achieve additional supplies of fresh water, has been to develop a whole new supply of water. Conservation measures, better water laws, vegetation control, water harvesting, and other similar techniques attempt to make better use of available supplies and hence make them go further. Desalination attempts to develop a new source of fresh water where no usable water existed previously.

### The Economics of Desalination

Economics plays an important role in the desalination of water since the cost of producing salt-free water must be attractive enough to compete with other sources of fresh water. Of course, if no other water is available and water is absolutely necessary (i.e., the users cannot move to another area where other sources of fresh water are present), then price is not much of a deterrent. However, in most situations, it is a matter of the comparative costs of one supply of water against an alternative supply so that economic factors will often dictate what type of water supply will be developed.

In the early 1970s, seawater desalination plants with capacities up to 30 million gal/day were producing fresh water at prices of about $1.00 per thousand gallons. Brackish water treatment plants were producing fresh water at costs of about 50¢ per thousand gallons in plants with capacities up to 1.6 million gal/day. As

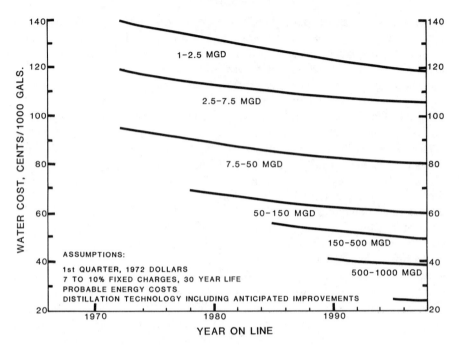

**Figure 6.6**   Costs for desalting seawater by range of plant size.
*Source.* Office of Saline Water (1973).

plants desalting brackish water with capacities of 6 to 8 million gal/day have
come on stream, this cost figure has dropped to less than 30¢ per thousand gallons.
While these costs are appreciably below costs 5 to 10 years earlier, they are still
high enough to limit widespread use of desalination for agricultural and industrial
purposes. Continued research and development is proceeding in the hope of reducing the cost of desalting seawater to 25 to 30¢ per thousand gallons. Figures
6.6 and 6.7 illustrate anticipated costs for desalting seawater and brackish water
over the next 25 years. The family of lines illustrate the economies of scale as desalting plants become larger in size, while the gentle downward slope to the lines
indicates cost reductions due to improved technology and equipment. These cost
figures are quite different from those suggested 15 years ago when it was anticipated that large-scale seawater plants would be able to produce freshwater at a cost
of about 25¢ per thousand gallons by the mid-1970s. Now, we realize that such low
costs may not be obtainable before the year 2000 and then only with the very
largest desalination plants. Figure 6.8 illustrates the main areas in the United States
where future desalting operations have a real potential for adding to the freshwater
supply. It is clear that the most immediate future for desalination lies with the
treatment of brackish water rather than with seawater desalting in the United
States. The Office of Saline Water (1972) anticipates that by the year 2000, the
U.S. need for desalted water will reach some 3 to 4 bgd of water.

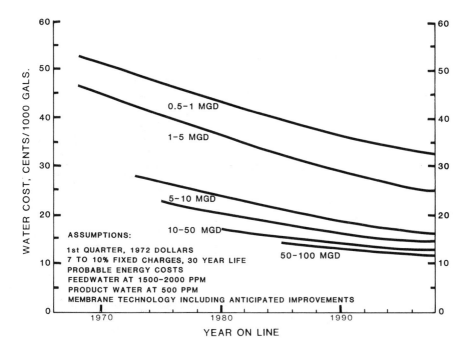

**Figure 6.7**  Costs for desalting brackish water by range of plant size.
*Source.* Office of Saline Water (1973).

Howe (1974) has provided a list of the various costs factors that must be considered in analyzing the economics of desalination:

(a)  costs of capital amortization of plant equipment;
(b)  costs of heat energy used for water production;
(c)  costs of electrical energy for operating pumps and other machinery;
(d)  costs of maintenance and operation—both labor and materials;
(e)  costs of site and buildings;
(f)  overhead costs.

He points out that costs of site and buildings now may have to include more than just the costs to purchase the land or to build the building because environmental considerations may force the construction of a facility that has sufficient aesthetic attractiveness to be acceptable in the area in which it is constructed. And, of course, included in the list above is the need to consider the costs inherent in getting rid of any waste heat and concentrated brine produced during the desalination process. In many locations, the disposal of these products of desalination can cause significant problems and increase costs appreciably.

Based on theoretical considerations, it is possible to determine the minimum amount of work necessary to remove a particular quantity of salt from water. It is assumed that the flow rates of seawater entering the system and concentrated

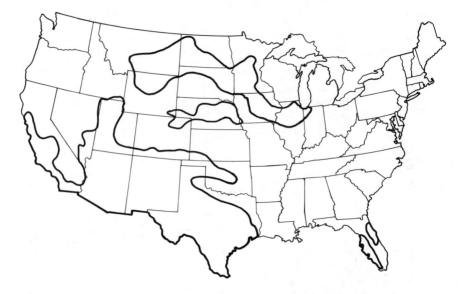

**Figure 6.8** Principal areas of the United States where desalting of brackish or sea water might produce a useful water supply.
*Source.* Office of Saline Water (1972).

brine leaving the system are very large when compared to the rate of production of fresh water so that the concentration of salts in the inflow and outflow streams is very nearly the same. Figure 6.9 shows how the effect of salinity of the water influences the work necessary to separate the salt from the water. With normal seawater having salinities of 33,600 ppm for the Pacific Ocean, 33,800 ppm for the Indian Ocean, and 36,000 ppm for the Atlantic Ocean (they approach 7000 ppm in the Baltic Sea and 43,000 ppm in the Red Sea), we find the minimum amount of energy necessary to remove the salt in average seawater to be about 2.9 kw hours per 1000 gal of water. This is a figure based on ideal efficiencies. In actual practice, the minimum energy requirements are usually two to four times greater to separate fresh water from seawater.

Since some 2.9 kw hours of energy must be used in the separation of 1000 gal of fresh water from seawater, this additional energy must ultimately be discharged from the treatment plant if temperatures are to remain constant. It is often necessary to incorporate a stream of cooling water into the plant design in order to keep the temperature rise in the effluent to within environmentally acceptable values. In a typical distillation facility, the combined waste brine and cooling water leaving the treatment plant may be of the order of 12 to 15°F warmer than the influent water and contain the concentrated materials removed from the influent water along with perhaps 0.15 ppm of copper due to the corrosion of copper tubing in the heat exchangers. This effluent water may be harmful to the aquatic life already established in the receiving waters. To minimize the effect of such effluent

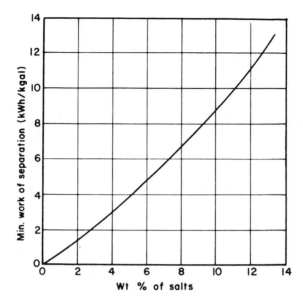

**Figure 6.9** Effect of salinity on minimum work of separation. Recovery of freshwater from seawater and its concentrates. Note: Work of separation based on infinitesimal production of desalted water.
*Source.* Reprinted from Howe (1974), page 76, by courtesy of Marcel Dekker, Inc.

streams, dilution is used extensively by treatment plants located on coastlines with large quantities of mixing water available.

For plants at more inland locations—using brackish water, for example—dilution is not feasible. Such plants now tend to dispose of the effluent by evaporation, pumping into deep underground strata, pipeline transport to ocean sites for dilution or further concentration to produce chemicals that can be marketed. Evaporation from open ponds seems to be most favored at the present time. Because of the slowness of the evaporation process, however, large ponds are often necessary to handle the effluent stream from just moderate-sized plants.

## Desalination Processes

The various techniques that have been evolved to separate fresh water from salt water may be grouped in several different ways. Following Howe (1974), it seems logical to discuss them in terms of (a) processes involving the phase change in water (distillation, freeze separation, hydrate separation); (b) processes utilizing surface properties of membranes (electrodialysis and reverse osmosis); and (c) processes utilizing ion-selective properties of solids (ion exchange). Distillation, electrodialysis, and reverse osmosis are techniques currently in use in large-scale,

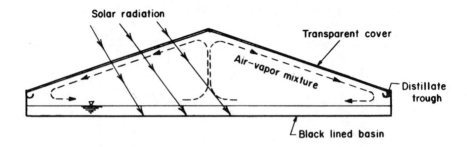

**Figure 6.10**  Schematic cross-section of a solar distillation still
*Source.* Reprinted from Howe (1974), page 76, by courtesy of Marcel Dekker, Inc.

practical, on-line systems—although the latter two techniques are much better adapated to the treatment of brackish water with lower salinities than seawater. Ion exchange, freeze separation, and hydrate separation are techniques with special problems that have limited current development. While they have been used in special situations, they are not in widespread use at present on a commerical basis. They may be considered as experimental.

### A. Processes Using a Phase Change of Water

1. *Distillation.* Nature's own way to provide the fresh water for precipitation is by means of distillation. The technique is straightforward, relatively simple, and can operate on the energy provided by the sun. It is slow, however, and if the extent of the oceans were not as great as they are, precipitation totals would be much lower than they are in many places around the globe.

Distillation involves the application of heat to a body of saline water to form vapor. Since most of the chemical materials found in saline water are nonvolatile, they will not be evaporated and will remain in the saline water. The vapor must be removed from the liquid surface from which it originated and then be recondensed again usually by contact with a cold surface.

a. *Solar distillation.* To illustrate the process in its simplest (but not necessarily cheapest) form, Figure 6.10 is a schematic sketch of a simple solar distillation still. In essence, there is a source of saline water—often a shallow layer of water—in a black-lined basin or tray to encourage heating as much as possible. The basin is covered by a transparent glass or plastic cover gently sloped so that water droplets condensing on the underside of the surface will roll down the surface to be collected in small troughs around the lower edges of the transparent cover.

Solar radiation will pass through the transparent cover, heat the water, and result in evaporation of fresh water from the saline water in the basin. The rising water vapor will contact the relatively cool inside surface of the cover, condense, and flow off in small droplets to be collected in the troughs. Air circulation is

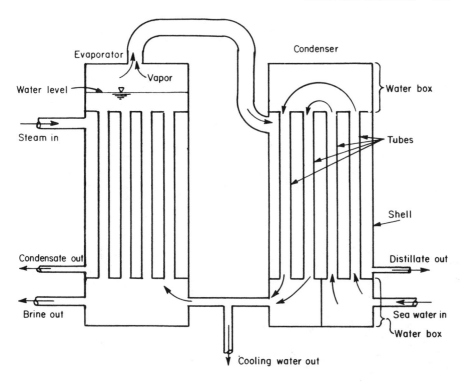

**Figure 6.11** Schematic cross-section of simple distiller.
*Source.* Reprinted from Howe (1974), page 62, by courtesy of Marcel Dekker, Inc.

by free convection. Clearly, the system needs a dependable source of solar radiation for most efficient operation so that the technique is most practical in cloudless coastal areas or desert areas with brackish water.

A standard production figure is 1 gal of distilled water from every 12 ft² of basin area per day. In spite of the fact that the energy source is free, the water so produced is anything but free. Because of the costs of the structure and especially the transparent cover, it is estimated that the cost of fresh water from a solar still is about $3.00 per thousand gallons, well above the cost for fresh water produced by most other techniques.

b. *Simple and multiple-effect distillation.* Figure 6.11 is a schematic sketch of a simple distillation process using a confined source of heat in place of solar radiation. In principal, it operates like the solar still. Two units are shown in the sketch: one, an evaporator, in which vapor is formed from the heated saline water (in this case, steam provides the source of heat); and the second, a condenser, in which the vapor is converted to a liquid form by contact with cooled surfaces. To provide some small economies in the process of heating the saline water, this water is first used to cool the vapor in the condenser. In this process it receives some heat from the vapor

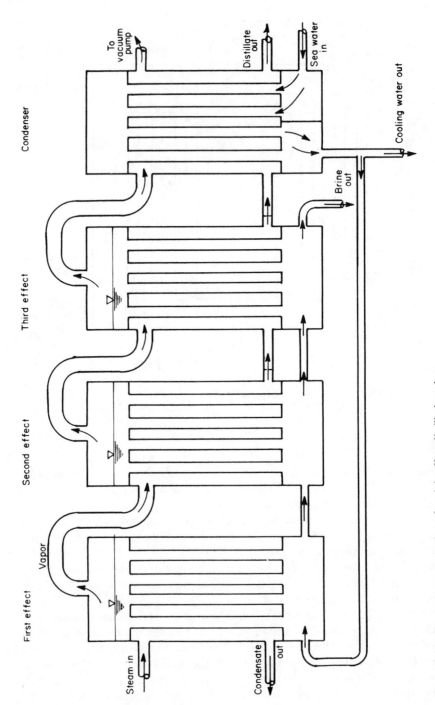

**Figure 6.12**  Schematic cross-section of a triple-effect distillation plant.
*Source.* Reprinted from Howe (1974), p. 64, by courtesy of Marcel Dekker, Inc.

and so passes into the evaporator with a temperature somewhat above its entering temperature.

Each of the two chambers consists of a bundle of tubes enclosed in a shell. There are water boxes at the ends of each container to speed the flow of saline water to and from the various tubes where heat is added or given up.

It is possible to increase the quantity of vapor produced by a given quantity of steam if several evaporators are placed in series. In Figure 6.12, three evaporators are connected in series to one condenser chamber. As in Figure 6.11, the first evaporator of this triple-effect distillation plant receives steam heat from an outside source. The steam must have a higher temperature than the boiling temperature of the saline water. The vapor produced in the first evaporator is condensed in the shell side of the second evaporator. This produces the heat necessary for the evaporation of some saline water in tubes of the second evaporator. To produce a flow of heat to the saline water in the tubes, the temperature of the condensing vapor must be greater than the temperature of the water in the tubes. The process is repeated again in the third evaporator. Since the condensation of each pound of steam will result in the production of a little less than 1 lb of water, it can be seen that the triple-effect distillation unit will result in the production of a little less than 3 lbs of fresh water with the initial input of 1 lb of steam. For this reason, the process is called multiple-effect distillation.

The sketches do not show needed auxiliary equipment. For example, a vacuum pump is required to remove air from the system and to establish a pressure gradient through the system so that the brine and the vapor will be able to move from one part of the system to another.

c. *Flash distillation.* "Flash" evaporation of hot saline water provides an alternative means by which vapor can be produced fairly economically. The flash of evaporation is produced by rapidly reducing the pressure in heated saline water. As the pressure is reduced, so is the boiling point of the liquid and a portion of it will flash into steam. Thus, flash distillation plants are designed so that the saline water is heated under higher and higher pressures to prevent the formation of vapor during the heating process. After the desired temperature of the liquid is achieved, the pressure on the liquid is reduced in a series of steps. At each step, a portion of the hot, saline water is converted to steam and is removed from the system.

As in the case of the multiple-effect distillation system described previously, certain economies can be achieved if a number of flash stages are included in the whole system. Figure 6.13 only shows three stages, but usually over 20 stages are included in the on-line system. The Chula Vista, California multiple-stage flash distillation system installed in 1967 with a capacity of 1 million gal/day has 68 stages, while a 4.8 million gal/day unit installed in Kuwait in 1970 has only 26 stages.

Each stage consists of a flash chamber located below a set of condenser tubes. Saline water enters the condenser tubes of the last stage and is heated by the vapor rising from the flash chamber of that particular stage. While some of this warmed saline water is removed as cooling water, the remainder is mixed with

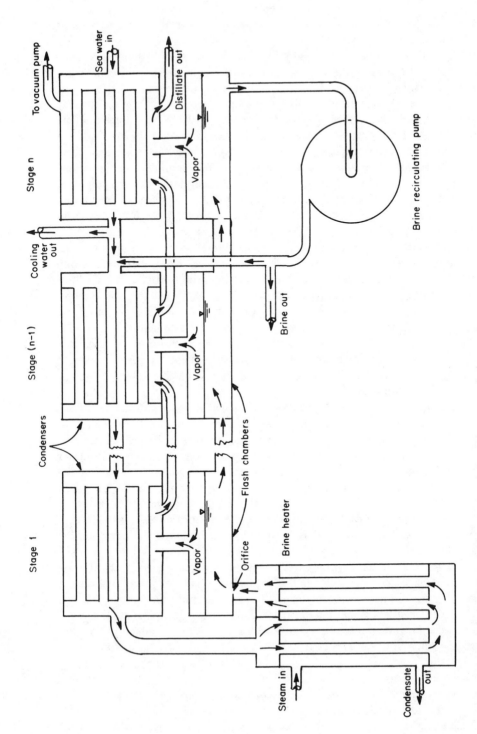

**Figure 6.13** Multiple-stage flash distillation plant (schematic cross-section). *Source.* Reprinted from Howe (1974), page 68, by courtesy of Marcel Dekker, Inc.

warm brine from the recirculating pump and is sent to earlier stages in the system. The final heating of the brine is accomplished in the tubes of the brine heater, with the source of heat being steam from an outside source. The hot brine is now sent to successive flash chambers where the pressure is reduced in steps with the production of a small amount of vapor in each stage. The vapor rises and is condensed on the relatively cool surfaces of the tubes in the condensers.

It is usually more convenient to have the flash units so arranged that a fixed small drop in temperature of the hot, saline liquid is accomplished in each stage. For example, for a 40-stage plant, the temperature decreases in each step might be 2.5°F so that the total temperature interval in the whole plant would be 100°F.

In the flash evaporation process, only a small amount of the influent saline water is evaporated (of the order of 10%). Thus, large quantities of saline water must be circulated through the system to obtain a relatively small amount of distillate. In the multiple-effect distillation plant described previously, the distillate constitutes about 50% of the influent saline water.

With large volumes of relatively cool saline water needed in the flash distillation process, the question might be asked whether it is better to circulate the water only once through the system or whether it can be recirculated a second time. Recirculation appears to provide certain economies because the saline water is already somewhat warmed as it starts its second cycle through the system. Also, to prevent scale build-up on the piping, the saline influent must be treated chemically. Reuse of this treated water decreases the costs of treatment to prevent scale. The salinity of the seawater can be increased by a factor of two before scale formation becomes a problem. One recirculation of the water will not increase the salinity by more than this amount. For these reasons, recirculation is practiced in most modern flash distillation plants. Recirculation is not utilized in multiple-effect distillation plants.

d. *Vapor-compression distillation.* The vapor-compression distillation process substitutes mechanical energy for steam in the vaporization process (Fig. 6.14). The process is relatively new, having achieved fairly rapid development in small systems during World War II in locations where large supplies of steam heat were not available.

As indicated by its name, the process involves compression of vapor produced in the evaporator under a high pressure. The condensation of this vapor will provide heat for the production of more vapor from the original water supply. A heat exchanger is also incorporated in the system to aid in the cooling of the effluent brine and, at the same time, to add heat to the influent saline water. The waste energy in a vapor-compression plant is kept to a minimum by the use of the heat exchanger, the temperature of the effluent stream being only a few degrees higher than the influent stream. As a result, the costs of fresh water from such a system have been estimated to be of the order of $0.50 per thousand gal if power costs are $0.01 per kw hour (Howe, 1974).

2. *Freeze separation (cold distillation).* This process utilizes the fact that ice crystals resulting from the freezing of saline water contain only fresh water. If these crystals can then be separated from the saline water from which they were

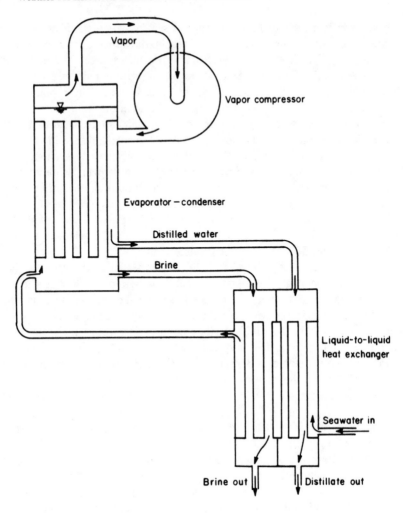

**Figure 6.14** Vapor compression distillation plant (schematic cross-section).
*Source.* Reprinted from Howe (1974), page 74, by courtesy of Marcel Dekker, Inc.

formed, rinsed to remove any salts that might adhere to the surface, and melted, fresh water will be produced. The process is attractive because the energy required to freeze water is very much less than the energy needed to bring water to the boiling point. However, the fact that less energy is required to freeze water is somewhat balanced by the need to use a relatively large amount of energy for the operation of a heat pump to dissipate the energy added to the system at temperatures well below ambient temperatures. Thus, the energy advantage over distillation systems is more imagined than real.

The freeze-separation process, while still considered somewhat in the experimental stage, has been utilized for one 250,000 gal/day plant in Eilath, Israel which is now on a standby basis due to the construction of a 1 million gal/day distillation plant nearby.

Because of the salt in the water, temperatures needed for ice crystals to form will be 4 to 6°F below freezing. Tests of conventional ice-making machines for the production of ice showed that they were not practical from an economic point of view. Use of water itself or some fluid with low solubility and low boiling point (such as isobutane) has been tried with better success. In these cases, a vapor compressor removes the vapor, and hence heat, from the freeze chamber resulting in some freezing of the saline water. Figure 6.15 indicates that the saline water is cooled before it is sprayed into the freeze chamber which is kept at a very high vacuum by means of a vacuum pump. The resulting vapor forms at the freezing temperature for the saline solution—about 7 lbs of ice for every pound of vapor.

The small plate-like ice crystals are moved into the wash column by means of a slurry pump. In the column, the ice crystals rise while the brine is pumped out the lower end of the column. The ice crystals are first washed with a small amount of fresh water to rinse off any brine and then are pushed down a sloping tube to the melter. Compressed vapor from the freeze chamber provides latent heat to melt the ice crystals.

Because of the use of the vapor compressor, the process might be considered to be a vapor-compression distillation process in parallel with a freeze-separation process. The water resulting from the freeze-separation process is not as pure as water from the vapor-compression process because of the difficulty of separating the ice crystals from the saline water. Because the vapor pressure of the water in the freezing chamber is of the order of 3 mm of mercury, rather sophisticated vacuum equipment is needed. Also, because of the very low density of the water vapor, a large vapor compressor is required to provide the proper weight of vapor to the melting unit. Use of refrigerants other than water eliminates many of the problems arising from the low density of the water vapor.

3. *Hydrate separation.* In concept, this process might be considered preferable to freeze separation because hydrate materials will crystallize at temperatures closer to ambient than those needed to form ice crystals. Crystals of a number of different hydrate materials (e.g., methyl bromide, propane) will contain a large percentage of water. The energy to form and maintain the crystal should be less than that needed to form ice. Pilot studies, however, have not shown that there is any real economic advantage to such a method and other problems involved in the process have made it less attractive than other freeze-separation techniques.

## B. Processes Using Membranes

These processes involve the removal of either salts or fresh water from a saline solution by passing the solution through selective membranes. Usually the process

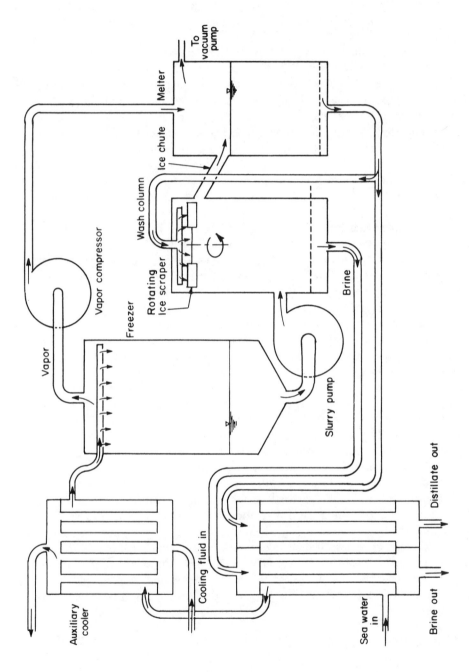

**Figure 6.15** Schematic cross-section of a freeze-separation plant.
Source: Reprinted from Howe (1974), p. 78, by courtesy of Marcel Dekker, Inc.

Auxiliary cooler

Vapor

Vapor compressor

Freezer

Rotating Ice scraper

Wash column

Ice chute

Melter

To vacuum pump

Brine

Slurry pump

Cooling fluid in

Sea water in

Brine out

Distillate out

is more effective on water with a relatively low salinity. Two different processes have been developed and tested—electrodialysis and reverse osmosis. Both processes involve the use of a membrane as a barrier between two water solutions having different concentrations of salts. In osmosis, water is transmitted through the membrane from the less concentrated solution to the more concentrated one. In dialysis, dissolved salts are diffused through the membrane from the more concentrated solution to the less concentrated one. The two processes have become more practical in recent years with the production of improved plastic membranes. For example, membranes for use in electrodialysis are now available containing either negatively or positively charged ions able to attract ions with opposite electrical charges. Synthetic membranes for use in reverse osmosis are now possible with the greatly increased mechanical strength and proper transmissivity necessary for use in desalination work. Further development work on membranes is currently under way.

1. *Electrodialysis.* A schematic diagram of an electrodialysis plant is shown in Figure 6.16. The saline water is divided into two streams. Ions are transferred from one stream (which then becomes the product stream) to the other (the waste stream) in the presence of a direct current electrical potential established perpendicularly to the water flow. The selective membranes stretched parallel to the stream flow prevent the transfer of the unwanted ions back again from the waste stream to the product stream. The membranes are actually set up in pairs, with one allowing only positive ions to pass through and the other only letting negative ions to pass.

In the electrodialysis process, the flow of electricity is by means of the migration of ions. With the use of the membranes in pairs, Na+ ions are transferred toward the negative electrode while the Cl-ions are transferred toward the positive electrode. The resulting water in the product stream is relatively pure but not as pure as it would be using a distillation process. Usually the salinity of the product water is 300 to 500 ppm. The latter figure is the maximum desirable value for potable water in the United States although water with 1500 ppm of salinity is in regular use in some parts of the world today. Removing more of the salinity would require the use of much higher voltages because of the significant electrical resistance of very pure water. Passage of saline water through one electrodialysis stack will reduce salinity by about 50% (Fig. 6-17 has ten such stacks). Since the cost of the plant depends on the number of stacks used, it is clear that costs can be greatly reduced if the process is used on low salinity water rather than on seawater. Electrodialysis is less expensive than distillation for water of salinity of 4000 to 5000 ppm but it is more expensive than distillation for use with seawater.

2. *Reverse osmosis.* In the reverse osmosis process, water passes from the more concentrated to the less concentrated solution under a pressure force which is greater than the osmotic pressure of the saline solution. Cellulose acetate membranes with a thickness of about 0.1 mm have been found to be the most effective in this application. Figure 6.17 is a schematic diagram of a reverse osmosis plant. The saline water flows through the inside of the tubes. The outer metal tubes are perforated with very small holes and lined with a porous material. Tubes of

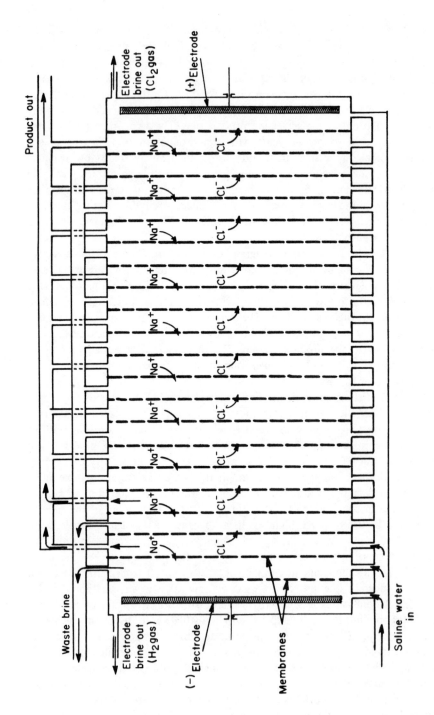

**Figure 6.16** Schematic cross-section of an electrodialysis plant.
*Source.* Reprinted from Howe (1974), page 82, by courtesy of Marcel Dekker, Inc.

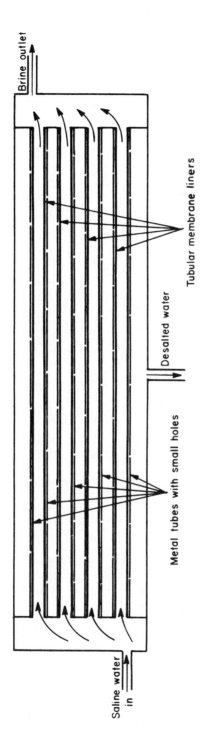

**Figure 6.17** Schematic cross-section of a reverse osmosis cell.
*Source.* Reprinted from Howe (1974), page 84, by courtesy of Marcel Dekker, Inc.

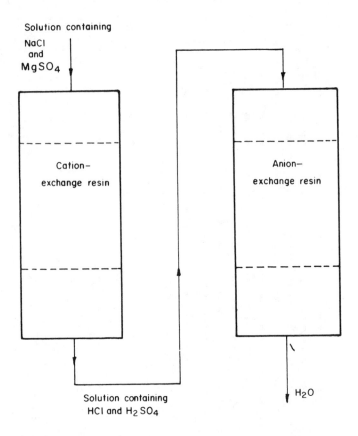

**Figure 6.18** Schematic cross-section of the ion-exchange polishing process.
*Source.* Reprinted from Howe (1974), page 88, by courtesy of Marcel Dekker, Inc.

cellulose acetate membranes are installed inside the porous lining of the metal tubes. Water flows through the membrane tubes at pressures of about 600 psi, sufficient to cause water to filter through the membrane where it will then move through the porous liner to the nearest hole in the metal pipe to be collected in the shell. With water of less than 6000 ppm salinity and a pressure of 600 psi, about 90 to 95% of the salt can be removed from the solution at a rate of 10 to 20 gal/day per square foot of membrane.

## C. Processes Using Ion Exchange

1. *Ion exchange.* Techniques used for water softening involve ion exchange. Water with magnesium and calcium ions is passed through a bed of natural sand-like

material such as hydrated sodium aluminum selicate. Sodium ions would be exchanged with the magnesium and calcium ions so the water leaving the bed would contain sodium ions and the bed would contain the magnesium and calcium ions. When the bed had lost all its sodium, it could be restored by reversing the process with the passing of concentrated sodium chloride through the bed.

As used in the desalination process, shown in Figure 6.18, two resin columns are used—one for cation exchange and the other for anion exchange. As indicated in the sketch, water with NaCl and MgSo$_4$ is passed through the cation exchange resin where the sodium and magnesium ions are exchanged with the hydrogen ions in the resin. This water, now high in hydrogen ion content, is passed through the anion exchange resin where sulfate and chloride ions exchange with the hydroxide ions. Combining the hydroxide ions with hydrogen ions forms water molecules free of all dissolved materials. The method can produce good quality water from seawater but, since the cost of regenerating the exchange resins is so great, it is preferable to use the process for only slightly saline water.

## Future of Saline Water Conversion

The actual desalination method used in any particular case will clearly depend on a number of factors which influence the cost per gallon of fresh water produced. Availability of power is, of course, a prime factor, as is the salinity of the influent water. In many cases, it has been found that some combination of processes may produce a final freshwater product at lower cost than is possible using only one process alone. This is especially true in distillation plants.

The three principal distillation methods are multiple-effect, multi-stage flash, and vapor compression. Extensive studies, in some cases resulting in actual construction of plants for test or production purposes, have been conducted on the following combined distillation plants:

(a) multiple-effect with multi-stage flash distillation,
(b) multiple-effect with vapor-compression distillation,
(c) multi-stage flash with vapor-compression distillation, and
(d) multiple-effect with both multi-stage flash and vapor-compression distillation.

Each has certain advantages over the others and all of the possible uses and limitations of each combination have not yet been fully explored.

While it is difficult to make direct comparisons of the costs of producing fresh water by means of each of the single- or multiple-process systems because of variations in power costs and influent characteristics, a few rough comparisons are possible. Some of the cost data are based on theoretical considerations while others are based on the results of actual on-line experience. Tentative comparisons as of the early 1970s are included in Table 6.3. The figures are, of course, subject to revision downward as further development of the process increases efficiency or reduces power needs.

It is clear that a variety of options are available for use in desalination plants. For example, combining a desalination plant with a power generating plant, such

**Table 6.3**  Comparison of Costs to Desalinate Water by Various Common
Techniques

| Technique | Cost Per Thousand Gallons |
| --- | --- |
| Multiple-effect distillation (20 effects) | $0.50 |
| Multiple-stage flash distillation | $0.25 |
| Vapor-compression distillation | $0.60 |
| Multiple-stage flash with vapor compression | $0.25 |
| Solar distillation | $3.00 |
| Electrodialysis (salinity reduced from 2100 to 440 ppm) 650,000 gpd plant | |
| at 48% capacity | $0.51 |
| at 98% capacity | $0.33 |
| Reverse osmosis (original salinity 2475 ppm–5000 gpd plant) | $1.16 |
| (estimated for 50,000 gpd plant) | $0.50 |

*Source.* Howe (1974), courtesy of Marcel Dekker, Inc.

as a nuclear reactor where considerable excess heat might be available for use in preheating the influent water at no expense, could lower cost figures significantly. Even so, costs of 50¢ per thousand gallons would be prohibitively high for agricultural uses and for many industrial uses. Thus, for a considerable time to come, use of desalination methods will be confined to those areas of the globe and to those particular demands where there are no economically competing water sources. Domestic water uses can be supplied by desalination along with certain low water-demand industries, but agriculture, except of the hothouse-type, will not be able to make use of desalination to supply its water needs. Even if further reductions in costs are possible, it must be remembered that most desalination activities will occur at sea level where water supplies are available while most uses of the water will occur at some higher elevation. Thus, pumping costs must be considered, especially if the water is to be moved any distance from the desalination plant. Desalination is certainly an important process to provide additional supplies of water and work must be continued in perfecting the various processes, but we should not conclude that the technique will be able to supply us with unlimited volumes of fresh water in the cost range to make it practicable for all possible uses.

## 7

MORE WATER THROUGH
VEGETATION CONTROL AND
WATER HARVESTING

## VEGETATION MODIFICATION AND CONTROL

Over much of the land area of the world, evapotranspiration dissipates almost two-thirds of the precipitation supply while streamflow removes the remaining third. In the United States, average annual precipitation is 30 in., evapotranspiration losses are 21 in., and streamflow is 9 in. Thus, any technique to reduce evapotranspiration losses can provide additional supplies of water for other, possibly more valuable, uses. Those evapotranspiration losses that are essential to the production of high value crops may only be able to be reduced slightly by improved cropping or irrigation practices, but those evapotranspiration losses that occur from noneconomic vegetation are susceptible to significant reduction. In the latter case, vegetation control can offer a means to make available new supplies of water. At least four different techniques have been tried in order to reduce evapotranspiration losses by farming and forestry practices with little or no change in the vegetation cover in the area. They include: (a) remove, replace, or thin the existing vegetation cover in the area; (b) apply transpiration suppressants to the existing vegetation cover or to the soil for uptake through the roots; (c) develop new strains of plants that transpire less without limiting yield; (d) grow plants within an enclosed structure so that transpired moisture is captured and reused.

The first technique has been studied for many years, not so much in an effort to make new supplies of water available but rather to learn the basic relationships between land use (or land-use change) and streamflow. Two common approaches have been utilized to study the effect of changing land use—the so-called "physical" approach and the "hydrometric" approach. Both methods have been used with varying degrees of success.

The physical approach to basin analysis involves understanding the effects of changes in specific hydroclimatic elements on streamflow. Muller (1966) suggests that while such physical method studies are being actively used at present and do provide much of our current knowledge about the relation between land cover and hydroclimatic processes, these studies are still incomplete and may possibly be misleading when the results of significant land-use changes are being evaluated since total water yields from the whole basin are often not measured. The hydrometric method, which developed early in the 20th century, involves the measurement of water yield from entire watersheds often under different vegetation covers. Several basic techniques are available for use. In one, two basins nearly identical in terms of size, climate, terrain, soil, and vegetation conditions are identified. One is used as a control and the other undergoes some particular change or treatment. Following a period of calibration (often 5 years or longer), an equation is developed that can be used to predict runoff from the watershed to be "treated" as a function of runoff from the control basin. After the basin is modified (generally through some form of vegetation change), the difference found between the observed and predicted water yield in the treated basin is accepted as the integrated effect of the land-use change. This is known as the "paired-watershed" technique.

A second technique is used if only one basin is available for study. Observations must be carried out on the single watershed over a reasonably long calibration period during which a statistically valid relationship between precipitation and basin yield is obtained. After land or vegetation treatment, a new relationship is found between precipitation and basin yield and the nature of the changes from the before-treatment period can be inferred from the previously developed statistical relationship.

Hewlett and Nutter (1969) point out that the hydrometric approach is costly, requires many years of observations, both before and after the land treatment, and is subject to various uncertainties, especially when there may be problems of leakage and underflow from the basin. Because of the need to eliminate problems resulting from changes in soil moisture storage, it is not possible to obtain short-term estimates of various hydroclimatic factors using this technique.

### Examples of Applications of the Physical and Hydrometric Methods

(a) *Physical method.* Muller (1969) has suggested the long-continued investigation of the effect of forests on precipitation as a practical example of the use of the physical method. Fernow, Chief of the Forestry Division of the U.S. Department of Agriculture, writing in 1892 (Fernow, 1902) suggested that forests would result in an increase in precipitation. This led others to conclude that deforestation would lead to a decrease in streamflow from the affected area, although Toumey (1903) concluded that forests in some areas would result in increased streamflow and in other areas would result in decreased streamflow. Using selected river-stage and discharge measurements, Chittenden of the Army Corps of Engineers argued (1909) that deforested areas would yield more water as streamflow although forests had no significant influence on precipitation.

Zon (1912) probably produced the most thoroughly documented review of forest-streamflow studies from North America and Europe. In it, he stated that forests increased precipitation—as most of the forest-streamflow investigations had found. Reasoning from this conclusion, he postulated that increasing deforestation throughout southeastern United States would result in decreased precipitation through the Great Plains and central United States. By 1945, increasing evidence to the contrary led Zon (1945) to express possible doubt concerning the real effect of forests on precipitation. Pointing out that rain gages in openings in forests usually registered more than nearby rain gages outside the forests, he accepted the possibility that these differences might be due to wind effects on "catch" in the gages rather than to any real influence of the forest itself. Any conclusions based on so-called physical methods must be carefully evaluated to insure that incorrect inferences have not been made.

(b) *Hydrometric method.* The original use of the hydrometric method in the United States began at Wagon Wheel Gap, Colorado, in 1910. The program, which covered 16 years, was conceived to provide information on the effect of changes in vegetation cover on water yield (Bates and Henry, 1928). Two nearly identical watersheds in the headwaters region of the Rio Grande River were selected. Each watershed was about 200 acres in extent and ranged in elevation from 9000 to 11,000 ft. Vegetation in both areas was Douglas fir mixed with aspen and grasses.

Precipitation and streamflow were measured on both watersheds for 8 years, during which time they were both maintained under as near normal conditions as possible. After the 8-year calibration period, all the trees on one basin were cut and burned. Hydroclimatic observations continued for 7 more years on each basin. The untreated basin served as the control while the treated basin provided quantitative values of the changes in streamflow due to forest removal. During the 7 years after cutting, no further clearing was attempted so that aspen quickly reseeded the area. Before treatment, mean annual water yield from the treated basin was 6.2 in. while after treatment it averaged 7.3 in. (the greatest increase, found in the third year, was nearly 2 in. greater than in the untreated basin). Water yield from the treated basin was only about 0.5 in. greater than from the untreated basin in the last 2 years of observations. Since streamflow on the control basin also increased 0.1 in., net increase in streamflow over the 7-year period was just 1 in. or 16% more than before treatment.

As it was recognized that the Wagon Wheel Gap data should not be extrapolated to other climatic or topographic areas, similar hydrometric studies were undertaken elsewhere. Among the best known of these studies were the ones at Fraser Experimental Forest in Colorado, Coweeta Hydrologic Laboratory in North Carolina, and Fernow Experimental Forest in West Virginia. The work at Fraser involved studying the effect of clearcut strips of varying widths (Love and Goodell, 1960). All told, 50% of the commercial timber and 40% of the treated watershed were cleared. Mean annual water yield increased 3.2 in. or 24% during the first 4 years after treatment.

At Coweeta (Hoover, 1944), the study was much like that at Wagon Wheel Gap although the area was considerably more humid. Precipitation ranged from 51 to 84 in. in the years of the study. All woody vegetation on the treated watershed was cut and left in place and sprouts were again cut the following year. In the year following clearcutting, streamflow increased 16.7 in. or 60%. In the next year, when sprouts were allowed to grow, streamflow was only 10.7 in. greater or about 30%.

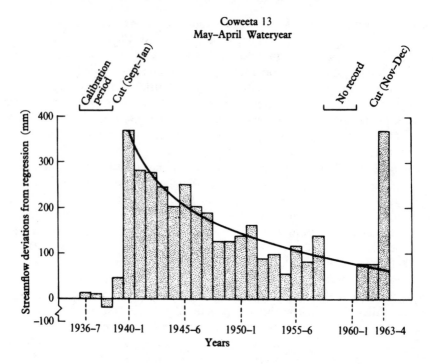

**Figure 7.1** Response of water yield to clear-cutting a forest and response to cutting again 23 years later, Coweeta Basin 13, May–April water year.
*Source.* Hibbert (1967), with permission of the Pergamon Press.

The study at Fernow involved five small but steeply sloped watersheds (Reinhart, Eschner and Trimble, 1963). Four were maintained under different degrees of logging while the fifth was the control. The basin with the most extreme treatment on which all of the commercial timber (about 85% of the gross timber volume) was removed showed a first-year yield increase of 5.1 in. or 19% over the before-treatment condition.

One interesting example of the response of streamflow to clearcutting was reported by Hibbert (1967). Two small forested watersheds near Coweeta were calibrated for 3 years at the beginning of the period of study. In 1940, one of the basins was clearcut. In the first year after clearcutting, annual streamflow from the treated basin increased some 373 mm over the value expected from the regression equation with the control basin. As the hardwood forest regrew during the following 23 years, streamflow from the treated basin decreased logarithmically (Fig. 7.1) until, at the end of this period, the flow was only 75 mm greater than the value predicted for a full forest cover. Clearcutting again at that time resulted in a second marked increase in streamflow from the treated basin, almost exactly equivalent to the response after the first cutting.

Pereira (1973) reports on a paired-watershed study of the water yields from

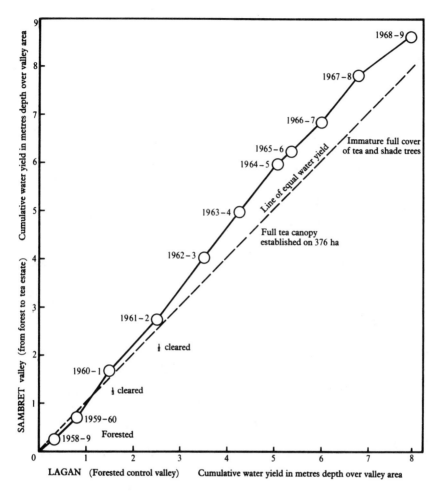

**Figure 7.2**  Ten-year cumulative water yields from forest and from tea plantation.
*Source.* Pereira (1973), with permission of the Cambridge University Press.

a forest cover being compared with water yields from a tea plantation in Kenya. While the two valleys, Lagan and Sambret, had forest vegetation at the beginning and approximately the same water yield, the result of clearing one-half of Sambret Valley and planting it with tea plants increased the water yield from that valley for several years (Fig. 7.2). From about 1964 on, after a full tea canopy had become established in Sambret Valley, there was little difference in water yield between the two valleys again. Thus, the water yield was little influenced by the completed land-use change although during the transition period when the tea estate was becoming established, an accumulated difference of about 1-m depth in water yield developed.

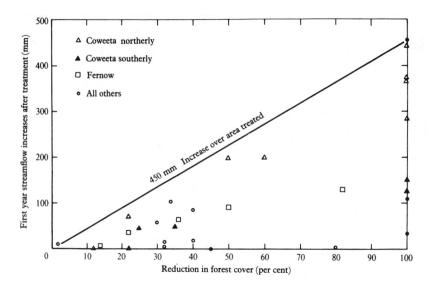

**Figure 7.3** Summary of experiments to study the response of forest cutting on water yield.
*Source.* Hibbert (1967), with permission of Pergamon Press.

Hibbert (1967) summarized nearly all available published material on water-shed experiments on land-use modifications involving some 39 experiments from North America, Asia, and Africa. Most were of the paired-watershed type. The streamflow response to vegetation modification in the first year varied significantly (Fig. 7.3) even in those experiments where complete clearcutting was practiced. Hibbert found greater increases in streamflow with cutting on north-facing slopes than on south-facing ones. The results suggest that in well-watered areas, stream-flow response is somewhat proportional to the degree of reduction in forest cover.

Pereira (1973), in reviewing the paired-watershed work, concluded that information on basin water yields resulting from planned vegetation changes is urgently needed since existing studies permit prediction in only relative terms. He felt that there was a growing recognition among experimenters that the period of paired-watershed or calibrated-watershed studies was about over, to be replaced possibly by more detailed physical studies of the individual components of the water budget.

Ward (1975) has also reviewed, rather critically, the results of previous hydro-metric and physical methods to study the effect of land-use change on streamflow. He points out that while most studies involving cutting of vegetation on a basin have resulted in an increase in streamflow (Fig. 7.3), some Russian experiments have actually indicated that removal of a forest cover decreased water yields. He suggests that lack of understanding of basin hydrologic mechanisms and processes may make it difficult to reconcile these apparently divergent results. For example, he suggests the more normal situation is for removal of a forest with its large transpiration demands or its appreciable interception losses to result in more precipitation going to runoff and streamflow. However, further information on the effect of snow blowing or drifting, snow accumulation, storage, and melt under specific forest covers and densities of stands might well explain why cutting the forest could lead to less rather than to more runoff. Ward further suggests that, while both interception and evapotranspiration losses from vegetation reduce water yield, it may well be that with some vegetation, interception losses are more significant than transpiration losses while with other vegetation the reverse might be true. The particular vegetation treatment to be used to increase water yield might be quite different depending on the dominant cause of the water losses.

Removal or replacement of a vegetation cover may, of course, have more far-reaching consequences than just changing evapotranspiration or interception losses. Ward (1975) suggests that changing a vegetation that produces a thick litter with one that does not may reduce water detention in the litter and so result in more overland flow or increases actual infiltration into the soil. Increased infiltration can result in more storage in the soil or movement to the groundwater table, or if impervious layers are present, more lateral movement as throughflow to nearby stream courses. Because of the particular relation among these factors, it may be extremely difficult to determine the exact reasons for changes in stream-flow as a result of a particular vegetation treatment.

Ward concludes his rather pessimistic but probably correct assessment of such watershed studies with a comment about the need for careful experimental control and data accuracy. The various hydrologic parameters of precipitation, evapo-transpiration, runoff, streamflow, and soil moisture storage are all most difficult to measure and are therefore subject to considerable error. At the same time, the scientist has essentially no control over the natural drainage basin. Given the nature of the possible errors in observations, Ward and others feel that it would be difficult to show statistically satisfactory quantitative data on the effect of land-use change.

**Effect of Conservation Practices**

Conservation practices, often no more than plowing on the contour, seeding pastures that exist on sloping land, strip cropping or, in some cases, terracing, have been shown to reduce runoff by 25 to 40% (Fig. 7.4). This reduction is more variable in wet years than in dry ones. The various conservation measures tend to allow precipitation to infiltrate and to be retained in the upper soil layers. As a result, greater evapotranspiration losses may occur so that water yield is less; but there will also be reduced danger from erosion and flash flooding.

A TVA study on Parker Branch watershed illustrates the results of good conservation practices on a small, badly-eroded area. Starting with 30% of the land area gullied and abandoned, the application of good farming practices and conservation measures over a 10-year period resulted in the following increases in yield: 60% in maize, 50% in tobacco, 25% in wheat, and 117% in alfalfa hay from

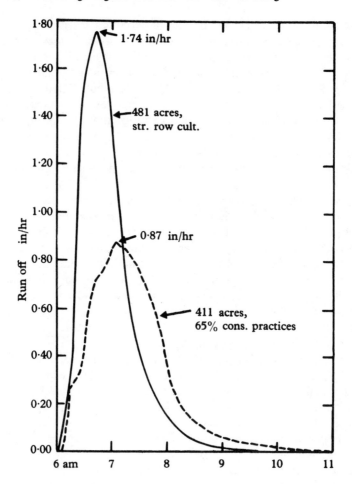

**Figure 7.4** Effect of conservation practices on runoff hydrograph at Hastings, Nebraska, from a storm of 2.75 inches.
*Source.* Pereira (1973), with permission of the Cambridge University Press.

the 43-ha farmed area. Also, there was a great reduction in peak rates of summer stormflow in Parker Branch Creek (Fig. 7.5). The average peak flows from summer storms were reduced by 50% when the soil was initially dry. When the land was near field capacity, stream runoff showed little difference as a result of the conservation measures. Erosion rates were reduced by two-thirds, more of the water infiltrated, and less ran off over the surface. The main effects, however, were on erosion rates and summer storm runoff.

A 30-year study at Coshocton, Ohio comparing two unimproved basins with two others under full soil conservation practices showed that the good soil conservation practices resulted in decreased stormflow peaks (although very large

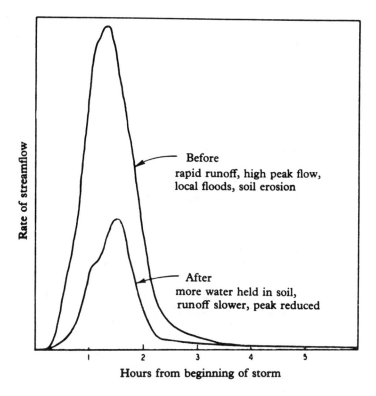

**Figure 7.5**  Hydrological effect of improved mixed farming on an eroded watershed.
*Source.* Pereira (1973), with permission of the Cambridge University Press.

storm peaks were not greatly affected), greater infiltration of precipitation, and decreased total stormflow. Most of these changes occurred during the first 20 years of the study with few significant differences existing during the last decade of the study. The changes in water runoff and stormflow generally resulted from deeper rooting depths for the vegetation on the basins with good conservation practices as well as greater water use by the more vigorous vegetation cover (Harrold, Brakensiek, McGuinness, Amerman and Dreibelbis, 1962; Ricca, Simmons, McGuinness and Taiganides, 1970).

### Effect of Vegetation on Snow Storage

The amount of snow stored in forested mountain watersheds usually bears a close relation to the water yield of streams draining those basins. In some respects, storage of water in the snowpack can be compared to storage in a reservoir. While the water in the snowpack cannot be so directly controlled, there is some indirect influence through treatment of the snowpack. With wise management, streamflow can be maintained at a reasonable level for longer than would be the case without snowpack management. Treatment of the forest should be directed toward increasing the depth of the snowpack on the ground, shading the snowpack to prevent evaporation or sublimation from the snow, protecting it from wind drifting, and reducing interception losses as much as possible. But at the same time, the timber resource must be managed to provide the highest possible yield.

Snow accumulation is clearly affected by the density of the forest stand, the type of trees in the stand (especially whether they are coniferous or deciduous), and the pattern of the stand (which involves the arrangement of openings). For example, studies have shown that interception is greater on the branches and needles of coniferous trees than on the bare branches of deciduous trees. Losses by evaporation from the intercepted snow can be higher than from snow stored on the forest floor. When winds blow the intercepted snow from the branches, it will usually end up on the snowpack, so high winds may actually reduce interception losses. Interception losses are greater in snowy winters than in dry winters because the opportunities for interception to occur are greater the more often snow falls.

Actually, interception losses are generally considered to be fairly small. West and Knoerr (1959) suggested a value of just 8% of the snowfall during the winter period might be intercepted in a dense coniferous woodland, while Geiger (1957) reported on work by Schubert who found a ratio of 10 to 9 in the snow depth outside a forest stand to the snow depth inside.

Evaporation losses from snow have been considered by a number of investigators although few definitive results have been achieved. On the basis of meteorological considerations, Linsley, Kohler and Paulhus (1958) suggested that evaporation rates from snow were much smaller than had previously been thought. They estimated no more than 5 mm/day (water equivalent) even during a strong foehn wind situation. The high albedo of the snow surface and the great amount of heat necessary to sublimate snow would argue more for melting than for evaporation. Linsley, Kohler and Paulhus (1949) felt that spring evaporation from snow in the west would seldom exceed 25 mm/month.

Jeffrey (1970) summarized existing knowledge about forest influences on snow accumulation and loss. Less snow accumulates beneath a dense forest canopy than a more open canopy, and less accumulates beneath forests than in openings in the forest stand. In the latter case, the maximum accumulation in the openings, which seems to be related to wind effects, occurs when the opening has a width equal to the height of the surrounding trees.

Forests will, of course, influence the snowpack on the ground through their effect on the rate of melting. Forests shelter the snow somewhat and reduce the radiation receipts. Advection of warm air in the spring will also be reduced since wind velocities will be less in forested areas. Radiation budgets in forested areas show that, as the percent of the forest canopy increases, there is a decreasing amount of radiation reaching the forest floor. The rate of decrease with increasing canopy cover varies with type of vegetation, being greater for balsam fir and lodgepole pine than for jack pine. Schomaker (1968) has discussed the relation of leaf formation to radiation received at the forest floor with a very pronounced drop in radiation receipts as brich leaves formed but only a minor change in radiation receipts under a spruce canopy. Even after the birch leaves were formed (and even with a greater canopy cover), more radiation penetrated the birch canopy than the spruce.

As the forest canopy develops, short-wave radiation (solar radiation) reaching the surface decreases. However, the trees will absorb solar radiation and they will, in turn, radiate to the snow surface. This long-wave radiation from the forest canopy increases fairly linearly with increasing canopy coverage. Figure 7.6 shows the relations between short- and long-wave radiation and the sum or net all-wave gain to the forest floor with increasing canopy. There is a rapid decrease in net short-wave and net all-wave radiation while the canopy cover increases from 0 to 20%. However, as the canopy cover increases from 20 to 100%, the more slowly decreasing short-wave radiation receipts are more than balanced by the increasing long-wave radiation downward from the canopy so that there is an increase in net all-wave radiation to the forest floor. At 100% canopy coverage, however, receipts of energy are only a little more than half of what they are with no canopy cover whatsoever.

Thus, Miller (1955) suggested that forest snowmelt was generally about 75% of the snowmelt in the open. One study in the Cascade Mountains showed that snow cover lasted 42 days longer in a forested area than in an open area. Melt rates are, of course, higher in small openings than in dense forest areas. The persistence of snow in forest openings is probably more a function of the increased accumulation of snow due to wind effects than the result of rate of melt, since melt rate will be greater in the opening than in the forested area. As a result, Costin, Gay, Wimbush and Kerr (1961), have suggested that a forest honeycombed with many small openings or with some cleared strips maximizes snow accumulation and results in greatest conservation of snow at the same time.

Increasing snow water supplies, while still favoring timber production and soil protection, is the continuing objective of current research studies. Establishing scattered openings in the forest cover may be good for water production but such a cultural practice is inefficient as a way to harvest timber. Thus, strip cutting has been used more extensively and it seems to provide a reasonable compromise. Strip clearing of trees prevents drifting of snow, reduces interception in the cleared areas, provides shade for the cleared areas so that radiation receipts are not too great. Lower wind velocities in the strip areas should lead to increased accumulation in the snow pack. Strip clearing, while possibly not as efficient as clearcutting, still protects the soil from erosion and may not increase cutting costs sufficiently to make harvesting a losing proposition.

## USE OF ANTITRANSPIRANTS

A second technique to reduce water loss from vegetation is to suppress transpiration. The use of antitranspirants on foliage, or their application to the soil itself

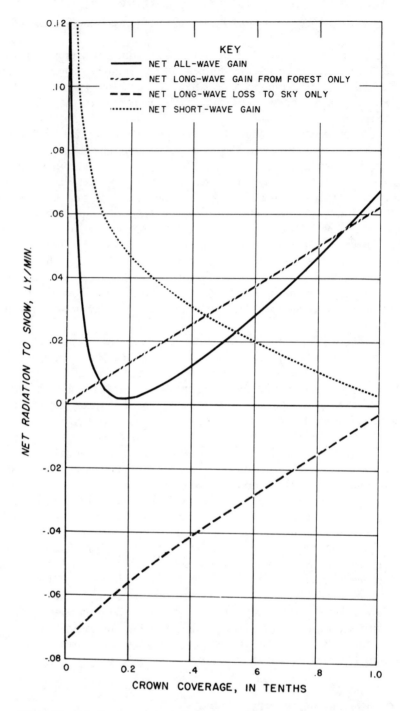

**Figure 7.6** Calculated net gains and losses of radiation to the snowpack in relation to canopy coverage for spruce-fir forest in Oregon.
*Source.* Reifsnyder and Lull (1965).

so that they are taken into the plant through the roots, are techniques still under-going extensive field testing. We need to know a great deal more about the role of transpiration in the life of the plant or tree. If reduction of transpiration will also limit plant yield or slow the growth of trees, water savings achieved in this manner may prove of little overall value. If, however, much of the transpiration of the plant is of little use in plant growth and if it can be reduced with little harm to the plant, then use of transpiration suppressants may prove to be of considerable value. Already nursery experiments have shown a higher rate of survival of seedlings treated with waxy materials or antitranspirants than of un-treated seedlings. This work has been related to problems of transplanting seed-lings when normal water uptake is impaired. Under those special conditions, the treated seedlings seemed to survive better.

Davenport, Hagan and Martin (1969) have identified three possible ways in which leaf sprays can be used to reduce transpiration: (a) use of reflecting ma-terials to decrease the heat load on the leaf; (b) use of film-forming materials to decrease the rate of water vapor loss from the leaves by decreasing their permea-bility; (c) use of materials that will increase stomatal resistance to water move-ment by closing or narrowing stomata.

The third technique is the one most frequently used at present. Both phenyl-mercuric acetate and decenylsuccinic acid have been shown to reduce stomatal openings under controlled conditions. In the field, aerial spraying is the only economic way to apply these materials. Such spraying is limited in effectiveness since stomata are generally on the underside of leaves and the sprayed material falls on the upper side. Also, unless repeated spraying is attempted, newly formed leaves may not be covered so that transpiration will only be partly reduced.

A 1974 study (Belt and King, 1977) of the use of a silicone antitranspirant, aerially sprayed onto a small (0.65-acre) basin covered with cedar and hemlock in northern Idaho, showed a 12% increase in streamflow from the treated basin over that from an untreated basin in a paired-watershed study. The differences were significant at a 97.5% level. The two watersheds had been compared for a 6-year period before treatment to develop the regression equations needed to compare the two basins. While soil moisture storage was also higher in the treated watershed, the differences between treated and untreated basins were not statisti-cally significant.

In spite of some success with the method of spraying antitranspirants, serious limitations still exist. Stomata serve as the entrance ways for carbon dioxide into the plant so that any chemical substance that will reduce transpiration loss by closing stomata may also interfere with carbon dioxide uptake and thus influence photosynthesis and plant growth. Reduction in carbon dioxide intake has been noted with the antitranspirants in use today. Thus, their use is dictated in only special situations where plant growth can be sacrificed in the interest of water conservation. Also, certain plant species have cells that are adversely affected by some of the antitranspirants in use today. Cell damage may result from their use on those plants. Finally, since transpiration serves to cool plant leaves, reduc-tion in transpiration will result in some additional heating of the leaf surface.

Whether this additional heat will adversely affect plant growth or development must be carefully determined for each plant species to be treated.

Work on reducing transpiration losses by use of antitranspirants is still in an experimental stage. Basic research so far has not been overly encouraging for the future widespread use of the technique. Where some retardation of plant growth is acceptable, and plant species are not damaged by their use, antitranspirants may have a limited usefulness. In addition to the successful practice in the early 1900s where nursery men and foresters used to dip seedlings to be transplanted in wax or wax-oil emulsions in order to prevent wilting, a National Academy of Sciences panel (1974) has listed the following situations where antitranspirants have been used in the United States:

(a) sprayed onto forested basins to increase streamflow;
(b) sprayed onto phreatophytes to reduce wasteful use of water;
(c) sprayed on highway plantings where irrigation would be expensive or hazardous;
(d) before harvest to increase the size of orchard fruits in those cases where size is closely related to moisture content;
(e) to reduce winter drying of plants where soils are often frozen.

While application of reflective materials to reduce the need for transpiration has not been widespread, some limited use with artichokes in Israel has been reported. In that situation, not only did the percentage of cuttings that rooted and the vegetative growth increase, but greater yields were also achieved.

Considerable research is still needed to understand the various problems with the use of antitranspirants. Especially attractive is the possible development of a systemic transpiration suppressant, a material that possibly can be applied to the soil and absorbed by the plant through the root system and then transported internally to the stomata. This technique of application would not only insure uniform coverage of all leaves new or old, but would direct the material to the stomata which occur primarily on the underside of the leaves and cannot always be reached by aerial sprays. So far, no real success has been achieved using this approach even though use of abscisic acid in a vase of cut flowers has been shown to reduce water loss from the leaves.

There is, of course, no guarantee that reductions in transpiration will be passed through the entire hydrologic system and be translated into increased streamflow. Whether this occurs or not, these important studies should continue. If such work will ultimately let us grow corn or wheat using 10 or 20% less water, the acreage of land in semiarid areas that would then become available to grow such needed crops would more than repay all the efforts going into antitranspirant studies. Using less water, so to speak, is almost as good as creating new supplies of water.

## WATER HARVESTING

Five inches (125 mm) of rainfall a year does not seem like a great deal of rain to most people. Yet, as some of the ancient scholars who wrestled with the problem of the source of water for streamflow finally realized, 5 in. of water falling over a good-sized basin can result in an appreciable volume of water. Five inches falling on 1 mi$^2$ of surface area, if all captured, will result in about 85,000,000 gal of water or 240,000 gal of water per day. If properly managed, this could irrigate 100 acres of farmland or supply domestic water for around 1500 individuals (10 mm of rainfall will produce 100,000 litres of water per hectare). Often, of course, the drier areas of the world may have soils which are quite rich agriculturally since they are unleached. In these areas, just small amounts of water applied at the right times can produce significant increases in food supplies.

Rainwater harvesting, as the technique to collect all possible water that falls on an area and to concentrate it for special agricultural or domestic uses is called, has been practiced in one form or another for thousands of years. It is quite possible that the reference to Canaan as a land of milk and honey in the Bible was in part, at least, acknowledgement of the successful application of rainwater harvesting techniques by the farmers in those water-short areas.

Rainwater harvesting is generally feasible in areas with 2 to 3 in. (50–75 mm) average annual rainfall. Some useful water will be made available during years when only 1 in. (25 mm) falls, but systems should not usually be planned in areas with such low average values of precipitation. Various techniques are available to collect precipitation. If the soil surface is fairly impervious, just creating a small pond in a surface depression may be sufficient to collect an appreciable amount of rainwater that might otherwise run off or sink into the dry soil unused. Where rainwater flows off over fairly impermeable hillsides, all that might be needed is the building of small ditches or rock walls sloping gradually across the hillside contours in order to direct the overland flow or sheet wash into specially prepared ponds or reservoirs. Clearing rocks and vegetation or compacting the hillside soil in various ways will increase the amount of overland flow and the amount of rainwater harvested by the hillside ditches.

Chemical treatment of soils to make them more impermeable is also successful. Addition of sodium salts to a clay soil will tend to make the clods of clay break down into small particles and thus seal soil pores and cracks. Other materials that have been spread onto soil surfaces to increase their ability to shed water are silicones, latexes, asphalt, and wax. Research is still under way using these materials, but they appear feasible for use on soils that do not swell with the addition of moisture.

Asphalt has particularly been used because it is quite effective, fairly long lasting, can be applied by spraying, and is relatively inexpensive, especially in areas where oil refining is carried out since the material is generally a low value by-product or pollutant resulting from the refinery operation. As used in the United States, asphalt, or a heavy oil mulch, has been applied by spraying hillslope or dune areas that have been cleared of vegetation, smoothed, and treated with a soil sterilant. Two coats of asphalt are used for rainwater harvesting—the first coat seals the pores while the second protects the surface from weathering. Such material has been found to last 4 to 5 years.

Granulated paraffin wax that melts and flows into the soil pores, or melted

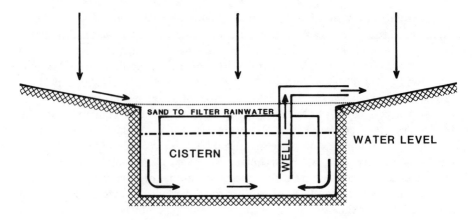

**Figure 7.7** Experimental sand-filled rainwater catchment with two water storage tanks. Sand acts as filter and reduces evaporation of stored water.

wax that is sprayed on the prepared surface are also effective soil sealants. Fink, Cooley and Frasier (1973) have shown that wax-treated plots permitted collection of 90% runoff of the rainfall as opposed to only 30% of the available precipitation as runoff from untreated areas.

Plastic sheets, butyl rubber, and metal foil have also been used to cover the soil in water harvesting operations. Often the plastic film is covered with gravel to protect it from wind damage and radiation effects. Use of these materials is considerably more expensive in both labor and materials than other methods but the rainwater harvesting area may last for 20 years or more without the need for further improvements.

The appropriate water harvesting method to use in a particular situation must depend on the characteristics of the site, the availability of the materials, and the experience of the labor involved. Among the soil characteristics to be evaluated are the water holding capacity of the soil, its tendency to erode, the presence of gravel or stones, and the permeability of the surface layers. Topography or slope of the land is also important since the system should be designed to transport the collected rainwater by gravity to the point of storage or use. Climatic factors of importance include the characteristics of the precipitation as well as other climatic factors such as wind, sunlight, and temperature. Storage must be a basic part of the rainwater harvesting system since precipitation is often quite infrequent and water demand occurs almost daily. If storage is in an open reservoir or pond, it is quite likely that much of the water collected in the rainwater harvesting scheme will be later lost by evaporation, or by seepage out of the bottom or sides of the pond. Thus, great care must be exercised in the design of the storage facilities to reduce evaporation and seepage to a minimum. One possible experimental design of a sand-filled, rainwater storage tank is shown in Figure 7.7.

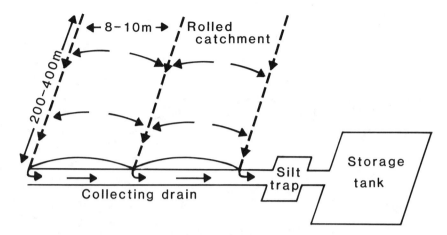

**Figure 7.8**  Rainwater harvesting scheme developed in Western Australia. Catchment graded and rolled to shed water. Water quickly flows to drain between catchments and to collecting drain at end of catchment, then through silt trap to large storage tank.
*Source.* Adapted from Anonymous (1956); Carder and Spencer (1971).

There are many advantages to a rainwater harvesting system. Once properly installed, the system should provide extra water without cost for fuel or power. The extra water, if used for drinking water for animals on ranges where food is available but water is not, may lead to significant increases in the carrying capacity of the land. The rainwater harvesting approach will also permit some smoothing of the great year-to-year variations in precipitation, so common in dry areas, and thus stabilize range carrying capacity. This will prevent some of the severe economic oscillations now experienced by rangeland managers in dry regions. Use of wax, reinforced-asphalt membranes, or plastic sheets covered with gravel have been installed in the southwestern part of the United States and provide water for less than 5¢ per cubic meter in an area of 12 in. (300 mm) of precipitation.

There are, of course, limitations to the rainwater harvesting technique. First, since the method depends on rainfall, if it fails there will be no rainwater to harvest. Second, poorly designed systems can lead to erosion inasmuch as the water falling on the harvest area is concentrated to flow to the storage area. Some water treatment or (at least) sediment traps may be needed to make the water ready for use. The systems generally have a limited lifetime and must be replaced or renewed periodically. While the technique may have begun in the Bronze Age, nearly 4000 years ago, it is still a relatively new technique in many parts of the world. One of the most extensive uses of the technique is in Western Australia where several thousand hectares of compacted earth catchments are used to supply water both for domestic and livestock use (Fig. 7.8; Anonymous, 1956; Carder and Spencer, 1971). Some 240 ha of asphalt-concrete harvesting areas have also been built to augment the water supplies for 32 small towns in Western Australia (Kellsall, 1962).

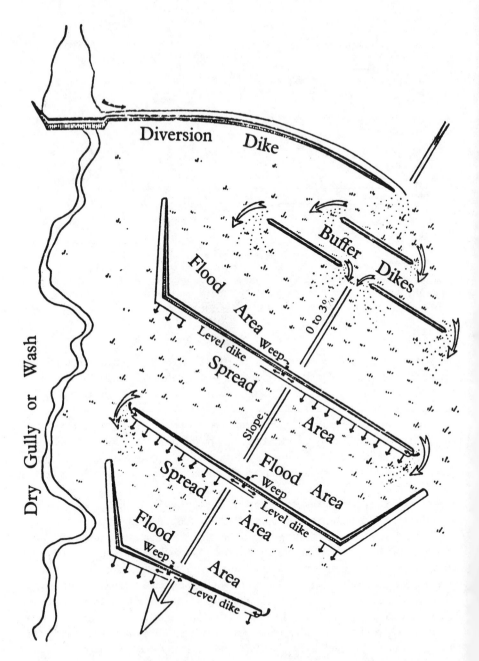

**Figure 7.9** Suggested spreading system for water flowing in wadi or small stream. *Source.* Branson (1956), with permission of the Society for Range Management.

Runoff agriculture is closely related to rainwater harvesting since it combines the system of harvesting of rainfall on unused areas with its immediate use as irrigation on a specific agricultural field or plot. Plants used in runoff agriculture systems should either be rapid growers, able to mature with possibly one irrigation and able to survive fairly long dry periods, or deep-rooting crops that will be able to tap water supplies stored more deeply in the soil and protected from evaporation losses. Deep-rooted, drought-resistant fruit trees, or some short-growing grains, such as millet and barley, show great promise for use in this type of agriculture.

Usually the catchment area, while extensive enough in size to provide the needed irrigation water, is divided into a number of small catchment areas in order to permit the farmer to control the runoff water, to direct it to selected portions of the irrigated· tract, and to prevent uncontrollable amounts of water from flowing through the collection ditches or overflowing the irrigated tract. The cultivated tracts are often terraced and have stone spillways that allow the farmer to direct the excess water from one irrigated field to the next lower one without damage.

One simple irrigation technique is to place stone barriers across wadis or intermittent stream channels. When the rains come and water flows in the wadi, it will be slowed and diverted from its course in the stream channel to flow over the rather broad, flat floodplain bordering the wadi. Strategic placement of rock barriers and crops (Figure 7.9) will allow maximum use to be made of the flood waters with minimum damage to land and crops. The technique involves directing the flow of the water in fairly thin sheets and low velocities to prevent erosion and to allow the water to infiltrate the tracts to be irrigated. But careful design and layout are necessary to withstand floods and prevent erosion.

Microcatchment farming is a particular variation of runoff agriculture. Here the agricultural field is divided into small basins, anywhere from $10 \text{ m}^2$ to $1000 \text{ m}^2$ in area. The basins are surrounded by low walls and sloped just enough so that the water falling in the basin will drain to one corner to irrigate the vegetation planted there (Fig. 7.10). Usually a small basin or pit is dug at the low point. The pit is filled with soil and manure and the area is kept loose to promote maximum infiltration and storage of the water draining from the rest of the microcatchment. The size of the microcatchment depends on the water requirements of the plant or tree growing on the catchment and the annual runoff volume expected from the catchment. Construction costs are very low—of the order of $5 to $20 per hectare depending on the size of the microcatchments. They are more efficient than larger water harvesting schemes since the water is used directly on the catchment and there is no need for transportation or long-term storage of the water.

Strip farming is a modification of the microcatchment farming technique. Rather than having individual catchments, the farmed area is divided into a series of sloping terraced strips from which water will flow off into the farmed strips on either side. The technique is similar to that which develops with a paved road. The vegetation on either side of the road will receive more moisture than will vegetation further back from the road.

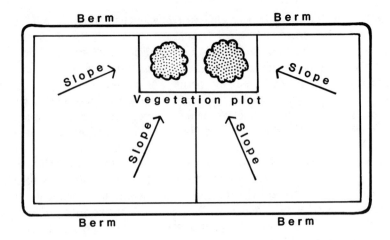

**Figure 7.10** Plan view of two adjacent microcatchments with drainage to cultivated plots in upper center. Size of catchment depends on annual rainfall and water needs of vegetation growing in cultivated plots.

The record for runoff agriculture is quite promising. For example, at "Conney-bar," Byrock, New Smith, Australia, where there is an 80-ha runoff agriculture area, it was found that the average carrying capacity of the region could be increased from 0.18 sheep per hectare on unirrigated range to 2.66 sheep per hectare on the irrigated tract. Seasonal feed shortages may still develop in very dry years but these shortages are usually not as severe.

Water harvesting and runoff agriculture are old farming techniques which have seen a rebirth of favor in recent years with the great need to increase agricultural production in water-short areas. The techniques have great merit for they will permit a flourishing agriculture in dry areas where the annual rainfall would argue against any agriculture. The techniques require wise use of the land with careful terracing and ditching in order to move water from the collection area to the area of use; but the techniques do not require any heavy earth-moving equipment or any engineering skills not possessed by the ordinary, careful farmer. The techniques should be applied more widely in the future as their clear advantages become more widely known.

## FRESH WATER FROM ICEBERGS

Human imagination has been aroused by the possibility of towing large tabular ice masses from the Antarctic to subtropical desert areas for use in agricultural or industrial development. It is the type of engineering undertaking that can be planned and discussed for years; but it contains so many unknowns that it will probably have to be attempted on (at least) a modest scale before we can be certain of the actual results. The whole project received a considerable boost in 1978 when the First International Conference and Workshop on Iceberg Utilization was held at the Iowa State University in Ames, Iowa. Proceedings of that conference have been published and they make for fascinating reading, even though they do not provide all the enlightenment one might wish (Husseiny, 1978).

There seems to be general agreement that the flat tabular icebergs from certain regions of Antarctica are the most usable since they possess the desired shape for towing. Icebergs should be roughly rectangular in cross-section to minimize rolling or calving during towing.

A minimum usable iceberg was described as having a thickness of between 170 and 280 m. Using a value of 250 m, this would mean that the berg should have a width of 375 m and be at least 1.5 km long. The surface area would be about 2 million $m^2$ and the water equivalent would be 0.12 $km^3$ (32 billion gal). An iceberg of this size would probably have to be encapsulated to prevent undue loss in transit. It could well represent the minimum amount of water delivery that would be price competitive with alternate sources. Clearly, larger bergs are available and might be considered. However, they might pose additional problems of power for towing, possible break-up on route, and the need for insulation or encapsulation. Icebergs with lengths over 20 km are not uncommon in the Antarctic area and, since annual production of icebergs is of the order of 1,000,000 million $m^3$, there would seem to be a wide variety of shapes and sizes from which to choose (Hult and Ostrander, 1973). Satellite imagery capable of high resolution would be extremely valuable in locating icebergs of the correct size and shape for towing.

Several papers at the Ames meeting addressed the problem of towing of icebergs. There was general agreement that it should be a fairly slow tow (approximately one knot) so that travel times to reach destinations in Chile or Australia might be 6 months while an Arabian destination might take over 9 months. With high sea temperatures, evaporative and melting losses could be great, as would be losses due to calving or sloughing off of side walls.

There was little general agreement on the feasibility of the iceberg utilization project. For example, Bader (1978) suggested that while considerable attention had been given to towing problems and the thermodynamics of the melting of ice in contact with moving water, very little information was available on how to convert the iceberg into fresh water delivered to the land at the end of the journey and next to nothing was known about the glaciology of icebergs. He felt that the simple project of moving a small iceberg to a protected harbor in Chile for a local agricultural development might be a worthwhile venture inasmuch as it would not involve problems of encapsulation, off-shore mooring of the berg, or large losses at sea. He did not believe that a scheme to tow much larger bergs greater distances or to shallower coasts where off-shore mooring is needed would be practical. Not only would super tugs capable of developing large amounts of power for many consecutive months of operation be needed, but problems of insulating the berg would have to be faced—especially the significant problem of calving along the sides. At the destination, slurry pipelines or augmented melting

**Table 7.1**  Relative Volume of Water in Icebergs, Lakes, and
Reservoirs

|  | $(\times 10^{12}$ gallons$)$ |
|---|---|
| Lake Erie | 128 |
| Lake Winnipeg | 84 |
| Lake Mead (Hoover Dam) | 10 |
| Grand Coulee reservoir | 3.1 |
| Large iceberg ($2.8 \times 11.2 \times 0.25$ km) | 1.7 |
| Shasta reservoir | 1.5 |
| Flaming Gorge reservoir | 1.2 |
| Hungry Horse reservoir | 1.1 |
| Small iceberg ($0.75 \times 3.0 \times 0.25$ km) | .12 |
| Hetch Hetchy reservoir | .12 |

techniques (use of waste heat from floating power plants) would have to be developed or the berg would have to be cut into pieces in order to bring it into protected but shallow harbors. This latter operation could result in considerable loss of fresh water.

Several scientists at the meeting in Ames were more optimistic about the feasibility of iceberg towing and utilization although they did agree with Bader on certain conclusions. For example, Weeks and Mellor (1978) agreed that an unprotected iceberg, regardless of initial size, would not be able to survive the losses incurred in the long slow trip to tropical latitudes. It might, however, be possible to move unprotected icebergs to Australia or Chile without too great a loss. They felt that the size of the towed iceberg would be a problem and argued that a berg greater than 2 km in horizontal dimension would be likely to experience mechanical breakup due to long wavelength swells. For efficiency in towing and to prevent capsizing of icebergs, the length/width ratio should be considerably greater than unity and the width/thickness ratio should be at least 1.5. They strongly recommended a pilot project with a small iceberg to test some of the many models and theories that have been advanced.

Two other studies reported on techniques for protecting the iceberg during transit by the use of foamed insulation and the use of floating solar collectors in the melting of the iceberg. Hussain (1978) pointed out that foam has many features that make it preferable to plastic wraps. It is easier to apply, has better insulating properties, and greater stability in transit. Polyurethane foam has the smallest thermal conductivity of all known foam products. It can be sprayed on both the top and bottom of icebergs and will conform to rough surfaces. The sides of the iceberg would have to be protected by preformed panels held in place by clamps or other methods of joining. Since the submerged vertical walls of the iceberg will be most subject to meltback and the formation of an underwater shelf, the sidewall insulation will be most critical and also the most difficult to install. Hussain estimated a cost of about $8 million to insulate a small iceberg 1200 m × 300 m × 260 m thick in polyurethane foam. Of this total, about $350,000 would be for insulating the top, $2.6 million for insulating the bottom, and $5.1 million for insulating the sides.

To melt the ice, Cluff (1978) suggested the use of waste heat from a floating tracking platform equipped with parabolic collectors to concentrate solar energy on photovoltaic cells. Use of waste heat from conventional power plants would improve the efficiency of the power plant but would also involve changes in power plant design, which power companies might be loath to undertake given the present state-of-the art in iceberg utilization.

It is difficult to grasp the magnitude of the icebergs that are being considered for towing and especially to have any feel for the amount of water actually stored in an iceberg. A large iceberg 11.2 km long by 2.8 km wide and 250 m thick contains about 5.3 million acre-feet of water (1700 billion gal). A small iceberg with a length of 3 km, a width of 750 m, and a thickness of 250 m would contain some 380,000 acre-feet of water (120 billion gal). For comparison purposes, this means that the large iceberg could supply *all* U.S. water need in 1980 (442 billion gal per day in Table 1.3) for a total of 4 days. The amount of water stored in large and small icebergs has been compared in magnitude with the volume of water storage in different U.S. lakes and reservoirs in Table 7.1.

Iceberg utilization is an intriguing idea whose time may not quite have come. However, as water scarcity continues to grow and the technology to move, protect, and melt icebergs develops, it appears likely that an attempt will have to be made to incorporate iceberg water into the water resource picture of at least one water-short place.

## 8

## WASTE WATER RENOVATION
## AND INTERBASIN
## TRANSFERS

## USE OF RECLAIMED WATER

Obtaining additional water by renovating sewage waste is a technique whose time may have come. It is not a new technique and, in fact, has been practiced in many areas of the globe for thousands of years. Where water supplies have been limited, sewage wastes have long been used to irrigate agricultural crops—the technique having a double advantage in that it supplies both moisture and fertilizer to the crops. The serious disadvantage is, of course, the possible transmission of viruses and other disease organisms.

For health and aesthetic reasons, the use of renovated waste water has grown only slowly in Western cultures, and then only in selected situations. With improvement in the technology for renovating waste water, with the greatly increasing need for water in our growing urban areas, and with the ability of industries to use water of different quality for different purposes, more water managers are turning to use of reclaimed water as a source for additional supplies or as an alternative to transporting water from distant sources.

While the direct reuse of municipal sewage for domestic purposes has not been accepted by a majority of people, the reasons are not entirely clear. Already, more than 40% of the population of the United States use water that has been previously used for some domestic or industrial purpose, while some 60% use stream water that has been used at least once upstream (Gloyna, 1966). A number of water systems currently accept water in a polluted condition, treat it, and supply it to customers for domestic and industrial purposes. This water is evidently accepted by the public because it comes from a river or lake and not directly from the sewage treatment plant of a city.

The recent spate of regulations requiring treatment of waste effluents, and the increased use of secondary and even tertiary treatment, will result in reclaimed water that may quite easily be of even higher quality than the original water source. Increasingly, the economics will be such that it will be less costly to utilize reclaimed water than to develop a more distant water source and to transport it to the user. Within a short time, it seems that social acceptance may be the major, if not the only, limitation to the widespread use of such water in municipal water systems.

Water reclamation is really only feasible in urban or industrial situations where large volumes of water are needed and similar volumes of waste water are generated. In such a situation, it may be economically feasible to build the necessary treatment plants and piping systems, as well as to find the variety of uses that will make recycling of reclaimed water worthwhile. In a few situations, reclaimed water from communities in farming areas might be sold to farmers for irrigation water, but the cost of distributing the water and the need for some way to get rid of the effluent in those seasons of the year when irrigation is not needed make such a reuse of water less feasible. Thus, in considering reuse of reclaimed water, we are primarily concerned with the use of such water in municipal water systems for domestic and/or industrial purposes.

If, for the time being, we disregard the matter of public acceptance, the question of use of reclaimed water hinges mainly on the economics of one source of water vs. another. If both sources are equally pure and usable, then the accepted source should be the least costly one. Since the present conventional treatment systems—primary and secondary treatment—do not provide water of potable quality, some form of tertiary treatment will be necessary.

## Waste Treatment Processes

Present municipal waste treatment processes are usually considered to be primary (mechanical), secondary (biological), and tertiary (chemical). Primary and secondary treatment processes are described as conventional while tertiary treatment processes, which are employed only selectively, are considered to be an advanced form of treatment.

Primary treatment involves removing the gross pollutants from the effluent by mechanical methods. The water flows through various screens which capture suspended solids (or through a comminutor or grinding device to cut the solids into smaller pieces), into grit chambers where separation of grit from organic solids and oils and scum occurs, and then into a settling tank where some of the suspended solids will settle out. Finally, chlorine is added to the waste, if it is to go to a stream, to kill or reduce the number of bacteria. Primary treatment will remove 50 to 60% of the suspended solids and 30 to 40% of the biochemical oxygen demand (BOD) from the wastes (Johnson, 1971).

Secondary treatment begins where primary treatment ends. The water passes through various biological processes to remove the oxygen-demanding materials and other solids from the waste water. A standard treatment uses a bed of rocks or crushed stones over which the waste water is sprayed (trickling filter or high-rate biofilter). Microorganisms, which feed on the organic materials in the waste, grow as a slime on these rocks. As the water trickles down through the rock bed, these organisms remove much of the organic load in the waste water, reducing the BOD of the effluent by about 85%.

Sometimes an activated sludge process is used to reduce the BOD loading of the waste water. Aeration of waste water results in the growth of microorganisms (activated sludge) which consume the organic matter in the waste. After leaving the aeration tank, the sewage and sludge mixture—called mixed liquor—flows into a sedimentation tank where solids are removed. Some of the sludge removed in this step is returned to the aeration tank to help treat the incoming sewage. The activated sludge process has certain advantages over the trickling filter process. The tanks required are smaller than those needed for the trickling filter, the process does not result in the development of odors and flies, and it is slightly more efficient than the trickling filter technique, removing some 89 to 95% of the BOD loading. However, it can be upset by significant changes in waste loading resulting from the discharge of varying industrial effluents, as can the trickling filter process. Chlorine is often added to the clarified liquid as the final step in the secondary treatment process.

Lagoons or stabilization ponds are sometimes used in place of these other two techniques as secondary treatment. The effluent from the primary treatment is allowed to flow into large storage ponds where sunlight, algae, and oxygen, all working together, will break down and stabilize the organic matter in the water. All that is needed is a large (or preferably a series of large) shallow ponds and sufficient time for the process to work. Odors will often develop, especially if the lagoon is so heavily loaded that anaerobic conditions develop. Large areas are needed due to the fact that storage times can be appreciable. From time to time, the bottom of the pond must be cleaned to remove collected sediments and a place must be found to dispose of such sediments. Lagoons cannot be thought of as merely ponds to hold the effluent until it is removed by evaporation for, in many areas, annual precipitation exceeds annual evaporation so that the depth of liquid in the pond would increase as a result of the excess precipitation if there were no bottom seepage or surface flow from the pond. In the case of some toxic wastes, storage ponds with impervious bottoms are being used with mixed results.

While primary and secondary treatment remove up to 95% of the BOD loading of the effluent, the waste material remaining in the effluent may still be sufficient to cause degradation of receiving streams. Further treatment (tertiary treatment) is often necessary to remove the remaining pollution and to render the effluent safe for stream disposal or human use. Such specialized treatment is often expensive.

## Types of Pollutants

To understand the problems involved in waste treatment, it is necessary to consider the range of pollutants that must be handled in a sewage treatment plant. Eight major categories of pollutants can be found in our sewage streams: (a) oxygen-demanding wastes, (b) disease-causing agents, (c) plant nutrients, (d) synthetic organic chemicals, (e) inorganic chemicals or mineral material, (f) sediment, (g) radioactive materials, and (h) heat.

**Table 8.1**  Relative Concentration of Biochemical Oxygen
Demand in Typical Domestic and Industrial Wastes

| Source | Mean BOD (mg/ℓ) |
|---|---|
| Domestic sewage | 200 |
| Poultry plant | 480 |
| Synthetic fiber plant | 520 |
| Brewery | 610 |
| Milk processing plant | 1000 |
| Tannery | 1100 |
| Meat packing plant | 1100 |
| Potato processing plant | 1340 |

*Source.* U.S. Federal Water Pollution Control Agency
(1969).

The oxygen-demanding materials are the normal organic wastes and ammonia
found in domestic sewage and in the plant and animal wastes from food processing
plants. Typical concentrations of BOD in domestic and industrial wastes are given
in Table 8.1. These organic wastes are usually consumed in the secondary treatment
process or by the addition of oxygen to the waste. If they are allowed to enter
a receiving stream or lake untreated, they can result in a rapid lowering of the
oxygen content of the water and possible fish kills.

Disease-causing agents, infectious organisms, viruses, and other bacteria can
be easily transmitted through water treatment systems from one source to another.
Such agents can generally be treated effectively by modern disinfection techniques.
Use of chlorine has greatly decreased the danger from disease-causing agents but
care must still be taken to insure that new strains of viruses or bacteria do not
develop. As a result of improved treatment techniques, the typhoid death rate,
which in 1900 was 100 per 100,000 of population, had dropped by 1950 to less
than 1 per 100,000 people in the United States.

Plant nutrients consist primarily of carbon, nitrogen, and phosphorus, which
are often present in large quantities even in water that has gone through conventional treatment processes. They contribute to the rapid growth of algae and weeds
in streams and lakes. These may be harmful directly because of odor and scum
problems and through possible clogging of water intake lines. They become an
even more serious indirect problem because when they die they create a high
secondary oxygen demand which may cause massive fish kills and the well-known
eutrophication of lakes. These waste materials are present in sewage wastes, in
the waste from certain industrial processes, and in drainage from fertilized fields.
Biological (secondary) treatment not only does not remove phosphorus or nitrogen
from the waste but it may convert the organic forms of these materials into mineral
forms which are more readily available to plant life.

Synthetic organic chemicals include detergents and other cleaning substances, synthetic organic pesticides and industrial chemicals, and the wastes resulting from their manufacture. These materials may be toxic to both fish and humans or result in significant taste and odor problems. They generally resist conventional waste treatment processes. The long-term effects of small doses of many of these materials on humans are still to be determined.

Inorganic chemicals and/or mineral substances include a wide variety of acids, salts, chemical compounds, and solid materials that result from various industrial and agricultural processes. Leaching of water through fertilized fields or from mining operations may result in the movement of considerable amounts of minerals and acid wastes into surface streams. Not only will these substances kill fish life and degrade the streams, but they cause hardness of water and result in corrosion of water treatment or distribution systems and so increase the cost of water treatment.

Sediments are generally thought of as the particles of sand, silt, and clay eroded from land or washed from the paved areas of urban communities. Various construction operations (housing developments, highway building, land remolding) are excessive contributors to the sediment problem; poorly designed farming operations can also contribute to the sediment load of streams. Sediments fill stream channels and reservoirs, cover fish spawning grounds and food supplies, damage pumping or power generation equipment, and necessitate the continual dredging of harbors and channels. Though less of a problem than some forms of pollution, sediments are insidious in that they lead to deterioration of our water supplies and at the same time represent a great loss to our soil and agricultural resource.

Radioactive materials in streams result from the processing of radioactive ores, the use of radioactive materials in power reactors and in various industrial and medical facilities. Because so little is really known about the long-term effects of radioactivity—or the possibility of these materials accumulating in the sediments at the bottom of streams—this type of pollution may pose serious future problems for humans. At the present, such problems are quite localized because of the restricted use of radioactive substances.

Heat is a very different type of pollutant. Since heat in water reduces its capacity to absorb oxygen, the disposal of industrial cooling water in streams can lower its ability to degrade oxygen-demanding waste material and, thus, to support fish life. Changing the temperature of the receiving water significantly may result in the development of an entirely new ecosystem in the vicinity of the heated water outfall. Periodic changes in the quantity or temperature of the heated effluent can result in significant changes in ecological conditions in the receiving stream and even massive fish kills locally. Steady doses of heat in controlled amounts have been shown to be quite helpful in speeding life processes of certain aquatic organisms such as oysters.

Most of our wastes are, of course, mixtures of several of the eight types of pollutants discussed above resulting in more complicated and expensive treatment needs. For example, normal municipal wastes, involving mixed domestic and industrial effluents, usually contain oxygen-demanding substances, synthetic organic chemicals, sediments, heat, and possibly inorganic chemicals, disease-causing agents, and plant nutrients. Until recent years, our prime concern had been the removal of oxygen-demanding substances from the waste in order to improve the BOD loading of surface streams. Many states have strict laws regulating the quantities of BOD that may be allowed to enter a surface stream from municipal or industrial outfalls. We now recognize that plant nutrients, especially because of their secondary oxygen demand, and persistent synthetic and inorganic chemicals are becoming even more serious problems than the normal oxygen-demanding substances. At present, the freshwater volumes needed to hold persistent chemicals to acceptable levels are, in most cases, lower than the volume of water needed to control the oxygen-demanding type of pollutants. However, as water treatment improves and less oxygen-demanding material is allowed to enter surface streams, chemical pollution may easily become our most significant problem.

**Tertiary Treatment**

The strict laws against water pollution and the increased need for reuse of waste water call for the development of new or improved tertiary water treatment facilities. Some of these are already available and in use but their widespread acceptance as part of "conventional" waste water treatment is still necessary if we are to achieve full control of pollution and at the same time make more water available for reuse.

One technique that has long been used (even though it is classified as an advanced or tertiary technique) involves coagulation-sedimentation. The addition of alum, lime, or iron salts to an effluent as it comes from the secondary treatment causes small particles of solids to floc or clump together into larger masses. Holding the effluent for a period in a sedimentation tank will allow the larger masses to settle to the bottom where they can be removed. This method removes essentially all of the settleable solids from the effluent and can reduce the concentration of phosphate by as much as 95%. Phosphate is possibly the critical plant nutrient to be removed from streams because municipal effluent is the primary source of phosphorus. Water vegetation can always obtain nitrogen directly from the atmosphere if they have need of it.

Some of the more persistent organic materials which resist normal biological treatment can be removed by an adsorption process. Here, the effluent passes through a bed of activated carbon granules which remove more than 98% of these persistent organic materials. The activated carbon granules can be cleaned by application of heat and reused. A newer variation on this technique involves the introduction of powdered carbon directly into the water flow. The organic materials attach themselves to the carbon which is later removed from the stream by means of coagulating chemicals. Powdered carbon is more effective in removing the persistent organics than the bed of activated carbon. Its future use will depend on whether the powdered carbon can be easily cleaned for reuse.

Waste water that has undergone secondary treatment followed by tertiary treatment involving coagulation and adsorption will have a chemical quality quite similar to rainwater except for the possible addition of salts. In municipal use, salts in water have been known to increase by 300 to 400 mg/ℓ. Electrodialysis, a process involving electricity and chemically-treated plastic membranes, is used to remove these salts. A mineral salt in water generally breaks down into ions— atoms or small groups of atoms having an electrical charge. In electrodialysis, the positive-charged ions are attracted through a membrane to the negative-charged pole while the negative ions are attracted through another membrane toward the positive electrode. The positive and negative electrodes will remove most of the salts and allow the clean water between the two membranes to flow off for further use or ultimate discharge.

We have just scratched the surface in water treatment techniques. New industrial operations will result in new materials being introduced into our water supplies and new challenges for water treatment engineers. New and improved combinations of existing treatment techniques will result in more efficient or cheaper treatment of effluents. The goal should always be to make the renovated water better than it was before use.

One new technique that may accomplish tertiary treatment at a fraction of the cost of more conventional methods involves the use of hyacinth retention ponds. Water hyacinths, long considered a weed of great nuisance by water treatment engineers, appears to have certain abilities to remove large quantities of nitrogen and phosphorus from water and so perform much of the work of tertiary treatment.

A large-scale field test was undertaken in Coral Springs, Florida during 1978 and 1979 in which some 100,000 gal of water, having undergone secondary treatment, was moved through five experimental ponds covered with water hyacinths over a 6-day period of retention (Anonymous, 1979). The five ponds were interconnected so that water was held in the first pond for a 2-day period and then in each successive pond for just a 1-day period. The inflow water only had domestic wastes which had been treated to remove 95+% of the BOD and suspended solids.

Hyacinths have a very high uptake rate of nutrients and they incorporate these nutrients into their biomass. To maintain high nutrient uptake rates, however, the hyacinths have to be harvested periodically and either used for fertilizer or to produce methane.

Early results of the field test suggested that the hyacinth-covered ponds lowered the nitrogen effluent level to 0.6 mg/ℓ from 10–30 mg/ℓ inflow. Federal standards are no more than 3 mg/ℓ. Phosphorus levels were reduced from 5 mg/ℓ in the inflow to about 3.5 mg/ℓ in the effluent while the Federal standards are 1 mg/ℓ. Effluent water from the hyacinth ponds also showed no total or fecal coliform counts although inflow water had 256 and 124 organisms/100 ml, respectively. Also chlorinated hydrocarbon concentrations were reduced by 50% and trace metals were also significantly reduced.

Tests were undertaken to see if reducing inflow volume to the ponds would improve phosphorus removal. However, it is quite possible since nitrogen removal

Table 8.2   Average Cumulative Removal of Different Water Pollutants by Means of Different Waste Treatment Processes (Expressed in Terms of Percent Removal)

| | 5-Day BOD | Chemical Oxygen Demand | N | P | Total Dissolved Solids |
|---|---|---|---|---|---|
| Primary sedimentation | 35 | 37 | 17 | 15 | 0 |
| Activated sludge (secondary) | 92 | 86 | 24 | 21 | 0 |
| Lime clarification | 97 | 91 | 28 | 90 | 20 |
| Rapid sand filtration | 98 | 94 | 28 | 95 | 22 |
| Granular carbon adsorption | 99 | 97 | 28 | 95 | 22 |

*Source.* Johnson (1971) as reported by Robert Smith, Robert Taft Engineering Center, U.S. Environmental Protection Agency.

was greater than needed that the final solution may be to increase total water inflow to the system (to reduce costs for the treatment ponds) and to use alum or ferric chloride to precipitate the phosphorus. Clearly, water hyacinths will not work everywhere or on all types of effluents but they give promise of an inexpensive form of tertiary treatment in certain situations.

**The Cost of Tertiary Treatment**

Cost, of course, is a major consideration in the use of advanced or tertiary treatment facilities. Conventional secondary treatment may remove up to 95% of the BOD loading in an effluent. Lime coagulation may raise this to 97% removal at a cost of about 7.5¢ per 1000 gal treated. The use of rapid sand filtration and carbon adsorption will increase the removal of BOD to about 99% but the cost will be about 13¢ per 1000 gal. Lime coagulation mentioned above will also remove up to 90% of the phosphorus (secondary treatment removes only about 21% of the phosphorus). Sand filtration, at 3¢ per 1000 gal, increases the phosphorus removal to 95%. Electrodialysis, to remove salts in the effluent, is a fairly expensive operation costing about 14¢ per 1000 gal treated. Since lime coagulation is relatively inexpensive and highly efficient in increasing the removal of phosphorus and oxygen-demanding wastes, it is felt that it will be one of the first advanced or tertiary treatments added to conventional primary and secondary treatments (Johnson, 1971). Table 8.2 provides a breakdown of the average cumulative removal of different pollutants for various conventional and advanced treatments. Table 8.3 summarizes the incremental cost of advanced treatment and suggests possible uses of the water at each stage.

**Table 8.3**   Incremental Cost and Utility of Treated Sewage Effluent

| Treatment | Cost/Dollars Per 1000 Gallons[a] | | | Application[b] |
|---|---|---|---|---|
| | 1MGD Plant | 10MGD Plant | 100MGD Plant | |
| Secondary | | | | Nonfood crop irrigation. |
| Coagulation-sedimentation | .135 | .075 | .047 | General irrigation supply; low quality municipal supply. |
| Rapid sand filtration | .063 | .028 | .012 | Short-term water recharge. |
| Carbon adsorption | .310 | .100 | .070 | High-quality irrigation supply; good quality industrial supply; body-contact recreation; long-term groundwater recharge. |
| Electrodialysis | .220 | .140 | .090 | High quality industrial supply; indefinite groundwater recharge. |
| Disinfection | .012 | .009 | .006 | Potable water supply. |

[a]Smith (1968); Smith and McMichael (1969).
[b]Stephan and Weinberger (1968).
*Source.* Johnson (1971).

### Examples of Municipal Water Reclamation and Reuse

The idea of reuse of renovated water is neither new nor radical inasmuch as many individuals in the United States already use water that has been used and cleaned several times. Chanute, Kansas, in 1956–57, provided the first recorded case in the United States where renovated waste water was used as part of a municipal water supply. The community water supply, the Neosho River, was dry and the city was forced to recycle its own waste water after a 17-day storage period in the river channel. A safe water supply was produced through water treatment although townspeople did not accept it for drinking purposes because of odor, color, and foam problems which made it unattractive.

The first known instance of a planned use of renovated waste water for municipal supply occurred in the town of Windhoek in Southwest Africa, Windhoek is in an arid area with very limited water supplies. To increase supplies of fresh water would require overland transport for a considerable distance. To reduce this expense, it was decided to build a waste treatment plant capable of providing high quality water from the town's waste effluent and to mix it in the ratio of about one-third waste to two-thirds fresh water to supply all municipal needs. The project was completed in 1968 and, after the first 3 years, no health hazards had been found and no strong objections had developed from the local inhabitants. Still, it is to be expected that it will be some time before extensive human use of renovated water will come about.

Communities that use groundwater for their water supplies, or are currently using fresh water to prevent salt water encroachment in coastal aquifers, have an opportunity to use renovated water to recharge groundwater supplies. The additional purification resulting from storage in and slow movement through groundwater aquifers, as well as the dilution afforded by mixing with the rest of the groundwater, can lower any chance of health hazards and will increase public acceptance of the renovated waste product. Public use would not be direct but rather indirect through the groundwater supplies; this should mollify public aversion to the idea of direct reuse of treated water.

In water-short areas, with urban populations often growing more rapidly than the ability to increase water supplies, water managers are being faced with difficult decisions. In a number of situations they are turning to advanced treatment of municipal sewage followed by water reuse to satisfy various low-grade demands such as irrigation of crops, golf courses, and cemeteries, and in creating water-based recreational areas. Swimming or other water contact activities have generally not yet been included in the recreational areas (McCauley, 1977). Water managers may have been overly cautious in the use of this reclaimed water, for some of it is exceedingly high grade and potable.

Lubbock, Texas, a town of around 146,000 people in 1970, has grown rapidly in recent years. Even so, water managers have been able to increase water supplies to 88 mgd while consumption averaged only 26 mgd in 1971. Thus, considerable extra capacity exists and there should be little need for extra supplies for almost two decades (McCauley, 1977). The city has upgraded its waste treatment facilities. Primarily, it had sent its waste water through trickling filters having a 17-mgd capacity. The treated effluent was then spread on city and private farmland to irrigate nonedible crops. The addition of a 12-mgd activated sludge system provided effluent of good quality except for a high nitrate concentration.

To keep effluent out of the Brazos River and to reduce the demand for fresh water, water managers have encouraged the use of this reclaimed water. About 2.5 mgd will be provided to the generating plant of the Southwestern Public Service Company. More surprising, some of the water was to be used in an aquatic park with recreational lakes. In all, eight lakes were planned, the first six within the city limits and the last two outside the city limits (Freese, Nichols, and Endress, 1969). Almost all of the recreation, including even boating and swimming, would be restricted to the last and largest lake outside the city limits. The lakes within the city were to exist as ornamental lakes with no body-contact water activities planned.

While the plan received favorable feasibility reports and public acceptance was quite good, financing was difficult to arrange. Four of the lakes within the city were completed by 1976 but a lack of money and a conflict with the development of an interstate highway caused the construction of the last two lakes within the city to be stopped. The two lakes outside of the city were also eliminated and with them the recreation portion of the program. The four lakes now in operation are largely ornamental (McCauley, 1977).

The Santee County Water District, located in a fairly dry area of southern California, has no water shortage. Santee now receives Colorado River water and may, in the future, obtain northern California water. Though water is available, its price has climbed appreciably and, in 1973, the wholesale purchase price of water was estimated at $75 per acre-foot ($230/million gal) (Houser, 1970). Santee was not part of the San Diego Metropolitan Sewer District because of the expense of entering the system and the feeling that it would be possible to treat and reuse waste water. In 1959, a 1.5-mgd sewage treatment plant was built to provide primary and secondary treatment. Two (and later other) oxidation ponds completed the water treatment. The effluent was used for golf course irrigation but other uses were sought. The development of recreational lakes was suggested by the geology of the area and the availability of lake sites without major excavation costs. Based on feasibility studies and preceded by a good community education program, the Santee Water and Sewer District in 1966 received money to construct a 2-mgd tertiary treatment plant. The system would remove nitrates and phosphates (Houser, 1970). Public support and acceptance of the project to use the tertiary-treated water for recreational lakes were strong.

By 1961, the effluent from the existing treatment plant was already being passed through a three-lake chain. The chain was increased to four and then later to six, and the lakes were improved with landscaping, stocking with fish, and the addition of picnic tables. The lakes were fenced off, however, to restrict public uses (McCauley, 1977). The Santee Water and Sewer District felt that it would be satisfactory to open the area for nonbody contact recreation in 1961 but the County Health Department refused permission because of possible health hazards. Approval would be granted with the introduction of additional treatment and with continuing analysis of the water for viruses. These steps were initiated and the area was opened to nonbody contact recreation. Fishing for fun was allowed at first but later fish could be removed and eaten. One controversial program that began in 1961 was the use of the reclaimed water in a swimming pool. At the beginning, the program was closely supervised and only 25 individuals at a time were allowed in for ½-hour periods. Careful health records were kept on individuals admitted but no waterborne diseases appeared. The program to use reclaimed water in the pool continued until 1968 when Federal funding stopped and the cost of maintaining the pool water in an aesthetically pleasing condition forced termination. Because of a chemical reaction, the pool water turned yellow-brown. This was unpleasing to users and raised the possibility of lawsuits if a drowning occurred.

In the 1970s, declining interest on the part of the operators, lack of Federal funds, the build-up of sludge in the lakes, and the increase in suspended and dissolved solids made continuation of the project a liability. Santee joined the Metropolitan Sewer District so that no more water was available for the lake project and the system was terminated with far less publicity than had accompanied its beginning.

To keep Lake Tahoe on the California–Nevada border just south of Reno free from contamination, residents banded together in a "Keep Tahoe Blue" campaign that resulted in the most advanced waste-treatment installation in the United States at the time (Culp and Culp, 1971). The water from this treatment plant was then piped out of the Tahoe Basin to nearby Alpine County where it formed the bulk of the water in Indian Creek Reservoir. Water quality standards established by Alpine County were more demanding than those set by Federal or state guide-lines and they have been met by the South Lake Tahoe Public Utility District. It would appear that the quality of the reclaimed water is so good now that it could be reused within the Tahoe Basin if future needs make it necessary. A water shortage in the basin may develop in 10 or 15 years if population growth con-tinues as expected (McCauley, 1977).

Possible reuse of water within the Tahoe Basin is somewhat limited since there are few nonpotable users and drinking of the effluent is not to be permitted. Discharge of the renovated water into Lake Tahoe would probably not harm the lake itself because of the high quality of the reclaimed water but there is still some opposition from the "Keep Tahoe Blue" coalition. More reuse possibilities exist in Alpine County where the water is presently being used for recreational purposes, including trout fishing and swimming. Indian Creek water is also used to supply some irrigation needs. So far, public acceptance of the use of reclaimed water for recreation and irrigation has been good—a tribute to the very high quality of the effluent from the South Tahoe treatment plant.

Several other examples of municipal use of reclaimed water, either actual or planned, could be cited but the experiences are quite similar to those already discussed. At Antelope Valley near Lancaster, California, for example, waste water is utilized in a three-lake recreation facility (26 acres of lake surface). Actually, only about 0.5 mgd of tertiary-treated effluent flow to the lake system to replace evaporation losses and the small amount of irrigation water which is removed in order to improve circulation in the lakes. Good local support for the project has developed because of the great need for recreational facilities and a well-executed public information program. The recreation area will have camping, fishing, picnicking, and boating. Water-contact activities are not envisioned.

Colorado Springs, which has undergone rapid population growth, is facing questions concerning the future adequacy of local water supplies. The present secondary sewage treatment plant can handle 13 mgd on the average and a small tertiary unit can treat 3 mgd. The tertiary treatment plant, which involves chemical coagulation and carbon filters to remove suspended and dissolved particles, pro-vides water to a power plant and for irrigation (McCauley, 1977). Approximately 5 mgd of effluent from the secondary treatment plant undergoes additional renova-tion by passing it through two pressure sand filters. This water is then pumped to storage reservoirs and made available to large nonpotable users. As a result, almost half of the water supplied to Colorado Springs is reused at least once.

Renovation and reuse of water are increasingly evident in water-short areas. To prevent shortages in even humid areas with large population concentrations, a form of nonpotable reuse may be necessary. As treatment systems improve and we become more certain of our ability to remove all viruses and other harm-ful materials from waste water, it is clear that direct reuse of reclaimed water will grow in importance and begin to contribute significantly to our overall water resources picture.

### Proposed Direct Potable Water Reuse

Previous discussions of water reuse have concentrated on cities in which actual reuse was or is being attempted. It can also be instructive to look at one large urban area in which direct potable reuse of sewage effluent is being planned on a large scale.

Denver, Colorado, located in a semiarid region with considerable agricultural, mining, industrial, and domestic demand for water, has been operating at the very limit of water supplies for some time. Rapid population growth in recent years has strained the city's water supplies even further. The Denver Water Department is operated as an independent water company supported by its own revenues rather than by tax funds. The company serves about 1 million people. In 1974, it pumped about 72 million gal of water. The per capita consumption was about 220 gal/day, one of the higher values in the United States.

The Water Department must develop additional supplies. This does not necessarily suggest bringing in more fresh water from distant basins or sinking additional groundwater wells (such sources are really not available). Other sources that have been considered include weather modification, vegetation control or watershed management, reduction of canal seepage or evaporation, and even reclamation and reuse of sewage effluent.

The Water Department has been considering possible reuse for the past decade and has concluded that (a) reuse for agriculture is not likely because of problems with various water rights decrees; (b) reuse for groundwater recharge is limited by the geology of the area and problems of ownership and control of such groundwater; and (c) reuse for industrial purposes is not feasible economically, in part because of the need to install a dual-piping system to suburban areas. Potable reuse appears to be the only practical way to supply additional water for the city (Heaton, 1978).

Direct and indirect reuse is possible. Direct reuse involves transferring the sewage plant effluent directly to treatment facilities and from there to the city supply system. Indirect reuse involves disposing of the sewage plant effluent in a local river to permit some natural purification to occur before removing the water at a downstream treatment plant. Indirect reuse is more pleasing to the public but it has resulted in several serious objections in the Denver area. Summer flow in the river is so small that the river would not provide any dilution—the stream would consist almost entirely of sewage effluent. Other contaminants in the stream from upstream areas might actually result in a streamflow of lower quality than the sewage effluent itself. Thus, the Denver Water Department has opted for direct reuse of sewage effluent as part of the potable water supply of the city (Heaton, 1978). Although this appears to be the most satisfactory solution to the problem of reuse in the Denver area, it has been dictated by the special constraints that exist in that area. It may not be the most practical solution in other areas.

The Water Department sampled public attitudes toward direct potable reuse and found a willingness to consume reused water. The public made it clear that

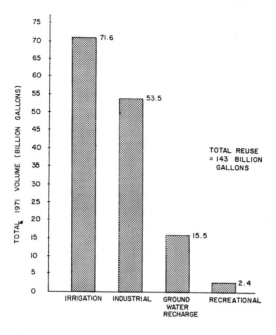

**Figure 8.1**  Relative reuse volumes in the United States.
*Source.* Schmidt, Beardsley and Clements (1973), with permission of the American Institute of Chemical Engineers and EPA.

the water must be equivalent in quality in every respect to the best available natural water source.

Plans are now under way for a small 1-mgd treatment plant at a cost of approximately $8 million. Water from this plant will be used primarily to study health effects. If this water proves to be entirely safe for human consumption, the city will undertake the next step of building a plant to reclaim up to 100 mgd of effluent. Testing of the treated effluent will be a most complex and difficult task because of the lack of consistent, acceptable limits for all chemicals and organisms. Those individuals who are against potable reuse feel that it is impossible to isolate and understand every potentially deleterious compound in the effluent. Since individuals who now use water from polluted rivers are already drinking water contaminated by such sewage effluents, it is still imperative that we obtain sound answers to the question of what constitutes a safe water quality (Heaton, 1978).

**Industrial Use of Treated Waste Water**

Use of treated sewage effluent is growing. A survey in 1971 showed that 144 municipalities in the United States supplied reclaimed waste water for reuse. A total of 143 billion gal of treated sewage effluent were utilized: 71.6 billion gal

**Table 8.4**  Industrial Reuse Operations in the United States

| Location | Producer | User | Purpose |
|---|---|---|---|
| Bagdad, Arizona | Bagdad Copper Corp. | Same | Process |
| Morenci, Arizona | Phelps Dodge Corp. | Same | Process |
| Burbank, California | City of Burbank | City Power Generation Station | Cooling |
| Colorado Springs, Colorado | City of Colorado Springs | City Electric Division Martin Drake Plant | Cooling |
| Baltimore, Maryland | City of Baltimore | Bethlehem Steel Corp. | Cooling and process |
| Midland, Michigan | City of Midland | Dow Chemical Co. | Cooling |
| Las Vegas, Nevada | City of Las Vegas | Nevada Power Co. | Cooling |
| Law Vegas, Nevada | Clark County Sanitation District | Nevada Power Co. | Cooling |
| Enid, Oklahoma | City of Enid | Champlin Refinery | Cooling |
| Amarillo, Texas | City of Amarillo | Southwestern Public Service Company | Cooling |
|  |  | Texaco, Inc. | Cooling |
| Big Spring, Texas | City of Big Spring | Cosden Oil & Chemical Co. | Boiler feed |
| Denton, Texas | City of Denton | Municipal Steam Electric Plant | Cooling |
| Lubbock, Texas | City of Lubbock | Southwestern Public Service Co. | Boiler feed and cooling |
| Odessa, Texas | City of Odessa | El Paso Products Co. | Boiler feed and cooling |

*Source.* Schmidt et al. (1973) with permission of the American Institute of Chemical Engineers and EPA.

(50%) were used for irrigation, 53.5 billion gal (37%) for industrial use, 15.5 billion gal (11%) for groundwater recharge, and 2.4 billion gal (2%) for recreational purposes (Fig. 8.1). The water volume for industrial reuse is quite high, but 44 billion gal of the total 53.5 billion gal are used each year by just one customer, the Bethlehem Steel Company plant in Sparrows Point, Maryland.

Only 15 industrial plants made use of some municipal waste water in 1973 (Table 8.4). Of these, 3 were city-owned power plants and 12 were privately-owned

**Table 8.5**  Major Industry Classifications Using Municipal Waste Water

| Industry | Number of Plants | Percent of Total Volume Reused |
|---|---|---|
| Basic metal manufacturing | 1 | 74 |
| Power generation | 7 | 20 |
| Petrochemical | 5 | 5 |
| Mining and ore processing | 2 | 1 |

*Source.* Schmidt et al. (1973), with permission of the American Institute of Chemical Engineers and EPA.

companies. The breakdown by industry of the total volume of 53.5 billion gal used by the 15 industries is given in Table 8.5. Two-thirds of the water is used for cooling purposes, with the rest equally divided between boiler feed make-up water and in manufacturing processes (Table 8.6).

The use of reclaimed water for cooling purposes requires special treatment of the water to prevent condenser tube fouling and other problems resulting from the composition of the water supplied. Water of excellent quality can normally be used with only an increase in chlorine and the addition of some acid and corrosion inhibitors. For most renovated water, clarification is also necessary to remove suspended solids and organic material before use.

The three plants using renovated water for boiler feed make-up water also treat the water substantially before reuse, the type of treatment depending on the operation of the boiler. With a low pressure boiler, the water must be clarified, softened, and lowered in phosphates. Use in higher pressure boilers requires that the water must also have the dissolved solids removed or undergo deionization.

**Table 8.6**  Type of Industrial Reuse in the United States

| Type of Use | Number of Plants | % of Total | Reuse Volume (mgd) |
|---|---|---|---|
| Boiler feed | 3 | 17 | 1 |
| Process | 3 | 17 | 1 |
| Cooling | 12 | 66 | 154 |

*Source.* Schmidt et al. (1973), with permission of the American Institute of Chemical Engineers and EPA.

So far, the use of reclaimed water in processing has been confined to mining and steel making. Each has a particular type of water that is needed but none of the processes require potable-quality water. Bethlehem Steel, the largest user of treated effluent, uses most of its water for cooling purposes. Other uses include gas cleaning, process temperature control, de-scaling systems, fire protection, air conditioning, mill hydraulic systems, and road equipment washing (Schmidt et al., 1973).

Cost is the primary factor that makes industry consider use of reclaimed water. Where fresh water of good quality is available at reasonable cost, there is little incentive to consider renovated water. However, where large volumes of water of nonpotable quality can be used and supplies are limited so that fresh water will be expensive, use of treated sewage effluent becomes a viable alternative.

Two costs must be considered in the case of treated sewage. First, the payment to the municipality for the water supply as well as the cost of any pipes or other facilities needed to deliver the effluent to the using plant. Second, it is necessary to consider the cost of any in-plant treatment necessary to develop the quality of water needed for the use contemplated. The cost of procuring reclaimed water is almost always less than for fresh water but the cost of treatment is usually more. Table 8.7 sums the information available in 1973 on costs for obtaining and treating sewage effluent for those plants willing to report such information. The range in costs is wide, from essentially zero up to $821 per million gal. The cost to treat water in-house contributes the largest part of the total cost of using treated water. An intangible benefit from using reclaimed water is, of course, its public relations value for most citizens would approve of industries that are not depleting resources that are in high public demand. It is not possible to put a dollar value on this goodwill, however.

**Human Considerations**

At present, renovated waste water is being used for irrigation (both on lawns and on certain crops); in a few, rather restricted municipal situations, it is being used for industry, aquifer recharge, recreation, and some domestic supplies. Such uses are more concentrated in the arid and semiarid portions of the country and are most widespread for irrigation and industrial cooling. Irrigation use might be considered the least efficient use of renovated water in terms of economic returns for volume of water used but, of course, it requires lower quality water. Most irrigation uses involve nonedible crops, such as cotton or fodder crops and certain grains (Law, 1968). Johnson (1971) does not believe that agricultural uses of treated water will increase significantly in the future although he feels that direct piping for lawn or golf course irrigation may indeed increase.

Freshwater supplies from groundwater, nearby rivers, and more distant reservoirs and lakes, as well as desalination of brackish water or sea water or the use of renovated water are the usual alternatives for municipal use. Clearly, if nearby freshwater supplies are available, they will be most economical and they will be utilized first for additional municipal supplies. However, in most parts of the

**Table 8.7**  Industrial User Costs for Reclaimed Water

| User | Cost to Procure Effluent ($/mg) | User Treatment Cost ($/mg) | Total Effluent Cost ($/mg) |
|---|---|---|---|
| Bagdad Copper Corp. Bagdad, Arizona | 0 | 0 | 0 |
| Phelps Dodge Corp. Morenci, Arizona | 0 | 0 | 0 |
| City of Burbank, California | 43 | 100 | 143 |
| City of Colorado Springs, Colorado | 351 | – | – |
| Bethlehem Steel Corp. Baltimore, Maryland | 1.33(avg) | N/A | N/A |
| Dow Chemical Co. Midland, Michigan | 3.33(avg) | N/A | N/A |
| Nevada Power Co. Las Vegas, Nevada | 25 | 200 | 225 |
| Champlin Refinery Enid, Oklahoma | 7 | N/A | N/A |
| Southwestern Public Service Co., Amarillo, Texas | 80 | 175 | 255 |
| Texaco, Inc. Amarillo, Texas | 90 | N/A | N/A |
| Cosden Oil & Chemical Co. Big Spring, Texas | 79(avg) | 742 | 821 |
| City of Denton, Texas | 80 | 100 | 180 |
| Southwestern Public Service Co., Lubbock, Texas | 144 | 160 | 304 |
| El Paso Products Co. Odessa, Texas | 125 | 550 | 675 |

*Source.* Schmidt et al. (1973), with permission of the American Institute of Chemical Engineers and EPA.

country, adequate supplies of water from groundwater are no longer available in the quantities needed for municipal systems. Thus, the use of renovated water must compete with saline or brackish water, polluted river water, and the storage and transport of more distant freshwater supplies. The cost for renovated water may be no greater and, in many cases, it can be less than the cost of producing fresh water from these other sources. Desalination of water has been running about

$1.00 per 1000 gal although estimates abound that it will be reduced to 25¢ per 1000 gal in the future. Even so, it will be limited to coastal or near-coastal use and would not seem to be able to compete economically with the use of renovated water. Treatment of brackish water might compete economically if pumping and transportation costs are not too great and the supply is adequate. Treatment of polluted river water supplies involves about the same costs as treatment of municipal wastes. Finally, the use of distant reservoirs and lakes may be a marginal economic proposition depending on the distance of transport, the need for pumping, and the cost of reservoirs or storage facilities in the source region.

Renovated water competes favorably on a cost basis with other available water sources but it is viewed less favorably from a health and aesthetic viewpoint. The health risk would appear to be real, largely because of our present lack of knowledge about viruses and our inability to test for them effectively at all times. Chlorination will remove viruses from water supplies but its effectiveness varies with the chemical quality of the treated water. Thus, if water quality varies, one may well question whether chlorination will continually be effective in removing viruses.

We are not certain whether we can even detect the presence of a virus in water. Present processes of detection involve the concentration of the virus and its innoculation into tissue cultures to determine density. Often the presence of fecal coliform in the water is used as an indicator of the possible presence of virus. However, since some scientists feel that coliform bacteria may be less resistant to chlorination than virus, the absence of coliform bacteria may not be indicative of the absence of virus as well.

Finally, the level of virus in water supplies that can be accepted as safe is still unknown. Does a low level of virus act as a form of immunization or will such low level amounts result in virus outbreaks in less resistant or infirm individuals? Clearly, the nature of the health risks and the general lack of knowledge concerning the dangers will limit the widespread use of renovated water for potable water supplies for some time. It should not, however, argue against the selective use of such supplies for nonpotable uses to free limited freshwater supplies for higher quality uses and thus make existing supplies go further.

Studies of human attitudes toward use of renovated water differ somewhat from what might be expected on the basis of other studies of human perceptions. Willingness to accept renovated water and even to use it seems to be related to education level and public knowledge of the source and its treatment. Water managers, however, almost universally agree that the public will not accept renovated water. This might be expected in view of the fact that the manager is (a) charged with marketing a product, (b) wants to keep volume of use high, and (c) is able to pass increased costs along to the user. Where renovated water is viewed as the only alternative supply, managers do show some interest in considering it. As the level of knowledge increases, and the cost of various alternatives become more clearly understood, acceptance of renovated water for municipal uses seems to increase.

## INTERBASIN TRANSFERS OF WATER

Water management means providing the right amount of water, of the right quality, at the right time for the purposes needed. But distribution of water in nature often does not accord with the distribution of people or the distribution of water need. For example, irrigation for agriculture is always needed in arid areas where water is in limited supply while cities are not always located on the basis of adequate water supplies. The water supplies of a large land mass such as North America or of even a portion of it, such as California, may be fully adequate for the needs of the area in question yet they may be insufficient locally for areas within North America or California. One way to achieve additional supplies of water locally is by transfer of water from an area of excess to an area of deficit.

### Examples of Water Transfer Programs

(a) *California.* One of the earliest large-scale water transfer projects in the United States involved moving 150,000 acre-feet of water per year from the Owens Valley on the eastern side of the Sierra Nevada mountains to Los Angeles. The first proposal in 1873 suggested transfer of water from the Sacramento Valley to the San Joaquin Valley but it was not until 1913 that completion of the Los Angeles aqueduct finally brought water from Owens Valley to the city. The cost of $24.5 million covered the building of the aqueduct as well as the purchase of land and water rights in the Owens Valley. Several extensions of the aqueduct to the Mono Basin have increased the amount of water that can be obtained to 320,000 acre-feet in 1940 and again to 472,000 acre-feet in 1968. Recent decisions to pump additional water from the groundwater in the valley have brought a long-simmering dispute between valley residents and city planners to a head.

The Owens Valley situation is fairly unique since much of the land with the accompanying water rights already belongs to the city of Los Angeles. One-quarter of the business establishments in the four small towns found in the approximately 2000-mi$^2$ valley are leased from the city while most of the small farmers in the valley are tenants of Los Angeles. When the water rights were obtained in the early years of the 20th century, it was claimed that Los Angeles needed the water. However, much of the water taken from the valley after the completion of the Los Angeles aqueduct in 1913 was actually used to irrigate the San Fernando Valley.

Prior to 1970, Los Angeles obtained about 70% of its water requirements from Owens Valley. A second channel from the valley to Los Angeles was completed at that time and the city increased its take of water from the valley by 50%, reducing the amount of water purchased elsewhere from 30% to 4%. This reduction in the volume of water purchased elsewhere provided great savings to the city. It is estimated that for each 60 million gal pumped from the valley daily, the city will save approximately $7 million a year. However, the recently completed second channel cannot be filled by surface water from the valley and so the city of Los Angeles is applying for permission to pump large amounts of water from

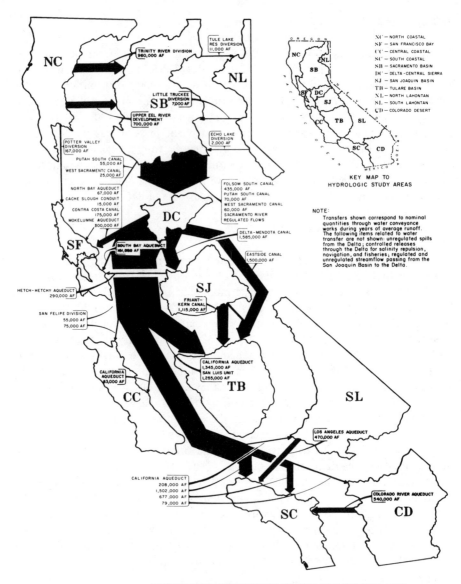

Within the figure:

NC — NORTH COASTAL
SF — SAN FRANCISCO BAY
CC — CENTRAL COASTAL
SC — SOUTH COASTAL
SB — SACRAMENTO BASIN
DC — DELTA-CENTRAL SIERRA
SJ — SAN JOAQUIN BASIN
TB — TULARE BASIN
NL — NORTH LAHONTAN
SL — SOUTH LAHONTAN
CD — COLORADO DESERT

KEY MAP TO
HYDROLOGIC STUDY AREAS

NOTE:
Transfers shown correspond to nominal quantities through water conveyance works during years of average runoff. The following items related to water transfer are not shown: unregulated spills from the Delta; controlled releases through the Delta for salinity repulsion, navigation, and fisheries; regulated and unregulated streamflow passing from the San Joaquin Basin to the Delta.

TRINITY RIVER DIVISION 960,000 AF
TULE LAKE RES DIVERSION 11,000 AF
LITTLE TRUCKEE DIVERSION 7,000 AF
UPPER EEL RIVER DEVELOPMENT 700,000 AF
ECHO LAKE DIVERSION 2,000 AF
POTTER VALLEY DIVERSION 167,000 AF
PUTAH SOUTH CANAL 55,000 AF
WEST SACRAMENTO CANAL 25,000 AF
NORTH BAY AQUEDUCT 67,000 AF
CACHE SLOUGH CONDUIT 15,000 AF
CONTRA COSTA CANAL 175,000 AF
MOKELUMNE AQUEDUCT 300,000 AF
FOLSOM SOUTH CANAL 435,000 AF
PUTAH SOUTH CANAL 70,000 AF
WEST SACRAMENTO CANAL 60,000 AF
SACRAMENTO RIVER REGULATED FLOWS
DELTA-MENDOTA CANAL 1,585,000 AF
SOUTH BAY AQUEDUCT 161,000 AF
EASTSIDE CANAL 1,500,000 AF
HETCH-HETCHY AQUEDUCT 290,000 AF
SAN FELIPE DIVISION 55,000 AF 75,000 AF
FRIANT-KERN CANAL 1,115,000 AF
CALIFORNIA AQUEDUCT SAN LUIS UNIT 1,345,000 AF 1,255,000 AF
CALIFORNIA AQUEDUCT 83,000 AF
CALIFORNIA AQUEDUCT 208,000 AF 1,502,000 AF 677,000 AF 79,000 AF
LOS ANGELES AQUEDUCT 470,000 AF
COLORADO RIVER AQUEDUCT 540,000 AF

PROJECTED INTRASTATE WATER TRANSFERS FOR
1990 LEVEL OF DEVELOPMENT

**Figure 8.2** Hydrologic study areas of California and projected intrastate water transfers between hydrologic areas of California for 1990 level of development: water from Northern California will constitute a significant supply for Southern California.
*Source.* California Department of Water Resources (1966).

the water table. The court has temporarily limited the city to 97 million gal/day from groundwater although the city is requesting the right to pump 204 million gal/day during dry periods of the year. All parties agree that such heavy pumping might exceed the safe yield of the groundwater and result in some "mining" of the groundwater reserves. Environmentalists have now entered the battle to try to protect the water table (and the overall water resources) of the valley. Economics figure greatly in the arguments since Los Angeles can obtain needed water elsewhere but it is cheaper to obtain increased volumes from Owens Valley. The court must decide whether the possible degradation of the water resource environment of the valley is worth the savings in water costs for the people of Los Angeles.

Other large-scale water transfer programs are also under way in California (Fig. 8.2). For example, a 240-mi aqueduct was built in the 1920s and 1930s to bring Colorado River water to cities in Los Angeles, Orange, and San Bernardino counties. Later expansion of the aqueduct, which includes some 92 mi of tunnels and 63 mi of lined canals, has provided a capacity of 1.2 million acre-feet per year (Nadeau, 1950). Some 30% of that water has been used in recent years for groundwater recharge to prevent salt water encroachment due to previous overpumping of coastal wells.

The Hetch Hetchy Water Supply and Power System carries water from the Yosemite National Park across the San Joaquin Valley to San Francisco while the so-called "Statewide Water Plan" of 1931, involving the Shasta and Keswick Dams on the Sacramento River and the Friant Dam in the Sierra Nevadas as well as the Contra Costa and Delta-Mendota canals, supplies municipal and industrial customers in Contra Costa County and agricultural interests in the San Joaquin Valley (Howe and Easter, 1971).

The largest of the California water projects, known as the Feather River project, involves the transfer of water from a large storage facility on the Feather River in northern California southward to the Los Angeles area. The system will be able to deliver 4.2 million acre-feet of water annually of which nearly half will go to the Metropolitan Water District of Los Angeles, Orange, and San Bernardino counties. The 1961 estimated cost of the project was $2.5 billion although it is certain that such a figure will be exceeded with rising construction costs.

(b) *Colorado.* More than 50% of the water supply of the city of Denver comes via interbasin transfers from watersheds on the western slopes of the Rocky Mountains. Two other Colorado projects, the Colorado–Big Thompson and the Frypan–Arkansas, were initially conceived to bring water primarily for irrigation purposes, the former involving 230,000 acre-feet per year and the latter 75,000 acre-feet per year. The presence of the water supply has, of course, led to the increase in municipal and industrial demands as well so that not all of the water resulting from these transfers is used as originally planned (Howe and Easter, 1971).

(c) *New York.* Not all large water transfer schemes are confined to the dry portions of the country. New York City has long obtained most of its water supplies by means of transfers from upland reservoirs and, finally, from the Delaware River. While the early reservoirs (begun in 1842) were in the Hudson River drainage and so technically they did not involve interbasin transfer, a Supreme Court decision in the early 1930s allowed the city to tap Delaware River water for additional supplies over the objection of the State of New Jersey. One million acre-feet of water annually are brought from the Delaware to New York City, somewhat over 50% of its total water supply.

## Problems in Interbasin Transfers

Thus, interbasin transfers of water have been practiced in various parts of the country for long periods of time. Cities, industries, and agriculture have all expanded because of the presence of this imported water. Yet there are many questions and problems involved in the whole matter of interbasin transfers which must be fully explored before further widespread use of the technique can be recommended. For example, if water is made available in an area by means of interbasin transfer, not only does it diminish the amount of water for development in the area from which the water is removed but it creates a growing demand for water in the area to which it is delivered. The history of interbasin transfers shows that this increased demand for water will almost always result in further need to increase the amount of water transfers in the future and, of course, it will lock the local, state, or Federal government into a moral obligation to continue to supply water to meet the needs so developed. Should these needs have been met in the first place or should the water users have been encouraged to move to areas where adequate water is available without transfers? Such questions cannot really be answered after the fact. Los Angeles already exists and it cannot be moved. Shortage of local water will not limit future growth as long as the local municipal users are willing to pay the increased costs of interbasin transfers. Certainly those costs are less than the costs of relocating large portions of the city with its services and industries elsewhere. But the question of the moral obligation to maintain adequate water supplies once they have been provided still remains. Interbasin transfer of water as a technique to provide additional supplies of water to an area of need is more insidious than some of the other techniques to increase water supplies. Interbasin transfers occur under strict engineering control. Once the storage and transport systems are in place, the quantity of water that can be delivered is fixed. Plans can be made to utilize this water and municipal and industrial developments will occur. With cloud-seeding, vegetation-control, or water-harvesting techniques, one may be able to make additional water supplies available but the quantities are often much smaller and they are certainly more variable over time. They cannot be counted on to produce X acre-feet of additional water each year and so expansion plans cannot be firmly based upon this additional water. Rather, those techniques promise a variable but hopefully larger supply. There is little moral obligation to continue with the water-producing technique inasmuch as firm annual increases in water volume were never contemplated at the beginning. Interbasin transfers of water must be treated quite differently from the other possible techniques for making additional water for they create an obligation to continue the practice regardless of cost once they have been started.

### Future Interbasin Transfers

Many water transfer plans are being planned for the future. Warnick (1969) has summed up information on a number of the plans formulated during the period 1950-1968 (Table 8.8). While most of these involve transfers of water in the arid and semiarid western portion of the country, they do give an indication of the range of ideas being explored to try to make additional supplies of water available to areas of great need.

No discussion of interbasin transfers would be complete without a word about one of the most grandiose of all transfer schemes—the so-called North American Water and Power Alliance (NAWAPA) suggested by the Ralph M. Parsons Company. The figures for the project are just staggering. In essence, it involves the transfer of surplus water from Alaska and northwestern Canada into other regions of Canada, into 33 states in the United States, into the Great Lakes, and into 3 northern states in Mexico (Fig. 8.3). To do so would involve:

(a) A dam on the Copper River of Alaska 1700 ft high—or 450 ft taller than the Empire State Building;

(b) Storage of water in a number of tremendous reservoirs, 15 of which would be larger than the largest North American reservoir at Lake Mead behind Hoover Dam;

(c) One storage area—the Rocky Mountain trench covering parts of Idaho and Alberta would have a capacity of over 518 million acre-feet (16 times greater than Lake Mead);

(d) Construction in a series of stages up to 20 years;

(e) Tranferring 110 million acre-feet annually with provisions to increase the amount to 250 million acre-feet annually;

(f) Providing 70 million kw of power (slightly more than the combined installed hydroelectric capacity of the United States and Canada in 1965);

(g) A cost of over $100 billion initially (Ralph M. Parsons Company, 1964).

In addition to providing a tremendous amount of power and irrigation water (enough to irrigate 56 million acres in the United States and Mexico), the plan (Fig. 8.3) would provide water to stabilize the water levels in the Great Lakes and increase power production at Niagara Falls, and would provide water for municipal purposes in southwestern United States. The Parsons Company suggests that the system would pay for itself in 50 years as a result of the sale of water and electric power.

While the plan is innovative and deserves a great deal of serious study, it is clear that it will not have easy sailing. Approval may never be achieved because of the serious international, Federal-state, and interstate problems that it would create due to the political and economic problems involved and because of environmental arguments that have already been voiced. The plan would flood thousands of miles of valleys, some of the most wild and scenic on this continent. It would cover rich farmland and valuable mineral deposits and the massive transfers of water might alter ecological systems to the detriment of many endangered species.

**Table 8.8** Summary of Information on Plans Proposed for Regional Water Transfer

| Code No. | Project Name | Agency Sponsor Company Sponsor Author of Plan | Approximate Date of Proposal | River Basin(s) for Source | River Basin(s) of Use | Countries Involved | States Involved |
|---|---|---|---|---|---|---|---|
| 1 | United Western | U.S. Bureau of Reclamation Rep. R. J. Welch– Calif. | 1950 | Columbia River North Pacific Coastal streams | Great Basin South Pacific Coastal Plain Colorado River | U.S. Mexico | 11 Western States |
| 2 | California Water Plan | California Depart. of Water Resources | 1957 | Northern California rivers | Central Valley California South Pacific Coastal Plain | U.S. | California |
| 3 | Pacific South-west Water Plan | U.S. Bureau of Reclamation W. I. Palmer | 1963 | Northern California streams Colorado River | Lower Colorado River South Pacific Coastal Plain | U.S. Mexico | California Arizona, Nevada Utah, New Mexico |
| 4 | Snake-Colorado Project | Los Angeles Dept. of Water & Power S. B. Nelson | 1963 | Snake River | Colorado River South Pacific Coastal Plain | U.S. Mexico | Idaho, Nevada Arizona California |
| 5 | North American Water & Power Alliance (NAWAPA) | Ralph M. Parsons Co. | 1964 | Alaskan & Canadian rivers, with Columbia River | Great Lakes Basin South Pacific Coastal Plain Colorado River Texas High Plains | U.S. Canada Mexico | Western states Texas Lake states |
| 6 | Yellowstone-Snake-Green Project | T. M. Stetson, Consulting Engineer | 1964 | Yellowstone River Snake River | Green River Colorado River | U.S. | Montana, Idaho Wyoming, Lower Colorado states |
| 7 | Pirkey's Plan Western Water Project | F. Z. Pirkey, Consulting Engineer | 1964 | Columbia River | Colorado River Sacramento River South Pacific Coastal Plain | U.S. Mexico | Oregon Washington California Utah, Arizona Nevada |
| 8 | Dunn Plan Modified Snake-Colorado Project | W. G. Dunn, Consulting Engineer | 1965 | Snake and Columbia rivers | Great Basin Snake River South Pacific Coastal Plain Colorado River | U.S. Mexico | Idaho, Oregon Washington Utah, Arizona Nevada California |

| No. | Project | Agency / Engineer | Year | Source | Service Area | Country | States |
|---|---|---|---|---|---|---|---|
| | Project | L. F. Miller, Consulting Engineer, Maryland | 1965 | Columbia River | Oregon valleys, Central Valley, California, South Pacific Coastal Plain | U.S. | Oregon, Nevada California |
| 10 | Undersea Aqueduct System | National Engineering Science Co. F. C. Lee | 1965 | North Coast Pacific rivers | Central Valley South Pacific Coastal Plain | U.S. | Oregon California |
| 11 | Southwest Idaho Development Project | U.S. Bureau of Reclamation, Region 1 | 1966 | Payette River Weiser River Bruneau River | Snake River | U.S. | Idaho |
| 12 | Canadian Water Export | E. Kuiper | 1966 | Several Canadian rivers | Western states (indefinite) | U.S. Canada | All western states |
| 13 | Central Arizona Project | U.S. Bureau of Reclamation | 1948, 1967 | Lower Colorado River Basin | Colorado River | U.S. Mexico | Utah, Nevada Arizona California |
| 14 | Central North American Water Project C3 NAWP | E. R. Tinney Washington State University, Professor | 1967 | Canadian rivers | Great Lakes Entire western states | U.S. Canada Mexico | Great Lakes western states |
| 15 | Smith Plan | L. G. Smith Consulting Engineer | 1967 | Liard River McKenzie River | All river basins of 17 western states | U.S. Canada Mexico | 17 western states |
| 16 | Grand Canal Concept | T. W. Kierens, Sudbury, Ontario | 1965 | Great Lakes and St. Lawrence River | Canadian river flowing to Hudson Bay | U.S. Canada | Great Lake states |
| 17 | Beck Plan | R. W. Beck Associates | 1967 | Missouri River | Texas High Plains | U.S. | South Dakota Nebraska Kansas, Colorado Oklahoma, Texas |
| 18 | West Texas and Eastern New Mexico Import Project | U.S. Bureau of Reclamation & U.S. Corps of Engineers | 1967 (1972 due) | Mississippi and Texas rivers | High Plain of Texas and New Mexico | U.S. | Oklahoma, Texas New Mexico Louisiana |
| 19 | Pacific-Mead Aqueduct Augmentation by Desalinization | U.S. Bureau of Reclamation | 1968 | Pacific Ocean | Colorado River | U.S. Mexico | California Arizona |
| 20 | Yukon-Taiya Project | Alaska Power Administration | 1968 | Yukon River | Taiya River | U.S. Canada | Alaska |

*Source.* Warnick (1969, pp. 345, 346). Copyright, 1969, AAAS and University of Arizona Press.

**Figure 8.3** Map of proposed North American Water and Power Alliance Water System. *Source.* Ralph M. Parsons Company (1982), with permission.

The fact that the scheme is feasible is a witness to man's technological ability. To build the system just because we have the technological skill might not be practical because, once constructed, vast amounts of labor and money would be necessary to keep the system operating. There is just no way it could be stopped without disrupting a large portion of our national economy. We would be trapped into a form of hydraulic civilization (Wittfogel, 1956) with possibly some of the same consequences of the former great hydraulic civilizations of the past (for example, Egypt, India, China, Iran).

## COST OF OBTAINING ADDITIONAL WATER

The foregoing three chapters have attempted to consider various ways in which more water can be made available for various national needs. It is clear that a variety of techniques are available and that they all have a role. Some are more effective in one climate than another, others are site specific (desalination) and cannot be considered where saline water is unavailable, and still others will work well if water quality is less important than water quantity. Economics plays an important role at all times in determining which technique might be most feasible, but costs are determined in part by the use to which the water is to be put, the area of the country, competing supplies, labor costs, and many other factors.

The analysis of benefits from additional water is extremely difficult because of the various uses to which water is put, and because the alternatives to the cost of producing more water are often not fully analyzed and understood. While the cost of adding X more acre-feet of water to a municipal supply might be possible to determine, the benefits derived from that much additional water are more difficult to ascertain since the water is used for a wide variety of purposes, each having a different value to different individuals. What is the benefit from a green lawn as opposed to a brown and withered one? If the community decision is to do nothing and bear the shortage of water, how can the economic loss be evaluated? If such a decision results in a change of water-use habits in a community and, therefore, in a more efficient use of water, a considerable benefit may result from no additional water. How can this unknown be offset against suggested benefits from having more water? At the present moment, costs of adding more water by the various techniques suggested in the previous chapters are about all that can be compared.

Primary concern has been for the addition of more water both for agriculture and municipal use in the arid and semiarid parts of the United States. The possibility of rainmaking, desalination, vegetation management, water harvesting, reduction in evaporation and seepage losses, waste water reclamation, as well as interbasin transfers have all been suggested. Improved water laws which can also result in a reduction in waste of water will be discussed in Chapter 9.

Cost figures for each of the various techniques of water conservation, water production, or water transfer are only approximate and depend on a wide variety of factors, but Howe and Easter (1971) have, at least, provided some suggested ball-park figures for comparison purposes. They conclude that, except for possible "crisis" situations, the cost of interbasin transfers would probably not equal the benefits gained nationally. They suggest that the cost of water to be made available from the major interbasin transfer schemes seems to run about $50 to $60 per acre-feet (15¢ per 1000 gal)—figures highly dependent on the cost of electrical power for pumping purposes.

In contrast to these costs, Howe and Easter suggest the simple technique of reducing water loss in the conveyance of irrigation water in canals could save as much as 9 million acre-feet a year at a cost ranging from $2 to $50 per acre-foot. They suggest the further development of surface sources (more reservoirs or impoundments), especially in the western Gulf and central Pacific areas, and estimate the cost of such work might be $3 to $41 per acre-foot of water provided. Though the water impounded might also require interbasin transfer to deliver it to the area in which it is needed, these transfers would be more local in nature and could compete quite favorably with the large-scale interbasin transfer schemes discussed above. Howe and Easter (1971) recognize that the use of impoundments involves certain loss of value in the land and resources being covered by the impoundment and they warn that the cost figures may not properly reflect the real loss in value caused by the impoundment. The same argument would apply, of course, in the case of the cost of the large-scale interbasin transfers given previously.

Desalination of water, while limited to coastal regions or areas of brackish water, so far has not been able to compete cost-wise with most other schemes for supplying additional water. At costs of $1.00 per 1000 gal ($325 per acre-foot), such water can only provide a feasible source for domestic or high-value industrial uses and then only in a situation where no other source is available. Even if the promised figure of 25¢ per 1000 gal is achieved ($80 per acre-foot), desalination techniques will still not be able to provide water for widespread agricultural purposes.

Renovation of waste water appears to be a cost-effective scheme, especially if potable water is not needed. Howe and Easter (1971) show how Los Angeles is currently producing 16 million gal/day of potable reclaimed water that is used as groundwater recharge for a cost of $18 per acre-foot.

The cost of weather modification—in this case, rainmaking—is very difficult to estimate since there is little statistical evidence whether the technique can actually produce more rain. Cost of the use of land-based silver iodide generators is quite small—farmers may be asked to pay 50¢ to $1.00 per acre of farmland where such generators are in use but, of course, their use may not produce 1 in. more rainfall than would have fallen without their use. We can obtain some suggestion of cost, however, if we use the figure of $1.00 per acre of farmland and the rainmakers' own claims of a 10 to 15% increase in rainfall. Assuming a humid area such as Delaware, where the June–July–August rainfall may total 13 to 14 in., a 15% increase in rainfall will amount to about 2 in. more during that 3-month period. This amounts to $1.00 for 2 acre-inches or $6 per acre-foot for the possible cost, an extremely attractive price if there is any guarantee that the water can be produced regularly.

Techniques to increase water supplies are available. They all need to be studied in detail and, in some cases, even to be tried in different parts of the country. Certainly, no one technique is far superior to all of the others at the present moment since each involves both positive and negative aspects. We need to increase our research on each of the techniques to understand better how costs may be reduced, how the water can be made to go further without adversely affecting our living standard, and how we may experience the smallest negative impact on our environment through plans to alter existing patterns of natural water supply. Such studies must consider water as a national resource rather than just as a local resource.

## PART D:
## MANAGEMENT OF WATER:
## LEGAL, POLITICAL, ECONOMIC

The world is not on the verge of running out of water. Because of the distribution of humans, their industry and their agriculture, we do have and will continue to have places where water is quite limited and additional supplies are needed. Techniques for making additional supplies available have been discussed in the previous section.

Probably more important in the overall problem of wise use of our water resources is the management of water—having the right quantity and quality of water present at the time and place of its need. Wise management of water will also lead to conservation and to restriction of wasteful misuse of water. Many of the actions we now take in the name of water management result in a waste of water. Some of our water laws, our interstate basin compacts, our economic policies with regard to water result in misuse or overuse of water. Thus, a review of the legal, economic, and political problems that have developed in relation to water resources is important if we are to achieve the wisest use of water supplies. Such an analysis may also suggest additional ways in which we can make our present supplies go further. The present section, therefore, should be thought of not only as a study of water management problems but also as an extension of the previous section in which we were seeking to achieve additional supplies of water.

## 9

# WATER LAWS

Legal control of water sources—water rights—has been one of the most compelling issues facing individuals as well as governments throughout history. Extensive and detailed water codes, water doctrines, water laws, or water policies have been developed; wars, skirmishes, as well as individual duels have been fought over control of water supplies; while nations have risen or fallen over the use and misuse of water sources. Water is as much a legal and political problem as it is an economic or physical problem.

Study of the legal aspects of water must include a consideration of existing water laws, how they developed, and how they influence the use and misuse of water. Understanding the development of our water laws may help to understand their current applications and why they may not be as rational under all situations as we would like them to be. But understanding basic water law, the rights of the individual to water, is only part of the problem since, in our modern society, most people live in towns or cities where their water needs are served by governmental agencies or by large private water companies. Thus, the important legal problems are solved not so much by the courts interpreting particular water laws as they might apply to two individuals but rather by legislatures and administrative agencies passing the necessary regulatory procedures and establishing the funding needed by large water management agencies to implement their operations. Thus, the householder in Cincinnati is not at all concerned with whether he has riparian rights or appropriative rights to the water in the Ohio River flowing a few blocks from his door, but he is concerned whether his local water department or water supply agency will be able to provide him with all the water he needs, in the right quantity, and whether, if he moves to the suburbs, the same type of service will still be available to him. Individual water rights have given way, in large measure, to collective or community water needs and the water doctrine or water laws, which used to resolve conflicts between individual users, are no longer as important or as applicable as the legislative-made rules governing the administration and operation of large public water management agencies.

## BRIEF HISTORY OF U.S. WATER LAWS

Most writers tacitly assume that the water laws of the United States developed from the English common law and the Code of Napoleon of 1804 (Thomas, 1969; Dewsnup, Jensen and Swenson, 1973). When those laws were taken from the humid part of the country to the dry and water-short regions of the West, they could not be applied as effectively. Thus, it is possible that a mixture of Spanish, Moslem, Mormon, and practical miner law led to the development of the water law that is now so common throughout the western United States. Certainly, looking at the similarities in the water laws, such a history could be surmised, although there are those who still question the actual origins of our various water laws. Some scholars suggest that while we did adopt English common law in the solution of other legal problems, English water law was unclear and far from straightforward. There is now even a suggestion that the riparian doctrine was first formulated in this country and then later taken to England and adopted into the English common law (Ellis, DeBraal and Koepke, undated). The first real exposition of the riparian doctrine as we know it occurred in 1827 in the decision of Mr. Justice Story in *Tyler v. Wilkinson*. Story seems to have drawn on English, French, and American sources in reaching his understanding of the riparian doctrine.

Regardless of its exact origin, it is clear that the water law of the eastern United States—known as the riparian doctrine—was fairly well adapted to the humid climates of France and England as well as the eastern portion of this country. Since the body of the doctrine really developed from judge-made common law decisions over a large number of cases, there is no reason to expect that English and American interpretation would be exactly similar.

The real break with the old riparian doctrine came in 1849 as thousands of miners poured into California seeking fortunes in gold. While they may have taken their ideas of the riparian doctrine with them, they were concerned with an entirely different water situation in California. They needed water to operate their sluice boxes but they did not own the land along the rivers. They were trespassers or squatters. In the East, they would have been entitled to no water since they were not landowners but, in California, they took the water anyway. The first man on the stream took what he needed and later individuals took what was left. Often the actual decisions were punctuated by rifles and shotguns, and justice was swift and abrupt, but it did not take long for a body of legal precedents to be developed quite different from the doctrine of the humid East. And so the doctrine of prior appropriation developed—partly modified by Spanish laws (and they, in turn, by Moslem precedents), since Spanish influence was strong throughout this portion of the country.

## THE RIPARIAN DOCTRINE

The riparian doctrine, which evolved in the early 1800s in the humid, eastern United States, is a set of principles governing private rights and responsibilities toward water. To understand the doctrine, it is necessary to remember that it developed in a geographical area where water was abundant for human needs, at a time when those needs were limited mostly to domestic and transportation uses, and cities were few. Irrigation and mining needs were generally unknown and the use of water in manufacturing was mainly to turn a millwheel for grinding or a waterwheel for power. The doctrine developed slowly on a case-by-case basis as judges made their decisions to settle disagreements between neighbors or individual landowners in the same watershed. Because of the multiplicity of decisions handed down, some in conflict with other earlier decisions, the riparian doctrine is often not clear-cut or easy to interpret, but the basic theory or theories of the doctrine are straightforward.

### Two Interpretations of the Riparian Doctrine

(a) *Continuous flow theory*: Under this theory, each landowner with flowing water running through or along one border of his property has the right to have that water remain essentially in its natural state, undiminished in either quality or quantity. Under this aspect of the doctrine, the riparian (the landowner on the stream or river) may use the water—but only on riparian lands—as long as his use does not affect in any material way the water available for riparians further down the watercourse. Lower riparians do not have to show any actual damage in order to stop upper riparians from a use that affects the flow of water in some material way. Use of water without some depletion or change in quality seems contradictory. Certainly the rule cannot be applied strictly in our present society with the population pressures we now experience and with the great agricultural and industrial demands for water that exist even in humid parts of the country.

(b) *Reasonable use theory*: This second theory, which may have developed later than the continuous flow theory, attempts to modify some of the contradictory aspects of that theory. It states that each riparian may put the water to use (on riparian lands) so long as the use does not interfere with the reasonable use of lower riparians. Note the change of emphasis on the effect the use has on lower riparians rather than on the streamflow itself. The lower riparian must, therefore, show damage to himself through the actions of upper riparians before a court case is feasible. This theory considerably liberalized the riparian doctrine used at the time in England.

### Ownership of Stream

The English system assumed that navigable waters and the stream beds under them belonged to the Crown rather than to the landowners on the banks of the rivers and that these waters were held by the Crown for the enjoyment and use of the public. Nonnavigable waters, however, were private and owned by the bordering landowners. Riparian land was the land bordering on the stream or watercourse as well as the land on which the stream was flowing. Each riparian owned the bed of the stream and, since the public could not gain access to the water (the stream was nonnavigable and thus to get to the stream, the public would have to trespass across private land), the riparian had all rights to the water. English law defined as navigable only those portions of streams that were affected by the ocean tides. All other parts of the stream were nonnavigable. This interpretation was rejected in the United States in 1851 when the Supreme Court decided that, since we had great rivers and lakes well above tidal influence which could be used for interstate and foreign commerce, a broader definition of navigability was needed. Thus, they defined any waters susceptible to navigation in interstate or foreign commerce as navigable. The definition was modified 20 years later (1871) when the Supreme Court said that waters that were susceptible to being used in their ordinary condition for navigation were navigable. They added that streams that were used for interstate or foreign commerce were navigable waters of the United States and thus under the control of the Federal government, while those waters that were only navigable within states were navigable waters of the states and not under the control of the Federal government. The states could make them public for the use of the people of the state or control them in other ways as they saw fit.

### Riparian Rights

Riparian rights to water arise as a result of ownership of land bordering a stream or water body and not because of any need or beneficial use of the water. There does not need to be any mention of transferring water rights when riparian land is sold because riparian rights are automatically transferred with the title to the property. The riparian right is actually a right to make a reasonable use of water on riparian land. Such rights are not to ownership of the water itself but to the reasonable use of such water. The riparian owner is governed by reasonable use standards in both quantity and quality so as not to damage a lower riparian but otherwise he is free to use or not to use the water. Originally, his right did not depend on use and he could let the water flow to the sea without use with no loss of his continued right to that water.

With the growth of water demands and with the increasing requirements by nonriparians for water, it has become apparent that the pure riparian system is a wasteful one. Needed water might flow unused to the sea while both riparian and nonriparian owners had great need for it but could not use it—the riparians because their use might exceed the reasonable use theory and the nonriparians because they did not own the right to the water.

In more recent times, there has been a movement toward modification of the riparian doctrine in favor of a permit system administered by a state agency. Permit systems have some features in common with the water appropriation systems now in use in the western United States but there are quite a few differences as well. Many of the permits are issued for a fixed time period, subject to certain conditions and limitations, and are often revocable. The rules applied to the permits vary widely from state to state in the eastern part of the country. In some states, the permits still function more as a systematic information gathering program than as a way to regulate and allocate water use.

In certain states, use of the permit system has legalized the diversion and use of water on other than riparian lands. It also licenses and regulates well drilling, limits the use of water within certain areas, serves to protect lake levels and minimum flows in streams, and provides the mechanism for supplying large quantities of water to various industries and municipalities in need of it. Of the 31 eastern United States, only 5 (Florida, Indiana, Iowa, Minnesota, and New Jersey) have what might be called strong water regulations for both surface and groundwater supplies. As demands increase, other states can be expected to introduce stronger regulations and greater centralized control over water use.

More recently, statutes have been passed in riparian states to establish large multi-purpose water conservation and water supply districts to supply water to municipal and industrial users as well as to satisfy recreation, wildlife, and various social and aesthetic needs. These districts take different forms and have different powers—some even the power to tax—but they usually have considerable authority to develop water from the river basin for human or industrial use. If their activities harm lower riparians, they are often legally bound to compensate those harmed. The interesting aspect is that the statutes creating these districts have, in a sense, authorized a form of condemnation procedure within a riparian legal system. The public, without riparian rights, may obtain rights to water at the expense of lower riparians. Even though compensation for damages may be required, it provides a way for so-called public rights to be given preference over the private rights of the riparian landowners.

In the traditional riparian doctrine, all rights are co-equal. No advantage should be gained by the riparian who begins his use before another but, in actual practice, earlier users are often protected at the expense of later users. Since each decision is court-made, there can be considerable variation from place to place and over time. In some cases, protection is achieved by means of statutory statements that certain uses will not be limited by subsequent uses or the courts have ruled that prescriptive rights have developed as a result of prior use of certain amounts of water for the prescribed period of time. The courts have generally ruled that it is not reasonable for later uses to interfere with earlier uses of the water—a tendency toward the "first in time is first in right" system prevalent in the western United States.

Again, traditionally, the riparian doctrine provides that a riparian cannot transport water outside the basin of the stream since the water would be lost to the lower riparians in that basin. Also, if riparian land is subdivided so that one of the new parcels is cut off from access to the water, it will lose its riparian rights to water. States disagree whether such land or any other nonriparian land purchased by a riparian can obtain rights to the water. Some states indicate that riparian water rights will be extended to such land newly incorporated with the riparian land while other states say no.

Individual court decisions have ruled that it is a reasonable use to impound water temporarily and then to release it for milling and power purposes but it is unreasonable to detain the water during the daytime and to release it at night when lower riparians cannot make as much use of the flow. It is also unreasonable to release the water suddenly to cause downstream flooding or to cause the water to back up and so to interfere with upstream mill or power operations.

While riparian rights cannot be lost through the nonuse of the water, it is still possible for a riparian owner to lose a certain amount of his right to water as a result of his own actions or those of others. For example, land titles, and the rights to water that may go with those titles, can be lost if the land is "possessed by another, adversely, openly, notoriously, and continuously" (Dewsnup et al., 1973) for a prescribed period of time as established by state statute. The most commonly prescribed period is 7 years. These regulations differ, but generally if an individual takes possession of or uses, in certain prescribed ways, a parcel of riparian land for a long enough period of time without interference or objections from the true landowner, the landowner will lose title to the land and to the water right as well.

If an individual makes use of his own property so as to infringe on the rights of another, or makes certain uses of another's property so as to infringe on his rights, and this infringement persists for a long enough time, a form of easement or a prescriptive right is acquired by the infringer. The adverse act of infringement must continue for some period of time—20 years in many states but less than this in other states—and the action must go unchallenged by the landowner who is being infringed upon for the prescriptive right to develop.

If the riparian owner consents to or otherwise permits the infringement of another, either formally by legal document or informally by oral consent, then no prescriptive rights can develop. If, however, a riparian takes an unreasonable amount of water from a river and his action is unchallenged by lower riparians who are injured or potentially injured by this action for the prescribed period, the upper riparian has achieved a prescribed right to that water in many riparian states. Construction of a dam that overflows lands of an upper riparian, if unchallenged for the prescribed time period, will give the operator of the dam the right to continue even if it does bring damage to the upper riparian.

Finally, riparian water rights may be limited or lost entirely by virtue of the common law doctrine of estoppel. An example best illustrates the workings of this doctrine. If a riparian owner assists another riparian owner in making use of significant amounts of water from the stream—such as by providing impoundments or canal facilities or other easements through his property—he cannot later complain about unreasonable use of the water since he assisted and agreed to the undertaking in the beginning. To allow him to cut off use of the water after he had helped in the original planning for the unreasonable use would be to work a fraud on the user. As a result, the assisting riparian may have his own use of water reduced or limited on the basis of the unreasonable use by the other riparian.

Under the newer permit system, permit rights may be lost through revocation or cancellation by the issuing agency or through failure on the part of the permit holder to live up to the terms of the permit. Almost all states require permits before dams can be placed in watercourses in order to protect public safety. Small facilities entirely on private property are usually exempted. As reasonable impoundments are permitted under riparian rules, the permits function more for information and safety purposes.

Diffused surface water is defined as water spread out over the surface forming no part of a natural watercourse. While such overland runoff may ultimately find its way into a watercourse, until it does it is considered diffused water. Most states allow the landowner to make whatever use he wishes of diffused water. Generally, he has the right to capture whatever he can and to put it to whatever use he wishes. At the same time, most states allow landowners to take reasonable steps to drain such waters from their lands or to stop or divert such waters from coming across their lands from higher-lying parcels. The older English rule, that the landowner had every right to defend himself in whatever way he wished against diffused water even if it harmed his neighbor, has now given way to a law of reasonablenss which allows the landowner reasonable actions to prevent such waters on his property or to drain them from his property, but he must make every effort not to cause too much inconvenience to his neighbor.

The rather inflexible riparian doctrine has begun to give way to a more flexible doctrine of reasonable use. The development of the permit system that allows for giving and taking away of water rights after fixed time periods is a step toward a more equitable distribution of water, possibly even to nonriparians. It does create a problem about the permanence of water. In some instances, developments that require water might not be undertaken unless there is some guarantee that a certain quantity of water will be available for a long period of time. The permit system has to be flexible enough to give assurance that developments needing water will have the permanent supply of water that they require.

The change to a rule of reasonable use recognizes that streamflow will be changed in both quality and quantity through use. The more flexible doctrine provides for this type of use while always keeping open the opportunity for review of individual uses to see that they are reasonable. As water becomes more limited and more highly prized in the humid part of the country, it is clear that the riparian doctrine will undergo even further changes, bringing it closer to the more legislative-sanctioned appropriation doctrine developed in water-short areas.

## THE APPROPRIATION DOCTRINE

By the mid-1800s, greatest demands for water in California and other settled areas in the West were from miners and farmers. Often these individuals had to divert water from streams to a more distant place of use. Since the land was just being opened to settlement, there were few courts and even fewer who wished to interpret the finer points of English or American riparian law. Often local rules or customs were sufficient and, if a water dispute did reach a court of law, it was usual for the riparian doctrine to be rejected in preference to local judgments. Riparian doctrine was just not suited to a dry climate area and to occupations which required the use of fairly large volumes of water. Miners and farmers needed to use certain quantities of water wether they owned land on a stream or not. The first person in the area—often trespassing on Federal land—took the water he needed, and the courts usually decided he had the perpetual right to that water whether a riparian landowner wanted it or not later. The riparian doctrine was rejected in California in practice, although not legally, for more than 30 years, after the miners arrived and instituted their own rule of "first come, first served."

The custom of the miners to appropriate whatever water was needed when they came to a stream gave rise to the appropriation system of water rights. The first individual to put water to a beneficial use, whether on riparian land or not, was recognized as first in right, and this right was protected in terms of quantity throughout time as long as a beneficial use for the water could be shown. Later appropriators of water would receive what quantities they needed as long as enough water remained in the stream. During dry periods, if it became necessary to cut back on water use, the later appropriators would be cut off entirely before the earlier appropriators would suffer any loss of water. Thus, the doctrine of appropriation developed where the right was determined by whether a beneficial use was being made of the water—and the status of the right was determined by the time the first appropriation was made. The appropriation was specific about the quantity of water that could be taken and little variation in quantity was permitted although the place of use, the place of diversion, and the use itself could be varied as long as the use could still be shown to be beneficial.

Water in the stream was treated as having no owner so that the first person to make use of it in a beneficial way had the right to continue to use that water. The time of the right would be determined by the time the appropriator actually began the diversion work—constructing a canal or ditch or otherwise making known his intention to use the water.

### Different Applications of the Appropriation Concept

(a) *California doctrine*: While the appropriation doctrine has long been considered to be the water law of the western states, it should be pointed out that the California legislature in April 1850 declared that English common law, as long as it agreed with existing state and Federal constitutions, would be the rule of law in the state. This included the riparian doctrine. Some 30 years later (*Lux v. Haggin*, 1886), the California Supreme Court reaffirmed the fact that riparian rights existed in California.

Thus, California had a mixed bag of water laws which might be called the California doctrine to distinguish it from the doctrine of prior appropriation as practiced in many of the other western states. The California doctrine provided that, as of 1850, rights to the unappropriated waters on nonnavigable streams on public lands belonged to the Federal government. When the Federal government gave or sold the riparian lands to a group or individual, the water rights owned by the government went with the land transfer. California did agree, however, that when there was beneficial use of water on public lands, the appropriator acquired from the government an appropriation water right. Thus California recognized both appropriative rights and riparian rights—the latter developing through the sale or transfer of land from the Federal government to private ownership. States following the same general doctrine—Hawaii, Kansas, Nebraska, North Dakota, Oklahoma, Oregon, South Dakota, Texas, and Washington—while accepting a dual set of water laws, have also passed statutes which recognize appropriation as applied to irrigation. The use of the dual set of laws has resulted in legal battles over water in a number of these states. Many so-called California doctrine states have found it necessary to modify their dual set of laws restricting the riparian doctrine. In California itself a constitutional amendment in 1928 did away with any idea of continuous flow theory and limited riparian rights to that water needed for the fulfilling of the beneficial use. It has been suggested that the many complex water law problems resulting from the dual water law system in California may well explain why that state seems to have a greater preference for public water projects than private ones (Clark, 1972).

(b) *Colorado doctrine*: The miners in California may have first instituted the appropriation system of water rights but it was not until 1882 that the Supreme Court of Colorado clearly set forth the theoretical background for the appropriation doctrine. In *Coffin v. Left Hand Ditch Company* in that year, a lower riparian lost his argument against an upper appropriator who diverted river water into another watershed for irrigation purposes.

The so-called "Colorado doctrine" states—Arizona, Colorado, Idaho, Montana, Nevada, New Mexico, Utah, and Wyoming—completely denied the riparian doctrine in favor of the priority of beneficial use to determine the acquisition of water rights. The doctrine stated that the riparian doctrine was unsuitable for areas of limited rainfall so that ownership of riparian lands carried no water rights. The only way an original water right could be acquired was by appropriation of the water to a beneficial use at a time when sufficient water was available in the stream to satisfy that use. Appropriations could later be transferred by purchase, inheritance, condemnation, or prescription.

The Colorado doctrine denied the Federal government any water rights although it did own public lands in Colorado. It also did not recognize any public interest in state control of water resources in the state. Both of these aspects have later been reversed by decisions of the U.S. Supreme Court.

To sum the appropriation doctrine, the following points can be specified:

(a) The rights of appropriators are not equal because priority of appropriation gives superiority of right.
(b) The appropriator obtains the right to a specific quantity of water if he can put it to a beneficial use and if the water is available in the stream.
(c) The water appropriated may be used on either riparian and nonriparian land. In fact, the appropriator does not even have to own land to obtain the appropriative right since beneficial use is the only criterion.
(d) The doctrine makes no distinction between so-called natural uses (domestic) and artificial uses (agriculture, manufacturing).
(e) The appropriative water right is quite specific in terms of quantity of water, season or period in the year in which the water may be taken, the nature or

purpose of the use, the place of use (although this may be changed in most states), and the priority date of the appropriation.

## Interpretation of the Appropriation Doctrine

One interesting aspect of the appropriation doctrine is that the date of beginning (rather than the completion) of the diversion for beneficial use becomes the date of the water right. Thus, if a farmer makes known his intention to divert water to irrigate 500 acres of land and works diligently (but slowly) at the project, the date of his appropriation is the date he started the whole project rather than the date he finished, which might be many years later. Other appropriators might have started projects later in time and finished them long before the irrigator finished his. They would, however, be junior appropriators to the irrigator who started his project sooner than the others. If the appropriator does not show due diligence in completing his diversion, his application may be judged as "lapsed," in which case he loses his priority application date.

Most appropriation states require that the appropriators file an application for water with some authorized state official, often the state engineer. He is charged with insuring that the use is truly beneficial and that sufficient unappropriated water exists in the stream channel. In recent years, this individual has been directed to recognize the need for a certain amount of unappropriated water at all times in the stream to satisfy fish, wildlife, and recreational needs as well as to maintain stream aesthetics. Thus, just because there is water in the stream, this is no indication that water is available for appropriation purposes.

Let us consider a situation where a series of appropriators have filed for various quantities of water from a given stream. The available water is all allocated and some of the applicants for water have had to be turned down because enough water is not available. Now, as a result of interbasin transfers or some other human activities which pump additional sources of water into the stream, more water becomes available. A new applicant might file to appropriate this new volume of water in the stream and be approved over the objections of those applicants who had been turned down previously. The earlier requestors for appropriative rights, whether approved or disapproved, did not anticipate additional volumes of water in the stream. Their applications were judged on the basis of what was available in the stream at the time. If more water later becomes available in the stream, new applicants have as much right to it as older applicants. Requests for the new water begin over again at the time when the new water is first available.

The question of efficiency in the beneficial use of water is somewhat difficult to resolve. If water is used so inefficiently that it is grossly wasteful, the use is illegal. Under this situation, the user may lose his rights to the water being appropriated (actually since his use was illegal, he never really had the right in the first place). If his use is only somewhat inefficient and he improves his efficiency of use, there may be a question concerning the use of the water so saved through increased efficiency. It might be argued that the appropriator could add the water so saved to his own appropriation since his action resulted in more efficient and beneficial use of a limited water supply which was already his own.

While appropriative water rights may be sold, problems can develop. A few states permit the sale of such rights but do not permit the transfer of the priority date of the appropriation. Rather, they establish a new date, the date of the transfer, as the date of the appropriation. Such action usually renders the water right worthless since most streams are fully appropriated and having a current rather than a much earlier priority date on the appropriative water right means that it is junior to all other demands and so it may not be able to obtain any water from the stream. Other states do not require such a change in date during transfer or sale.

Certain states prevent the sale of water rights held by irrigation districts or municipalities in order to protect water users in the districts against possible unscrupulous actions of managers of the districts. Such laws are usually harmful since the managers or directors of the irrigation districts are frequently local individuals who would not do anything to harm the users in their own districts. Often, the district may have more water than it needs and thus could obtain additional income by the judicious sale of water rights, but it is usually prevented by law from doing this.

Appropriation rights may be lost by the user as a result of abandonment, forfeiture, adverse use, prescription, and estoppel. Abandonment is sometimes difficult to prove since the owner of the right must discontinue beneficial water use with the intention of abandoning his water right. There is usually no specified time period during which nonuse must occur. It is difficult to determine the intent of the nonuse and consequently some states substitute nonuse for a specified period as tantamount to intent to abandon. In this case, forfeiture and abandonment are quite similar since forfeiture occurs when a user fails to make beneficial use of his water for a specified time period. These periods range from 2 to 5 years in different states. Forfeiture is not concerned with intent but only continuous nonuse during those periods of time when use was possible. Clearly, nonuse during a time when low streamflow conditions make it impossible for use to have occurred is not grounds for forfeiture. Intermittent nonuse is also not grounds for forfeiture as long as some use occurs within the specified time period. Some states, however, argue that if the user only takes water during part of the time period specified in his right, his water right might be changed to cover only the new period of use (if it is less than the older period) and the remaining rights to water returned to the stream for later appropriation to others.

Acquiring water rights by adverse use is now generally disallowed under most appropriation doctrines. As originally stated, a user could acquire such rights if he used the water for the proper statutory period (often 7 years) without the consent of or interruption by lower water users. This is the same doctrine that applies to land, but in the case of land it is much easier to determine if someone is illegally using or taking possession of a piece of land during the statutory period. If this same doctrine were applied to water, it would force all appropriators to check constantly all upstream users to insure that they were taking only the water they were authorized to take. Such diligence is too demanding and hence the doctrine of adverse use has been rejected in regard to water in most states. Still, water rights obtained by adverse use before the enactment of state laws eliminating such activities still pose some question. Normally, there is no record when the adverse use began since such information would not be publicized. Legal actions to confirm the propriety of water rights obtained under the adverse-use doctrine can occur at any time—even now, well after the time that the doctrine has been abandoned in most states.

Rights for water-related uses may be acquired by prescription in the West much as in the East. Actual water rights cannot be obtained by prescription since the rejection of the doctrine of acquiring rights by adverse use. However, prescriptive rights in land may be acquired as a result of use during a prescribed period— often 20 years—and this can include water-related rights such as rights to ditches, canals, and other similar facilities. The principles of estoppel are essentially the same under both the riparian and appropriation doctrines.

## FEDERAL RESERVED WATER RIGHTS

The Federal government owns vast amounts of lands, especially in the West. The courts have ruled in several important decisions that the Federal government has a reserved water right in them, the right having been created by withdrawing those lands from settlement and private use. It has been argued that the creation of an Indian reservation—or a national park or forest—restores the proprietary rights of the United States to the waters on that reserved land and thus establishes, for the United States, riparian rights unrelated to state law. Such an interpretation can create havoc with a system of appropriation which may already have assigned all available water in the watercourse to private beneficial users.

The Pelton Dam case in 1955 extended the right of the Federal government to control of water within states even further. The thrust of this decision was that the Federal government had the right to decide what to do with the water on its own reserved lands regardless of the wishes of the individual states. The authority to do so stems from the Property Clause in the Constitution which gives Congress the power to make whatever rules are needed to dispose of or control territory or property belonging to the United States. As a result, Congress has the authority to decide what it wants to do with reserved lands or the water on these lands (in this case to construct a dam) without regard for the wishes of the individual states. The Supreme Court has ruled that public lands are lands subject to disposal under public land laws and may be disposed of to private individuals. Water rights on these lands are, therefore, treated differently from water rights on reserved lands—national forests, Indian or military reservations— which are lands withdrawn from later settlement. Such reserved lands—and the majority of Federal land in the West is reserved for some purpose—have water rights belonging to the Federal government in a form of riparian ownership. Water rights on public lands having nonnavigable streams accrue to the states and may be assigned by the states as they see fit for the good of the public.

It is quite unfortunate that this interpretation of water rights in Federal reserved land was put forth so late in our development as a nation when other water laws were in operation and water from streams with limited flows was already fully distributed. The furor caused by this interpretation of water rights on reserved lands has not subsided yet. Especially in the case of Indian water rights, many new legal battles will be necessary before the relationship between appropriative rights and Federal reserved rights is clearly understood and necessary adjustments can be made.

## GROUNDWATER LAW

While we now understand that water in all parts of the hydrologic cycle is closely interconnected, legal concepts relating to groundwater were originally developed as if groundwater was physically separate from surface water. Also, the idea that there existed underground streams (aquifers), quite distinct from the water slowly seeping downward through the soil pore spaces, influenced the rules governing underground water. While such underground streams do exist, they are nowhere as numerous or as clearly identified as one would expect from the laws on the subject.

In developing the original common law on underground water, it was assumed that the groundwater beneath a parcel of land belonged to the landowner and could be used in any way the landowner wanted. The idea of the interconnectedness of the groundwater so that overpumping at one place would result in harm to the owner of another property in the same groundwater basin did not seem to be know by, or of concern to, the early jurists.

As more was learned about the movement of water underground and the quantity of annual recharge to the water table, groundwater law was modified to reflect a reasonable use doctrine. In many areas, the idea of correlative rights to use groundwater in common with others owning lands in the groundwater basin developed. In the western United States, written applications were required in order to appropriate water from the groundwater reservoir, a practice that allowed some control over the development of this resource.

Underground streams, if identified, are treated quite similarly to surface streams. In riparian states, the owner of the land under which the streams flow is entitled to reasonable use of the water as long as the use does not harm lower landowners. In appropriation states, the rights to the water in the underground stream depend on the beneficial use concept. Application is made to the state engineer or other authorized individual and, if unappropriated water is available, a permit is granted to make use of a designated volume of the flow for a specific purpose and time period. The real problem with underground streams is the great difficulty in identifying if one exists at a place or whether the groundwater has resulted merely from percolating waters. The rules for underground streams are not applied too often.

Most groundwater results from percolating waters and the laws to determine rights to percolating waters have changed over time from the concept of absolute ownership to reasonable use to correlative rights. As it became understood that underground water was merely a part of the hydrologic system, so that human actions in one place would influence the water table response in other nearby places, it was recognized that absolute ownership and uncontrolled use were no longer feasible. Reasonable use of the groundwater was needed to afford protection for all landowners against the willful or inadvertant action of just one or a few landowners. The determination of reasonable use had to be made on a case-by-case basis.

Refining the reasonable use concept resulted in the idea of correlative rights now used in a number of appropriation states. Literally, the idea of correlative rights gives each landowner in a basin the right to the percolating water in the basin in the same proportion that the surface area owned by the landowner bears to the total surface land area overlying the groundwater basin. Thus, if a landowner owns one-quarter of the surface land in a given groundwater basin, the landowner is given correlative rights to one-quarter of the developable groundwater. The rule is seldom enforced so rigidly but it does provide a good guide to determine water volumes to which each landowner is entitled.

One real problem with reasonable use of groundwater is the question of stability in the maintenance of the depth to the groundwater or of artesian pressures. Assume that a landowner drills a well to 100 ft and obtains an adequate source of water. Later appropriators install additional wells which, as a result of overpumping, lower the groundwater table to a depth of 500 ft. The original appropriator must either dig a deeper well and suffer the economic consequences of pumping from a greater depth or be deprived of water rights. The law does not provide a clear-cut solution to this problem, which is happening in many areas. The general concensus is that some lowering of the water table will occur with use (or some reduction in artesian pressure will occur) and that the early appropriators must expect change in the conditions of underground water. What is a reasonable change and how much mining of the water table can be permitted is settled on a case-by-case basis by the users in a particular area. Need for water and the potentialities for economic gain often obstruct the exercise of sound hydrologic judgment.

## A COMPARISON OF SOVIET AND AMERICAN WATER LAW

As major industrialized nations and as countries with areas of great climatic diversity, it is clear that both the United States and the Soviet Union face many of the same water problems. They have attempted their solution in different ways. American law (discussed above) has been concerned with resolving conflict between users who have rights to water and who need water to produce a product for private economic gain. The Soviet approach has been directed more toward comprehensive planning in order to achieve programmed economic growth for the State. Both countries seek to achieve efficient economic growth. The U.S. approach is that this will occur as a result of the interworkings of individual economic decisions, while the Soviet approach is that an efficient economy can only result from careful governmental planning of economic decisions. Water law in the Soviet Union clearly illustrates the government's significant role in all aspects of planning and management.

Foremost in Soviet water law is the fact that the water is owned entirely by the government—it was vested in the government after the 1917 revolution. The right to use water may be assigned to individuals or groups. Water users are only allowed to develop water resources if their use is in accord with the master plan for the national economy. While the law is supposed to give first priority to public and domestic uses, the major emphasis in recent plans for the national economy has been on industrial development, with less emphasis on water resource conservation. Comprehensive economic planning is a very difficult and complex problem in any nation so it is not surprising that the Soviet Union at times has had difficulty in keeping the priorities for water development balanced with the priorities for economic development.

Soviet water law, as it existed in the mid-1960s (Davis, 1971), seemed unable, at times, to establish proper priorities between uses or to resolve conflicts arising among users in a river basin. There was not a single, unified set of rules covering the range of water uses. Rather, there were a number of sets of water rules, one for each different type of use—industrial, municipal, hydroelectric power, navigation, etc. The rules were often quite independent of each other, as if water could be treated as a series of different resources. There is one general rule recognizing the unified nature of water, however. This states that in operating a water resources project, other uses of the water resource should not be modified or influenced in any way. Implementation of this principle has led to real problems in Soviet water law and possibly to the 1970 Principles of Water Law of the USSR. These Principles require, among other things, that in locating new facilities or in operating old facilities, other uses of the same water source must be considered. The various Republics that make up the Soviet Union have the right to implement the Principles by means of specific legislation.

The 1970 Principles recognize the need to resolve disputes among water users. Disputes are to be settled under administrative procedures legislated at both the National and Republic levels. One assumes that such legislation has or will be provided as a result of that decree.

Under Soviet water law, water is a resource to be employed in the national economic development. The particular relation between users and uses at a place must be determined by the economic needs of the area and by reference to the national economic plan. While it is too early to determine whether the 1970 Principles are effective in solving some of the problems resulting from the multiple uses of river systems and the inadequate communications among planning agencies, it is clear that the Soviets have recognized the problem and are seeking its solution.

Several points in the 1970 Principles may be of interest. They are summarized on the following page:*

*See Fox (1971, Appendix, pp. 221-239) for a full translation of the 46 Articles in the 1970 Principles of Water Law of the USSR and Union Republic's which appeared in *Izvestiya* on December 11, 1970.

(a) All the waters of the USSR constitute a single common state resource. These waters include rivers, lakes, reservoirs, canals, ponds, underground waters, glaciers, inland seas, and the territorial waters of the USSR.

(b) It is unlawful to put new or rebuilt factories, shops, industries, or communal or other facilities into operation if they have no equipment to prevent pollution or obstruction of water or its harmful action.

(c) Use of water bodies shall be permitted on an indefinite (permanent) or temporary (short-term, or up to 3 years, or long-term, over 3 years) basis.

(d) Water users can use the water bodies only for the purposes authorized. Users are obligated to use water bodies rationally and economically, to eliminate the discharge of any pollution, to work to improve the quality of the water, not to infringe on the rights of other users, and, if necessary, to keep a record of water use.

(e) Rights of groups or individuals to use water can be terminated when they no longer have need of the water, when their granted time period expires, or when there is a need to remove the water body from previous uses. The laws of the individual Republics may specify other reasons for termination. Losses caused by such terminations must be compensated for.

(f) The use of underground potable water for purposes other than drinking is usually not permitted. Other uses may be allowed where adequate reserves are available and surface supplies are limited.

(g) Groups using water for agricultural and industrial purposes must exercise care to reduce inefficiencies and unnecessary water losses. Waste of any type can only be allowed to enter water bodies with the authorization of proper agencies concerned with health, fish protection, and other interested groups.

(h) The operation of reservoirs shall be determined by the group concerned with the use and conservation of water in each particular reservoir.

(i) State record-keeping of water flows and water uses is an important undertaking to insure adequate information for planning purposes.

All told, the short document lists 46 useful principles and goals but without information on how best to implement those principles. Clearly, the methods of implementation in the United States would be quite different from those in the Soviet Union.

## CONCLUSIONS

While the subject of water laws has been discussed in a chapter following the section on ways to increase water supplies, it must be understood that improved water laws actually provide a way to make available additional supplies of water. Especially in the riparian states where needed water can flow unused to the ocean, it may be possible to increase our beneficial use of water appreciably without bringing significant damage to existing riparians. Laws relating to the discharge of pollutants into streams must also be modified and enforced if we are going to protect fresh water supplies from being further degraded. Even in the appropriation states, the permanent right to an appropriation of water may lead to some waste of water. Certainly, continued review of the water requirements of each of the beneficial users is needed along with regulations to encourage conservation and more efficient use of water. The useful water laws of the eastern riparian states can probably be merged with the more effective laws from the western appropriation states to provide a significant improvement in water use efficiency, greater water conservation, and increased supplies of available water. Water laws are not only a vital aspect of water resource management in their own right, but, if properly adapted to the area, they can be a technique by which more water may be made available for use where supplies are limited.

## 10

WATER RESOURCES LEGISLATION
AND THE MANAGEMENT
OF WATER RESOURCES

## A HISTORY OF FEDERAL WATER RESOURCES LEGISLATION*

Water resources development in the United States cannot really be understood without consideration of the history of Federal water resources planning and legislative programs. These programs reflect the changing relationship between the Congress and the Executive Branch, the social, political, and economic climates of their time, as well as the changing needs or demand for water as the United States moved from a primarily rural and agricultural economy to an urban and industrial economy. A study of changing Federal programs may give the impression of a series of random oscillations emphasizing different aspects of water. At times, it might even suggest a lack of planning; but, when the overall course of water resources development is traced, certain trends or recurring ideas and concepts emerge.

Water resources planning and construction programs are carried out by the Federal government under a number of different provisions of the Constitution. For example, certain water programs may be authorized and carried out under the constitutional power to make treaties (Art. I, Sec. 10; Art. II, Sec. 2) or to provide for the national defense (Art. I, Sec. 8). Other water projects may be authorized under the constitutional power to provide for the general welfare (Art. I, Sec. 8). Probably the most widely used constitutional authority over water programs is that giving the Federal government power to regulate commerce with foreign nations and among the states (Art. I, Sec. 8). It was generally understood that the term commerce included navigation.

While the law talks about navigable and nonnavigable streams, it is difficult to make a hard and fast distinction in practice. England had few inland streams

---

*A comprehensive survey of Federal Water Resources Programs from 1800 to 1960 has been prepared by Holmes (1972). It serves as the source for much of the material in this review section.

that had significant public uses, so that navigation and fishing in the tidewater areas were the major public uses of such watercourses. In the English system, navigable streams were defined as those affected by tidal fluctuations while streams above the head-of-tide were considered nonnavigable.

The early American definition of navigable and nonnavigable followed the English system, with the demarcation coming at the point where tidal fluctuations ceased to affect the stream. This interpretation was held until 1851 when the Supreme Court, reversing earlier decisions, rejected the head-of-tide as the boundary between navigable and nonnavigable streams. It did not make clear in that decision just what constituted a navigable stream but did allow that it could include inland streams far removed from tidal fluctuations.

In 1870, a partial answer to this question was achieved when the Court decided that all waters which were navigable-in-fact were navigable under the law. The phrase navigable-in-fact included watercourses that were being used for commerce or which were capable of being used for commerce in their ordinary condition. Watercourses that were only navigable intrastate would be considered waters of the state and subject to regulation by the state. A 1940 decision broadened the idea somewhat by stating that a stream was navigable-in-fact if it could be made navigable through reasonable improvement of the waterway. The type or degree of improvement might change over time; thus, an unreasonable improvement at one time might be considered, under law, reasonable at another time. Finally, a 1960 decision permitted Federal regulation of a nonnavigable stream since such an action would have significant influence on a downstream watercourse which was navigable. Congress, in recent years, has tended to put the broadest possible interpretation on the phrase "navigable waters" so that nearly all watercourses can be so considered.

Promoting commerce and transportation has long been one of the major concerns of the Federal government (witness the present emphasis on our interstate highway system). In the early 1800s, this interest was expressed in the government's desire to promote river improvements and canal developments to aid in inland transportation. It was clearly stated that navigable waters would be treated as public highways and would be maintained free for the use of all. The westward movement of the nation after the Louisiana Purchase resulted in the Senate asking Secretary of the Treasury Albert Gallatin to prepare a national plan for the development of roads and canals. The Gallatin Report of 1808 called for a nationwide system of canals and other river improvements that could be justified on the basis of the economic development of the West, national defense, and political unity.

While the Gallatin Report, and its later revision by Secretary of War Calhoun in 1819, suggested that Congress had the power to spend money on river improvements because of its concern for national defense and the general welfare, the famous *Gibbons v. Ogden* decision by the Supreme Court in 1824 acknowledged that the power of Congress to regulate interstate commerce also included the power to regulate navigation within each of the states insofar as that navigation was connected with commerce.

Legislation for channel clearing of the Ohio and Mississippi rivers was passed but the real question of whether the Federal government should be involved in such internal improvements remained a serious political issue during the pre-Civil War period. Because the U.S. Army Corps of Engineers was the only group with the technical abilities to undertake water-related programs, it became responsible for these activities beginning in the 1820s. The General Survey Act of 1824 empowered the President to employ civil engineers as well as the Corps to make survey plans of roads and canals deemed of national importance. Special congressional legislation, such as the first Omnibus Rivers and Harbors Act of 1826, provided separate authorization and appropriations for the Corps to undertake planning and actual work on rivers and harbors improvements under the control of Congress.

Following the Civil War and with the rapid opening of the western territory, the demand for Federal water resources undertakings increased. The 1874 Windom Select Committee called for a comprehensive program of waterways improvements that would provide midwestern farmers with cheaper transportation than that offered by railroads. Somewhat earlier (1849–1850), because of serious flooding in the lower Mississippi Basin where settlements were growing, Congress passed the Swamp Lands Act which granted Federal lands in Arkansas, Louisiana, Mississippi, and Missouri that were subject to flooding to those states with the stipulation that money from their sale be used for flood control and drainage works. Another Mississippi River flood in 1874 led to the appointment of a commission of engineers and in 1879 to the establishment of the Mississippi River Basin Commission, authorized to survey the river and prepare plans to improve navigation and to prevent floods. Post-Civil War sectional differences often limited appropriations to carry out action plans. At the same time, concern was expressed whether Federal funds should be used for local levee construction and such construction with Federal money was usually expressly prohibited.

As farmers moved into drier and drier areas, the need for irrigation became obvious. The Desert Land Act of 1877 authorized sale of 640-acre parcels of land in certain arid states and territories provided the purchasers would irrigate them within 3 years. Various land speculation scandals severely limited the implementation of this Act.

The period from 1900 to 1920, under the administrations of Roosevelt, Taft, and Wilson, marked a most significant and what might be called progressive period in Federal water resources programs and developments. The principles directing the water resources development during this period were (a) conservation of national resources for the use of present and future generations; (b) elimination of control and/or exploitation of the resources by monopolies; (c) elimination of "giveaways" of resources to special interests; (d) encouragement of individual, independent groups such as the family farm.

To aid in the development of the new programs, a number of study and action commissions were established—for example, the Inland Waterways Commission (which recommended the creation of a "National Waterways Commission" to coordinate the work of the Corps of Engineers, the Reclamation Service, the

USDA Forest Service and Bureau of Soils, and other concerned Federal agencies); the National Conservation Commission (which called for massive amounts of research to support ambitious plans for multi-purpose waterways improvements); and the National Waterways Commission (which urged legislation to prevent deforestation, to encourage water power development in the public interest, and to develop a Federal reservoir system for flood control to be justified by multi-purpose benefits).

A number of agencies were established to coordinate planning and development of water resources. For example, the Reclamation Act of 1902 established, among other things, the Reclamation Fund with money made available from the sale of public lands in 16 western states (Texas was added later as the 17th state—all states west of the Mississippi River except Arkansas, Iowa, Louisiana, Missouri, and Minnesota were included). The Secretary of the Interior was authorized to use the fund to plan and construct irrigation works in those states. The right to use the water so developed was limited to tracts not exceeding 160 acres and occupied by the landowner. Water users were also required to repay the estimated cost of construction (into the Reclamation Fund) without interest within 10 years. Later, the Secretary of the Interior was permitted to develop water power at reclamation projects and to lease surplus power or the right to develop power for 10-year periods if such action would not adversely influence the irrigation program. He could also sell water to towns in the vicinity of the projects. By 1920, he was authorized to sell water for purposes other than irrigation if such action would not limit the irrigation project or other earlier water users in the area.

The Forest Service was transferred from the Land Office of the Department of Interior to the Department of Agriculture in 1905 under the direction of Chief Forester Gifford Pinchot. The Weeks Law of 1911 permitted a national forest reserve program and made it possible to purchase land in basins of navigable streams if needed to regulate water flow.

The Waterways Commission, which existed only from 1917 to 1920, was authorized to coordinate the work of all Federal groups concerned with water resources and to prepare nationwide multi-purpose plans. The Commission was never able to function as proposed and was abolished by Congress.

The Federal Power Commission was started as a Cabinet-level committee of the Departments of War, Interior, and Agriculture to license non-Federal development of water power on navigable streams and to sell surplus government power when in the public interest. The Commission also had the authority to study water resources development, water power potentialities, and the water power industry.

Along with the increase in interest in water resources planning, the role of the Corps of Engineers in such activities also increased. The General Dam Acts of 1906 and 1910 required that applications to build dams be submitted for approval to the Corps and that approval be granted on the basis of multi-purpose use criteria. Major floods in 1915 and 1916 led to the Flood Control Act of 1917, giving the Corps responsibility for planning and building control works (not including reservoirs) on the Mississippi and Sacramento rivers. The law also provided that at least one-half of the cost of levee construction should be paid by the state or locality.

During the 1920s (until 1933) under Republican administrations, the general trend was to deemphasize the antimonopolistic policies of the previous two decades and to eliminate, as much as possible, government competition with private enterprise. For a while, work was stopped on the nearly completed Wilson Dam near Muscle Shoals while a private purchaser was sought. Hoover, as Secretary of Commerce, recommended national planning of public works as well as multi-purpose planning by the various governmental drainage basin commissions. But the matter of public vs. private control of power continued to be one of the most sensitive political issues in the nation for years.

In 1925, Congress asked the Corps of Engineers and the Federal Power Commission to prepare a list (with cost estimates) of navigable streams and their tributaries which might be surveyed for possible power development. This list was submitted in 1927 and printed in House Document 308. The 1927 Rivers and Harbors Act empowered the Corps to undertake these surveys (known as the 308 reports). And in 1928, following a most severe flood on the Mississippi River (see Chapter 11), the Corps was authorized to do what was necessary for flood control (including the construction of reservoirs on tributaries); the Federal government would cover the entire cost of the project in view of the fact that there had been large local expenditures for flood control previously. The Bureau of Reclamation meanwhile acquired control over all regional multi-purpose planning in the basin of the Colorado River.

While the Corps of Engineers still depended on congressional authorization and appropriations for projects, the general authority granted to undertake surveys as part of the 308 reports gave them considerable freedom to make general plans for all river basins (except the Colorado River basin) in the United States, setting, in most cases, their own priorities for action.

The depression of the 1930s and the new administration under Franklin Roosevelt resulted in the formulation of many construction plans in the form of public works projects in order to stimulate the economy. Multi-purpose water resource development with public power programs was promoted to provide regional economic growth and to benefit the largest numbers of people. Planning the development of "national" resources, which included both human and natural resources, became a key to the programs proposed. Roosevelt enlarged the role of the Executive in the proposing of legislation and Congress accepted its role to consider and modify the programs initiated in the Executive Branch.

A number of new water planning agencies were established to push new concepts of regional resources development. One cornerstone of this concept was the creation of the Tennessee Valley Authority in 1933, an agency authorized to exercise all Federal functions of development and management of water and land resources within a large geographic region. As such, it had the power to plan, build, and operate dam and reservoir projects for navigation, flood control, and power. Even more uniquely, it had to obtain only the approval of the House and

Senate Appropriations Committees before undertaking any proposed work. The TVA could also make and sell electric power from thermally-driven generators, build transmission lines, direct soil conservation, recreation development, and fish and wildlife improvements.

Four different National Resources Planning Boards (under different names) were established between 1933 and 1943 to prepare comprehensive programs of public works for a large variety of purposes including water resources uses. The President was given blanket authority to build or finance any public works project (except river and harbor improvements) in order to increase employment rapidly. The Bonneville, Central Valley, Fort Peck, Grand Coulee, and Parker dams were all begun under this authority and construction of the Wheeler Dam of the TVA was accelerated.

Two important actions of the National Resources Committee (as it was called in 1935) were the initiation of a nationwide study of drainage basin problems and detailed studies of particular river basins. The Committee also took over the responsibility of reviewing and evaluating annually the 6-year construction plans of the various Federal agencies called for under the Employment Stabilization Act of 1931.

The Public Works Administration was established to make loans and grants to both state and local governments for construction of municipal water and sewage works, for drainage, irrigation, and flood control projects, and for water power development. The PWA also helped supply planning consultants and provided technical assistance to state and local planning agencies.

While, in earlier years, flood control activities had been limited to specific basins or portions thereof, the 1936 Flood Control Act initiated a national flood control program and gave the Corps of Engineers the authority over all Federal flood control activities. This act also gave the Corps authority to initiate many preliminary surveys as well as to build a number of reservoir projects for the purpose of navigation, flood control, and recreation. The Corps was quite conservative in its views of including power production in its dam projects even after a 1938 directive authorized installing facilities adaptable to the later production of power in new dams.

Interagency coordination in planning of water resources activities was encouraged in the 1933-34 report of the President's Committee on Water Flow. The Committee, consisting of the Secretaries of Agriculture, War, Interior, and Labor, worked through six technical subcommittees, while coordination and administrative work was done by the National Planning Board. In spite of this overt effort at coordination, much of the legislation of the period contained no provisions stressing interagency cooperation. The Flood Control Act of 1936, while dividing flood control work between the Corps of Engineers and the Department of Agriculture on the basis of size, function, and geography, did not require any consultation on programs between the two agencies or with any other agencies.

Two major efforts in multi-purpose river basin planning occurred during the "New Deal" days. The first was the aforementioned "308 surveys" that the Corps of Engineers carried out. These surveys were to include data on navigation, water power, flood control, and irrigation. The second planning effort involved that of the interagency teams headed by the different national resources planning groups. In addition to the data obtained by the 308 surveys, data on pollution and pollution abatement, soil erosion and reforestation, recreation and wildlife conservation, and land drainage were also collected. Though the TVA had been established as a unique agency with comprehensive planning and operating powers, a 1937 proposal by President Roosevelt to create seven similar regional planning agencies was defeated in Congress.

President Roosevelt tried many times to have the National Resources Board established as a permanent board but he was never able to convince Congress. The President's attempt to subordinate the Corps' planning function to a National Resources Planning Board was undoubtedly considered an attempt to usurp congressional power. The Corps of Engineers was responsible to the individual congressional member who sponsored local improvements in his/her district. The congressional member was permitted to testify at public hearings, was informed first of any modifications, and could request a review resolution if unsatisfied with the Corps' preliminary report. This treatment resulted in Congress having a special interest in the Corps. In 1943, Congress not only appropriated no money for the National Resources Planning Board but stated categorically that it should be abolished and its functions not be given to any other group. The defeat, accomplished in spite of the President's personal appeal to retain the agency, was the result of efforts by a bipartisan coalition including the Rivers and Harbors Committee favorable to the Corps and conservatives who were concerned about the human resources planning also under active consideration by the Board.

With the end of the National Resources Planning Board, the Executive Branch lost much of its ability to prepare or to evaluate comprehensive water resources plans. New water programs became more the concerns of congressional committees responsible for legislation and appropriations. The various construction agencies—the Corps of Engineers, the Bureau of Reclamation, the Soil Conservation Service, TVA—developed their own strong liaison with Congress and their own independent programs based on history, mandates, interests, and areas of operation. The programs of the four construction agencies expanded more rapidly during the 1940s and 1950s than they had during the period of more publicized public works programs in the 1930s. The 1944 Flood Control Act authorized a record number of projects due to the fact that it was felt the end of World War II might bring a great surge of unemployment. Even though that was not the case, and there was little national interest in water resources programs, water development construction projects were rapidly pushed ahead. Only toward the end of the 1950s did one water resources related issue begin to attract more than local interest, namely the environmental issue which led, in part, to the passage in 1958 of the Outdoor Recreation Resources Review Act. That Act provided for a nationwide inventory and assessment of outdoor recreational facilities and needs.

At the same time, city water supply problems were emerging in both humid and arid parts of the country. The Water Supply Act of 1958 included storage for future municipal and industrial water supplies as of equal importance in the planning of multi-purpose reservoirs of both the Corps and the Bureau of Reclamation. The value of such storage could be included in the economic justification for the reservoir projects.

The important Flood Control Act of 1936 (and as amended in 1937) had empowered the Department of Agriculture to make preliminary surveys in the basins of each of the rivers (except the Colorado) the Corps of Engineers was authorized to study. Over the years, the Department made little use of this power since by 1954, when the original authorization was superseded by a small watershed program (Public Law 566), they had sent to Congress only 26 detailed basin studies along with a Missouri Basin Agricultural Plan.

## Recent Water Resources Legislation

In 1954, the Small Watersheds Act gave the Secretary of Agriculture permission to help local organizations plan and construct flood prevention works on watersheds not exceeding 250,000 acres in size. The plans finally achieved could result in no single structure having a storage capacity of more than 5000 acre-feet. Local interests were to pay for a fair share of the construction costs. Because of problems with the cost-sharing aspect, the Act was amended in 1956 to permit Federal payment of all costs allocated to flood prevention and to allow the Secretary of Agriculture to make loans of up to $5 million to local groups to finance their share of any costs. Works for industrial and municipal water supplies were authorized by the 1956 amendments but they had to be fully paid by local groups. Permissible structures were increased to 25,000 acre-feet of capacity as long as not more than 5000 acre-feet were allocated to flood protection. An amendment in 1958 permitted the Secretary of Agriculture to pay a fair share of the cost of works allocated to fish and wildlife development.

In this same vein, a Small Reclamation Projects Act was passed in 1956 which permitted the Secretary of the Interior to loan up to $5 million for projects to construct or rehabilitate small irrigation systems which might also include multipurpose functions. Loans were available for irrigation of single tracts greater than 160 acres but interest would be charged on the portion allocated to irrigation of acreage over the 160 authorized acres. The Bureau of Reclamation also became more involved in planning for both municipal and industrial water supplies, especially in the Southwest.

By 1953, the TVA had completed 20 multi-purpose dams on the Tennessee River and its major tributaries. Although this activity greatly reduced the heights of major floods, it still had not eliminated damaging floods on smaller streams in the area. In the mid-1950s, the TVA, therefore, began a program of land-use planning at the local level to try to help local communities mitigate flood damage where they could not completely prevent floods. By 1959, some 21 communities had begun floodplain planning studies while 9 communities had adopted programs of floodplain regulation. The apparent success of the program led the agency to recommend its use across the nation and the 1960 Flood Control Act permitted the Corps of Engineers to begin a similar program.

Several new planning agencies have been brought into the water resources picture in the last two decades. Even though the Public Health Service has been authorized since 1912 to provide information on health aspects of water pollution, concern over water pollution did not really become significant until after World War II. The 1948 Water Pollution Control Act gave the Surgeon General authority to cooperate with Federal, state, and local agencies in preparing comprehensive pollution control programs for interstate rivers and for the collection and release of information and technical aid. The Act made no provision for enforcement without the consent of the state in which the pollution occurred. This weakness was corrected in the 1956 Act which permitted the Surgeon General to call conferences with state pollution control officials to deal with interstate problems and to take action against alleged polluters without the consent of the state, if remedial action did not occur as a result of the interstate conferences. The Act also authorized 30% incentive grants to help subsidize construction of municipal waste treatment plants.

In 1946, a Fish and Wildlife Coordination Act required consultation between Federal agencies (or non-Federal agencies operating under a Federal permit) and the Fish and Wildlife Service in order to prevent loss or damage to fish or wildlife in the impounding, diverting, or controlling of waters. A later act in 1958 took a more positive position by requiring that Federal construction agencies include planning to enhance wildlife and fish resources as part of their own construction plans.

In 1947, the National Park Service was given full responsibility for recreation planning at reclamation projects. The Service had been doing this for Corps of Engineers projects since 1945 but only when requested to do so. Since the Service viewed the importance of reservoirs, drainage measures, and flood control activities somewhat differently from both the Corps of Engineers and the Bureau of Reclamation, it frequently opposed projects by these construction agencies on the grounds that they would despoil scenic or historic sites or wilderness areas.

Current Federal water laws and programs are quite involved and complex. More than 40 Federal agencies have some water programs or statutory responsibilities and the programs keep changing. Reorganization of the various agencies makes it difficult to determine specific responsibilities. The National Water Commission (1973) undertook a notable effort to compile an up-to-date summary of the various Federal water resources programs but even they encountered great difficulty because of changes enacted during the time in which they were making their summary. A brief discussion of a few of the more important current water resources programs is provided below:

(a) *Water Bank Act of 1970.* This law directs the Secretary of Agriculture to develop a program to conserve and improve the wetlands of the nation. Payments may be made to landowners to encourage them to keep their wetlands in a condition for migratory waterfowl, to protect hunting and fishing conditions, or to allow free public access for controlled recreational activities.

(b) *Council on Environmental Quality.* Created by the National Environmental Policy Act of 1969, the three-member Council, appointed by the President with the consent of the Senate, analyzes and interprets environmental trends, draws up possible policies to promote improvement of the quality of the environment, and seeks to understand and be responsive to the scientific, socioeconomic, and cultural needs of the nation in environmental matters.

(c) *Flood Control Act of 1970.* This Act authorized the Secretary of the Army through the Chief of Engineers to study the operation of all projects constructed by the Corps to determine whether modification in their structure or operation could improve the quality of the environment.

(d) *Environmental Protection Agency.* Reorganization Plan No. 3, 1970, transferred a number of programs dealing with pollution control to the EPA. The Pollution Control Act Amendments of 1972 greatly expanded the scope of the EPA's program with respect to setting stream quality standards and regulating discharges to streams. The Act defines the conditions for the issuance of discharge permits in terms of water quality standards in the receiving streams. State water standards already adopted and approved by the EPA Administrator are not to be changed by the 1972 Act. However, if the standards were waiting approval and were not in conformance with the provisions of the 1972 Act, the Administrator can require that they be changed within 90 days. If the state fails to submit standards that are acceptable, the Administrator can prepare his own regulations for water quality standards. If the state does not adopt its own standards that are acceptable within 180 days, the Administrator will promulgate his own water quality standards. The Act also calls for a progressive reduction in the quantity of pollutants discharged to navigable streams. Certain limitations had to be achieved by July 1, 1977, and a second level of reduced discharge must be reached by July 1, 1983. All discharges are to be eliminated by 1985.

The Act of 1972 is quite comprehensive in establishing four permit systems to regulate discharges and other human activities that influence water quality, to make it unlawful to discharge toxic substances, to limit thermal pollution of streams or nearby ocean waters, and to regulate ocean dumping of wastes. Enforcement procedures detailed in the 1972 Act include permitting the Administrator to bring suits against persons contributing substantially to pollution or introducing new pollutants into treatment works. The Administrator may also undertake enforcement measures within states that do not respond rapidly enough to repeated violations. The Administrator has the authority to insure that conditions for discharges stipulated by permits be rigidly enforced and may bring suit to see that the conditions are met. Violators, if convicted, are subject to fines not to exceed $10,000 per day while second and subsequent convictions may bring fines not to exceed $50,000 per day and/or imprisonment up to 2 years.

The Act also authorizes grants (a) to states to pay up to 50% of the administrative expenses of planning agencies to prepare comprehensive pollution control plans; (b) to finance Federal participation in demonstration projects aimed at control or elimination of acid mine drainage; and (c) to states or municipalities to pay for construction of publicly-owned treatment works.

(e) *Housing and Urban Development Act of 1965.* The Act authorizes the Secretary of HUD to make grants to finance projects for specific public storage,

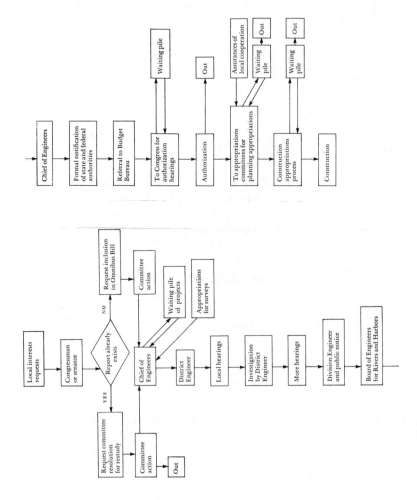

**Figure 10.1** Procedures for evaluating projects for Federal funding.
*Source.* Reprinted from Ferejohn (1974), with permission of Stanford University Press by the Board of Trustees of the Leland Stanford Junior University.

treatment, purification, and distribution facilities. Systems for collection and discharge of liquid wastes and storm water are included too. The Secretary can also make grants to states and local public groups to acquire and develop urban land for various public purposes and the Act authorizes the Secretary to establish a national flood insurance program. The Secretary must collect and publish information about both floodplain and coastal areas of high flood hazard, establish flood risk zones, and make estimates of the rate of probable flood-caused loss in these zones.

The foregoing examples are only a sample of the wide range of recent Federal programs dealing with water resources development and management. Clearly, many overlap and conflict. It is certain that even the agencies involved do not know of all possible conflicting or supporting programs in other agencies or, even possibly, the full ramifications of the programs they have been authorized to establish or enforce within their own agencies.

## A "HOW-TO" GUIDE FOR DEVELOPMENT OF A WATER PROJECT

Since Congress has long had the responsibility to support and encourage the development of water resources for the good of the nation, it might be instructive to review the steps that must be followed in the evaluation of a water project for Federal funding. Figure 10.1 is a flow chart of the procedures that are followed. Maass (1951) determined that each project must go through 32 distinct levels of consideration, although not at all of these levels are decisions made that affect whether the project passes on to the next level or not. At some levels, it must be determined whether the project meets certain "objective" criteria that automatically determine whether it passes or not. At still other levels, more formal decisions as to the worth of a project are made by committees or groups exercising independent discretionary power over the project.

A project normally begins with a local group sending a request for a study of a particular situation to a Congressman. If no study has already been completed, the Congressman will ask for the study request to be included in the next Omnibus Rivers and Harbors Bill. If a study has already been made, only a committee resolution is needed to authorize another study. Many of the Corps of Engineers studies in any given year have been authorized by committee resolution.

Once the study has been authorized, the project has to undergo a feasibility study. Here, the Corps of Engineers makes its own decision as to which requests will receive funds for feasibility studies in any given year. Preliminary investigations are carried out by the Office of the District Engineer with hearings at which local interests can express their views on the proposed project. The District Engineer forwards his report to the Chief of Engineers. At this point, benefit-cost figures are quite approximate for still the Chief of Engineers has the prerogative of deciding which projects should be studied further and which do not appear to be economically justified and can be eliminated.

Once a favorable report is given by the Chief of Engineers, the sponsors of the project must now obtain appropriations to carry out a more detailed study of the proposed project. Further hearings are held, a detailed plan of improvement is prepared, and a more realistic benefit-cost relationship is worked out. If this benefit-cost analysis shows a favorable situation (benefits exceed costs), the District Engineer will usually recommend that the project be constructed. There will, of course, be further reviews of the project by the Division Engineer, by the Chief of Engineers, and by the Board of Engineers for Rivers and Harbors. At this point, other agencies and groups are notified of the proposed construction. These groups include the Bureau of the Budget and Government agencies in the state in which the project will occur. These agencies usually make their own recommendations and send representatives to the authorization hearings before the appropriate public works committees of the Congress. All of the reports generated are submitted to Congress. Further approval of the project is now based solely on a benefit-cost ratio greater than unity.

In addition to hearings before the public works committees, other hearings will be held before Rivers and Harbors and Flood Control subcommittees. When the various project-by-project hearings are concluded, one of the public works committees will prepare an Omnibus Rivers and Harbors and Flood Control Bill. Some small additions or deletions may still be made on the floor, although when the project gets to this stage it will almost always be authorized as long as it has a favorable benefit-cost ratio. Ferejohn (1974) tells of one project included in the 1968 Omnibus Bill in which the Representative from the district involved argued strongly against the need for the construction. Since it was in the middle of his district and would not affect other districts, his wishes would have seemed to be paramount. However, the Corps recommended the project, it had a benefit-cost ratio of 6.0, and it was ultimately passed.

In recent years, Congress has required that projects have a preauthorization review by the President. The Budget Bureau of the President or his staff may object to some projects at this point, as clearly happened in the 1977 review process when President Carter requested no further funds be authorized for a score of water development projects in various stages of completion. Here, the authorization decision becomes crucial, as it was in 1977, with the Congress often wanting to authorize and the President wanting to delay. Usually some compromise is achieved that may not really be satisfactory to anyone.

After authorization, each project needs a "definite project report" and Congress will approve separately the funds needed to draw up these reports. This is a critical step because this is the time that the cost of the project is considered in relation to the overall availability of appropriations for the year. Authorized projects may have to wait here for some period before they receive approval for funds to be spent on the project report.

Lack of sufficient funds might not be the only reason for the failure of Congress to approve monies for the definite project report. In drawing up such a report, it is determined what lands will be taken out of private ownership, what local governments will contribute, what roads or bridges will be moved or modified.

Local residents now receive, essentially for the first time, detailed knowledge of who will benefit and who will suffer from the construction project, perhaps leading to stronger or more concerted local opposition than heretofore encountered. Repeated local hearings to resolve the problems often result in endless delays and ultimately could result in the killing of a project at this stage.

When the definite project report is completed, the project is then eligible for appropriation of funds for construction. The Corps must consider the make-up of its budget for the year and the need for funds for other construction projects. It goes to Congress (House Appropriations Subcommittee on Public Works) with its budget. After various hearings, the final appropriations bill works it way through Congress (usually requiring some conference committee work to iron out differences in the House and Senate versions of the bill). If the project is in the appropriations bill and funds are authorized for it, it will be built. Even though the Budget Bureau can now delay projects somewhat, they are rarely eliminated from the budget once they are included in it.

It is clear from the foregoing that many of the important decision-making stages in the life of a project involve the budget process once the benefit-cost ratio has been found to be favorable. Economic and, in a large measure, political decisions, therefore, influence the final approval of a project rather than environmental considerations, the rationality of the project, or even the wishes of the majority of the people.

## PROBLEMS OF BENEFIT-COST ANALYSIS

The use of so-called benefit-cost analysis to justify water development projects has probably resulted from the wording of Section I of the Flood Control Act of 1936 where it was stated that "... the Federal government should improve or participate in the improvement of navigable waters or their tributaries including watersheds thereof, for flood control purposes if the benefits to whomsoever they may accrue are in excess of the estimated costs ..." (Sec. 1, 49 Stat. 1570, 33 U.S.C. 701a). While the statement clearly applied only to flood control improvements by the Corps of Engineers and the Department of Agriculture, it was quickly accepted as a guideline by most water planning agencies for other water resource developments. The statement required that benefits should exceed costs but it did not spell out the nature of either the benefits or costs; it thus provided a loophole through which it has become possible to pass all manner of water projects whether economically justifiable or not. At the same time, different agencies employ different criteria for estimating both costs and benefits so that it is not a simple matter to compare the economic feasibility of projects across agencies (Blackwelder, 1976).

Benefits are usually identified as either tangible or intangible, Intangible benefits are those that are not easily evaluated in terms of dollars—for example, the saving of lives, aesthetic improvements, or the development of recreation facilities. Tangible benefits are often separated into both direct and indirect (or secondary).

Direct benefits involve the product or service that the water development project is intended to supply and indirect benefits include such things as construction jobs for unemployed workers or jobs for workers in other service industries at the construction site.

Estimating the benefits of a water development project is difficult because of uncertainties about future demands and prices. For example, in a navigation project, some estimate of kinds and quantities of materials to be hauled by water are needed as well as an estimate of the possible charges for such transportation. But these factors will depend on the availability of the navigation project and will change once it is developed and brought into use. The benefits from an irrigation project depend, in large measure, on estimates of agricultural yield, type of crops to be grown, and product prices in the future. The benefits from increased recreation in an area depend on the number of visitor-days that the presence of the project will develop and how importantly the recreation facility is evaluated by the users. Then too it is difficult to assign a dollar value to the recreational use of a facility.

In order to evaluate benefits, it is also necessary to know the value of the benefits given up. For example, in the construction of a dam and reservoir, a considerable amount of land will be taken out of production as a result of the filling of the reservoir. It has been estimated that some 300,000 acres are lost each year beneath the waters of new reservoirs. The benefits that might have been derived from the use of that land must, logically, be subtracted from the estimated benefits of having the reservoirs if a true benefit-cost analysis is to be performed. The use of the stream valley for recreation activities before the reservoir was constructed must have provided certain benefits. These too must be subtracted from the benefits accruing as a result of the construction of the reservoir.

Critics of benefit-cost analysis suggest that there is always a tendency for the construction agency to overestimate benefits while neglecting to list all possible costs. Their arguments do seem to carry significant weight. For example, a prime benefit of flood control projects is the value of "damages prevented" as a result of the existence of the control works. Damages prevented not only are difficult to estimate, but they also depend, in large extent, on the subsequent development of the floodplain below the dam. Clearly, many of the structures built on the floodplain would not have been built there had the dam not been constructed, providing as it does some measure of security from future flooding. Thus, according to this argument, almost any flood control work can be economically justified by the damages prevented as long as a sufficient number of valuable structures are built on the floodplain below the flood control work. In spite of criticism about the "damages prevented" argument, the Corps of Engineers relies extensively on this technique to justify new flood control works. For example, in the 1940s only about 10% of the benefits estimated by the Corps for flood control projects were assigned to preventing damage to future floodplain developments. Two decades later, an average of 40% of flood control benefits were based on the protection offered to future floodplain developments. In the case of the authorized Days Creek Dam in Oregon, some two-thirds of the flood control benefits accrue from protecting future floodplain developments. It would seem that to justify flood control works on the basis of protection of structures that will be built on a floodplain because it is supposedly protected by the flood control works is a weak and circular economic argument indeed.

Aside from the circular nature of the argument, there may, of course, be a fallacy in it as well. Dam construction will not always prevent floods. For example, the Johnstown, Pennsylvania floods of 1889 and 1977 both resulted from dams giving way, as did the Teton Dam flood in Idaho in 1976. And, of course, much of the rain that caused the Rapid City flood (see Chapter 11) fell in the unprotected portion of the watershed between the Pactola Dam and the city itself. Thus, assigning values for damages prevented is a risky business. Flood control work may prevent 9 out of 10 or 99 out of 100 of the floods, but floods will still occur on most rivers and the damages, when they come, can be extremely severe if unrestricted building has occurred on the floodplain.

Recreational enhancement has often been listed as a benefit of a proposed water development project. This would seem to be reasonable provided estimates of use are accurate and are balanced by a consideration of the loss of recreational value in the area as a result of the construction. For example, the New Melones Dam now being built on the Stanislaw River in California will cover a white-water recreation area used by some 20,000 raft and kayak enthusiasts as well as by other visitors each year. The Corps of Engineers estimated that 4 million visitors would annually use the new reservoir area although the Environmental Policy Institute suggests that this is a very high estimate in view of the fact that the Don Petro Dam and reservoir just 7 mi away and of similar size has only about 250,000 visitors a year.

In its plans for the Spewrell Bluff Dam in Georgia, the Corps estimated the visitor usage would be as great as the most heavily-used reservoir already constructed in the nation. Such a heavy usage may actually develop but it would seem hazardous to predict a record usage in drawing up benefit estimates.

Use of reservoir areas for recreational fishing may well be overestimated by construction agencies. Free-flowing streams are much better able to keep themselves free from pollution than reservoirs. Accumulation of nutrients in reservoirs can make them eutrophic, increase algal blooms, and cause health problems—thus making them less available for water-based activities. Such losses of recreational opportunities are seldom considered in benefit-cost analyses.

The channelization of streams and the building of other works to improve navigation are often justified on the basis of increased transportation usage as well as the secondary benefits resulting from the additional industry that will be attracted to the area. But estimates of such increased economic activity are often not borne out in actual practice. The Environmental Policy Institute points out that one is not justified in using the difference between rail rates and barge rates for the commodities that would move over the navigable stream as benefits that would accrue due to the navigation works. Rail rates generally overstate the real costs of rail transportation while barge rates usually understate the real costs. Thus, benefits should be determined not from a consideration of rates but rather by using the real resource costs to the nation as a result of using each method of moving goods. As an example, the Missouri River navigation project to provide a channel from St. Louis to Sioux City, Iowa was justified, in part, on the basis that it would save shippers $1.00 per ton shipped. However, the annual costs of operation and maintenance are now over $5.00 per ton shipped, which might lead one to question whether it is economically worthwhile to keep the project operating. The Corps also estimated that 12 million tons would be shipped annually if the project were authorized. This estimate was reduced after authorization of the project to 5 million tons. Actually, the traffic load reached 2.5 million tons in 1964 and has remained at about that level ever since.

If ones compares the fuel used to move a ton of goods by rail and by barge, barges appear to be the most energy efficient. But ton-mile comparisons do not provide the whole story. River routes are generally 25 to 30% longer than rail routes, and the use of unit trains for bulk commodities—the type that would be moved by barge—reduces rail costs considerably. Sebald (1974) concluded that, considering all aspects, railroads were 10 to 23% more energy efficient than barges. The question seems very much in doubt still.

### The Estimation of Costs

The foregoing has dealt mainly with possible problems related to estimation of benefits from a water development project. Other examples of overly optimistic estimation of benefits could be cited but the main point to be made is that it is extremely difficult to obtain realistic estimates inasmuch as so many unknowns are involved in any construction program. The same conclusion applies to the cost side of the analysis. While it is perfectly feasible to estimate immediate costs for labor and materials to build a dam or a levee, and to determine the cost of land, rights-of-way, and services that will be needed, these costs may be entirely unrealistic 5 to 10 years later when the actual project gets under way. The time lag between proposal of the project and actual construction is often so great that realistic cost figures are difficult to estimate.

Possibly more important are the costs of modifications, repairs, or remedial work that might be necessary as a result of poor initial planning or the occurrence of unforeseen contingencies. It is, of course, impossible to anticipate all possible negative developments that might occur with the building of a dam or the straightening of a river channel. Careful planning will certainly forestall some adverse consequences but not necessarily prevent all problems. When the cost to correct problems caused by the water development project are added to the cost of the project, many of them would no longer be economically feasible. For example, Stockton Dam on the Sac River in Missouri was built to provide flood protection for agricultural fields below the dam as well as to provide power. However, the carrying capacity of the river was not correctly figured; consequently, whenever water is released to run the hydropower turbines, the river floods downstream. At present, warning horns are sounded before water is released so that cattle and workers can be removed from fields adjacent to the river that will be flooded. Erosion is rapidly scouring the agricultural fields the dam was designed to protect. Either government purchase of the downstream floodplain or more extensive (and expensive) remedial work will be necessary to correct for this miscalculation (Anonymous, 1974).

Other dams have raised water tables in their vicinity, necessitating drainage works, while some reservoirs have permanently flooded more upstream land than the dam would protect from periodic flooding downstream. Still other dam-reservoir projects have resulted in the loss of prime fishing and hunting areas because of the drowning of valleys and the development of eutrophic conditions in the reservoirs. Reservoirs in karst or limestone areas subject to severe underground leakage or in areas with old oil wells, oil seepage, mine drainage, or toxic metal deposits may be subject to problems of contamination or loss of water and may require extensive remedial work if they are to be useful multi-purpose reservoirs.

Sedimentation problems plague all reservoirs but some much more than others. If the stream feeding the reservoir flows through areas in which the soil is easily eroded, some mechanism to remove the sediment before it settles out in the reservoir must be incorporated into the project. In benefit-cost analysis, seldom considered are the costs of keeping the reservoir free of sediments or the costs of relocating activities dependent on the reservoir (or supplying a new source of water) that must be incurred when the reservoir fills up. This is unfortunate since they are important costs that will develop as a result of the initial construction of the reservoir. Dams and reservoirs are monuments to our engineering capabilities but they may prove to be very costly monuments indeed—and possibly monuments to our lack of real understanding and foresight.

## The Question of Discount Rate

Another problem of considerable significance in any benefit-cost analysis is the choice of a discount rate to use on the money required to fund the project. In an ideal situation, competition for money among entrepreneurs should result in a market rate of interest which, in equilibrium, equals the marginal productivity of capital to each entrepreneur. The actual rate of interest that must be paid by each entrepreneur depends on the risks involved in investing the particular funds.

When the government attempts to decide whether to invest in a particular project, it must first select a rate for discounting both the projected costs and benefits in order to arrive at a decision as to whether the project is economically feasible. Should it use the market rate of interest or some other rate arrived at in a more complex manner which might be called a "social rate of discount?" The U.S. Army Corps of Engineers in the 1950s and early 1960s used a rate of discount of 2-5/8% (it has now been raised to 5-1/8%). Fox and Herfindahl (1964), in a review of Corps of Engineers projects for 1962, report that 178 projects were authorized with benefit-cost ratios greater than unity based on a discount rate of 2-5/8%. If, instead, the discount rate had been set a 4, 6, or 8% of the initial gross investments, they determined that 9%, 64% and 80%, respectively, would have had benefit-cost ratios less than unity and hence would not have been authorized. Thus, the discount rate is extremely important in benefit-cost analysis in those situations (such as water development projects) where a large capital expenditure is needed early in the project before any benefits develop. Usually costs are almost entirely confined to the first few years of a project while benefits may accrue slowly over the entire life of the project. Accordingly, the project becomes quite sensitive to the discount rate.

But what discount rate should be used? Ferejohn (1974) made a careful review of the subject, studying what might be called the riskless bond rate, the social rate of time preferences without consideration of risk, as well as the market rate of interest. He concludes that the actual choice of a discount rate is a most controversial and unsettled question among economists, mainly because it is more a political question than an economic one. Those agencies that use benefit-cost analysis, however, usually choose discount rates that are well below market interest rates and apply them uniformly across projects and equally to benefits and costs.

It is clear that in preparing benefit-cost estimates, both benefit and cost figures are only crude estimates. Comparisons that show a benefit-cost ratio close to unity must certainly be viewed with suspicion in view of the unknowns involved. Since there is always the tendency to overestimate benefits and underestimate costs, it might be desirable to set a benefit-cost ratio considerably greater than unity as the breakpoint to determine feasibility of construction.

## A Retrospection Study of Benefit-Cost Analysis

A recent analysis of benefit-cost estimates derived from projects in the Cumberland River basin of Kentucky and Tennessee corroborates some of the previous conclusions concerning overestimation of benefits and underestimation of costs (Martin and Thackston, 1980). The authors compare the benefit-cost ratios estimated at the time of planning with ratios derived from ex-post performance information on four projects. Four reservoirs on the Cumberland River—Wolf Creek Dam (in Kentucky), Dale Hollow Dam, Center Hill Dam, and J. Percy Priest Dam, all in Tennessee—were studied. The first three were authorized in the 1930s and begun in 1941 or 1942, although they were not finished until after World War II because the funds were needed for the war effort. Funds for the Priest Dam were not appropriated until 1963. The Wolf Creek, Dale Hollow, and Center Hill projects were not considered to have recreational benefits when initially planned and so benefits were estimated for only flood control and power. The Priest project included recreation as part of its initial benefits in estimating benefit-cost ratios.

Table 10.1 provides preconstruction estimates of the benefit-cost ratios without recreation benefits for the three projects not originally constructed for recreational use as well as the actual ex-post construction benefit-cost ratios both with and without recreation benefits considered. Since the Priest project considered recreation originally, these data are only given with recreational benefits included.

**Table 10.1**  Estimated and Actual Base Year Dollar Benefits and Costs
(Calculated at 3% Interest Rate) As Well As Benefit-Cost Ratios for
Four Reservoir Projects on the Cumberland River

| Component | Average Annual Base Year Dollars | |
| --- | --- | --- |
| | Estimated | Actual |
| *Dale Hollow Project* | | |
| Cost | 505,252 | 790,425 |
| Flood control | 206,583 | 284,503 |
| Power | 530,000 | 273,814 |
| Recreation | 0 | 419,754 |
| B/C ratio with recreation | — | 1.24:1 |
| B/C without recreation | 1.46:1 | 0.71:1 |
| *Wolf Creek Project* | | |
| Cost | 2,676,300 | 1,410,315 |
| Flood control | 1,162,000 | 555,969 |
| Power | 2,882,000 | 538,531 |
| Recreation | 0 | 884,097 |
| B/C ratio with recreation | — | 1.40:1 |
| B/C without recreation | 1.53:1 | 0.78:1 |
| *Center Hill Project\** | | |
| Cost | 870,587 | 1,002,494 |
| Flood control | ˙464,812 | 306,217 |
| Power | 1,296,000 | 368,970 |
| Recreation | 0 | 519,482 |
| B/C ratio with recreation | — | 1.17:1 |
| B/C without recreation | 2.02:1 | 0.66:1 |
| *J. Percy Priest Project* | | |
| Cost | 1,519,500 | 2,458,429 |
| Flood control | 989,000 | 1,389,634 |
| Power | 495,000 | 458,119 |
| Recreation | 989,000 | 1,714,025 |
| B/C ratio | 1.6:1 | 1.45:1 |

*Combining the first three projects, the overall B/C ratio is 1.29:1;
without recreation, 0.72:1.
*Source.* Martin and Thackston (1980) with permission of the American
Water Resources Association.

The table shows that the Corps of Engineers underestimated the construction cost of the Dale Hollow, Center Hill, and J. Percy Priest projects although they over-estimated the cost of the Wolf Creek project by some 47%. Estimated benefits from flood control were underestimated for Dale Hollow and Percy Priest projects al-though they were overestimated for the other two projects. Power benefits were overestimated for all four projects. As a result of these over- and underestimations of costs and benefits, the Corps of Engineers came up with strongly positive benefit-cost ratios for all four projects ranging from 1.46 to 1 for Dale Hollow to 2.02 to 1 for Center Hill even without considering any recreational benefits in three of the projects.

Martin and Thackston recomputed the benefit-cost ratios on the basis of the actual performance of the projects at a later time and found significant differences between estimated and actual results. When recreational benefits are considered at the four projects, they still all end with benefit-cost ratios greater than unity, suggesting that the projects were justified. However, the benefit-cost ratios have dropped, ranging now from 1.17 to 1 for Center Hill to 1.45 to 1 for Percy Priest. Without the inclusion of recreational benefits (Dale Hollow, Wolf Creek, and Center Hill were not originally built for recreational purposes), the ex-post con-struction benefit-cost ratios are below unity for these three projects (ranging from 0.66 to 1 for Center Hill, to 0.71 to 1 for Dale Hollow, and 0.78 to 1 for Wolf Creek). The breakdown of the figures in Table 10.1 suggest the type and magnitude of the problems that result from any benefit-cost estimates.

Martin and Thackston do not indicate whether the cost of benefits lost due to the construction of the reservoirs (land taken out of cultivation, homes moved, loss of river use) were first removed from the estimated benefits of the projects to obtain a truer picture of benefits. They do discuss the problem of the discount rate, however, and show that all four projects could be justified if recreational benefits were included and the internal rate of return (interest rate on money borrowed for the projects) was below 3-3/4% for Center Hill, 4% for Dale Hollow, 5% for Wolf Creek, and 7% for Percy Priest. Considering the time when these projects were constructed, the discount rate of 3% (which was actually used) seems reasonable, except for the Priest project which was not funded until 1963. Figure 10.2 shows how interest rate or internal rate of return influences benefit-cost ratios for different projects within the same basin and how sensitive project costs are to the selection of the proper interest rate.

An argument against supporting only projects with benefit-cost ratios greater than unity is possible. The argument hinges on the fact that some projects with benefit-cost ratios less than unity are still worthwhile since they could lead to a movement of money into needy hands. Thus, they might have socially de-sirable goals even though they might not be justified on the basis of strict eco-nomic considerations. Ferejohn (1974) argues that there is no evidence that such inefficient projects do, in fact, aid socially desirable goals nor is there evidence that it is these considerations which make such projects pass in the Congress. He concludes that there is a pork barrel and that it exists, in large measure, because of our system of representative government, majority rule, and the committee system procedure for review and approval of projects.

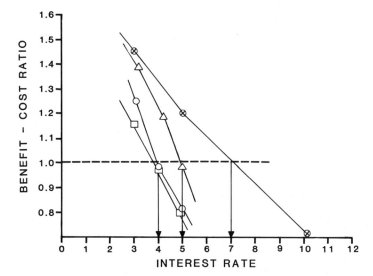

Figure 10.2. Internal rate of return at Dale Hollow, Wolf Creek, Center Hill, and J. Percy Priest Reservoirs on the Cumberland River.
*Source.* Martin and Thackston (1980), with permission of the American Water Resources Association.

## THE PORK BARREL

In considering the term pork barrel, layman usually do not utilize the significant benefit-cost argument. Rather, the lay view of pork barrel involves just the appropriation of money for projects in some other part of the country or in some other congressional district. The view is that as a result of logrolling—the trading of votes from one block or one individual to support a bill in which the first block or individual has little interest in order to receive in return the votes of others to support a bill favored by the first block or individual—a legislator can achieve the authorization of projects and improvements in his district at the expense of the taxpayers in all districts. Even if such projects have benefit-cost ratios greater than unity, such excess benefits seldom filter far from the district where the project is built. Thus, to unaltruistic taxpayers in districts far removed from the district of construction, the benefit-cost argument is less than compelling and the fact that some taxes must go to support projects that bring no immediate benefit to the local taxpayer is sufficient evidence of the existence of a pork barrel.

Several possible models were developed by Ferejohn to explain why the pork barrel phenomenon might develop and continue to exist under our type of government. While each of the models has certain limitations, he feels one suggested by Niskanen (1971) provides a very useful understanding of at least portions of the decision-making process. In his model, Niskanen divides the legislature into three equal-sized groups on an issue such as a water development project. The high-demand group includes congressmen whose districts have many rivers and harbors on which they want improvements. The medium-demand and low-demand groups have fewer rivers and harbors in their districts, respectively, and have less desire for improvements. Niskanen assumes that the high-demand group will dominate the committee that writes the water development legislation and that the medium- and low-demand groups will not be moved to form a coalition to change the legislation on the floor. While each district will share equally the cost of the program, the high-demand group receives the largest benefits since it has the largest number of projects while the medium-demand group receives the next largest benefits. The high-demand group will opt for the highest rate of expenditures with the low-demand group wishing for the lowest rate of expenditures on the program.

Since the high-demand group controls the Appropriation Committee, it will select a budget figure so that a majority of members in the medium-demand group will have their projects approved. The bill should easily pass inasmuch as the members of the medium-demand group, not feeling it is worth their while to defeat or amend the bill, will vote with the high-demand group. If the final budget figure is too small and does not result in approval of a majority of projects for the medium-demand group, they will vote against the bill along with the low-demand group and the bill will fail. Thus, the largest possible budget to insure sufficient projects with positive net benefits to the medium-demand group will be proposed. It is not necessarily clear whether economically unsound projects will be built under this model. Ferejohn (1974) concludes that this type of model can produce pork barrel behavior. The pork barrel, in part, results from the need to put together a winning coalition of legislators. The appropriations bill must have enough in it to attract at least a majority to vote for it. Inefficient projects can be included since the costs are shared by all and not borne solely by the residents of the district in which the construction occurs. Thus, costs appear relatively small for each individual and most taxpayers have no real way of knowing whether they exceed benefits or not.

There have been a few but largely ineffective efforts to stop the pork barrel. The suggestion to give the President power to veto selected portions of the Omnibus Rivers and Harbors Bill is predicated on the hope that he might be able to rid the bill of the most inefficient projects. Congress is not ready to give the President that authority. Of course, if the taxpayers themselves were willing to join in a strong stand against the economically least justified projects (even if they were in their own districts), changes could probably be accomplished but it would have to be clearly demonstrated to Congress that significant changes in the pork barrel approach were necessary. The system will be difficult to change without pressure from the electorate.

## RIVER BASIN MANAGEMENT

Over time, several different organizational structures have been evolved for the purpose of planning and managing river basin development. The differences are, in part, due to differences in the geography and geology of each basin, variations in the principal function demanded of the organizational structure, and the development of new ideas concerning the optimal structure. The principal organizational structures are (a) intrastate (state-created) river basin authorities; (b) joint Federal-state river basin planning and coordinating committees; (c) interstate or joint Federal-interstate compact commissions; (d) Federal-state governmental corporations.

### Intrastate Authorities

The earliest river basin planning and management organizations were established by states to exercise planning and/or regulatory powers within a single state. For example, the Wisconsin Valley Improvement Company, a corporation chartered by the state of Wisconsin in 1907, was owned by six paper mills and four power utilities. It operated dams and reservoirs on the Wisconsin River designed to supply both power and water to the industrial users. The Miami Conservancy District was set up by the Ohio legislature in 1914 as a result of a serious flood which damaged the Miami River Valley and the Dayton area in 1913. It was originally established for flood control purposes but later was broadened to regulate streamflows, to improve regional water quality, and to conserve and develop needed water supplies. A number of river basin authorities have been established in Texas, starting with the Brazos River Authority (1929) and the Lower Neches Valley Authority (1933). These authorities have quite broad powers to regulate basin developments, including construction projects for flood control, power, water supply, navigation, pollution, recreation, and conservation. The various Texas river basin authorities have been a very positive force in developing the water resources of the state, generally by means of revenue bonds which do not add to the public tax burden.

Intrastate authorities have been found to be satisfactory organizational structures, particularly if they are given fairly broad regulatory power and independent financial support. They are especially useful in bringing together water resources activities in basins in which many different local groups are operating and in which closer coordination of planning and development is necessary.

### River Basin Planning Commissions

In 1939, the Corps of Engineers, the Bureau of Reclamation, and the Department of Agriculture entered into an agreement to consult in the preparation of river basin surveys. By 1943, the Department of the Interior and the Federal Power Commission also joined in this agreement to establish a Federal Interagency River Basin Commission to continue the coordination of basin survey activities begun earlier. This Commission had some problems in achieving effective coordination since it had no statutory standing, no budget, and its decisions were advisory only. Implementation of decisions were dependent on voluntary cooperation of the individual agencies. It was extremely difficult to reconcile the different agency plans and policies.

To find a more effective water planning structure, proposals were made to establish basin commissions combining Federal, state, and private interests. The 1965 Water Resources Planning Act allowed the President, by Executive Order, to establish a river basin commission upon the request of either the Water Resources Council or a state. The Council and at least half of the states in the basin must agree to the formation of the commission, except in the case of the Upper Colorado and Columbia basins where three-quarters of the states must agree. Commissions would not be established in areas covered by the TVA or the Delaware River Basin Commission. The commission would serve as the main agency to coordinate all water development plans for the basin regardless of source. It would be responsible for preparing an up-to-date comprehensive joint plan for water and land resources development in the basin. The commission's activities were limited to planning rather than to regulation, construction, or management and the authority of the commission could not limit any preexisting authority held by state or Federal agencies. As a result, such commissions have little authority and no way to enforce decisions. Rather they can only facilitate discussion or provide a stage for bargaining among the separate entities in the river basin planning picture. In such a role, the personality and ability of the chairman become important to the success of the commission.

### Federal-Interstate Water Compacts

Compacts among states to deal with water problems in basins involving more than one state require the consent of Congress if the national interest is involved. Interstate compacts were used originally under the Articles of Confederation to solve boundary, navigation, and fishing problems. In the Colorado River Compact of 1922, a new problem, that of making water allocations to the various states in the basin, served as the basis for an interstate compact. In the following 50 years, more than 30 interstate water compacts have been entered into to solve a variety of water resources problems. Chief among these are compacts for water allocation, pollution control, flood control and planning, and project development and coordination.

In regions of water shortage, compacts for water allocation are quite efficient in permitting the development of rivers without time-consuming legal battles among neighboring states for their respective water rights. Most of the water allocation compacts protect existing water uses and rights.

In 1921, the Supreme Court and the Congress approved the use of interstate compacts to solve pollution control problems. New York, New Jersey, and Connecticut entered into a compact in 1935 to improve water quality in New York harbor. Since then, 10 other compacts have been established to deal with interstate water pollution problems. The powers of the commissions resulting from these compacts vary widely. In the case of the Potomac River compact, the commission can only study and recommend remedial action, whereas the Delaware and the Susquehanna River Basin commissions have the power to set water quality standards and to enforce them. While the original water pollution compacts (New York harbor, Ohio River, Tennessee River, Potomac River, New England agreement) were single-purpose compacts, the more recent compacts include water pollution control as only one portion of the entire compact.

A few compacts deal primarily with flood control aspects of basin development (Red River of the North, Connecticut River, Merrimack River, Thames River, Wheeling Creek). These compacts resulted from the flood control program of the 1930s and were established to promote interstate cooperative action.

There has been a growing need for better coordination of comprehensive water resources planning and programs not only within the Federal government but also between the Federal government and the states. One possible suggestion is for the Federal government to be included as a partner in multi-state compacts in order to make clear its authority within such basins. Early attempts in the Missouri Basin Survey and in New England to establish Federal-interstate compacts were not successful. However, the Delaware River Basin Commission, which developed as a result of several decades of controversy over water allocation, received permission to begin operations in the late 1950s. The Delaware River Basin Commission is fairly unique in that the United States is a signatory party with the several states and the commission is given extremely broad regulatory and enforcement powers. No project that has "a substantial effect" on water resources in the basin can be undertaken by any public or private group without the express consent of the commission. The Basin Commission also has the power to allocate water to the various states in a fairly flexible manner under a doctrine of equitable apportionment. While it does not have any power to tax, it does have fairly broad financing power. Most important, all water projects planned by Federal, state, or local groups must conform to the Basin Commission's comprehensive plan.

The Federal-interstate compact on the Delaware River served somewhat as a model for the later Susquehanna River Basin Commission and aspects of the compact have been incorporated into more recent agreements now being drawn up. The Delaware River Basin Commission has achieved a number of worthwhile results. It has established basin-wide water quality standards, encouraged regional solutions to waste disposal problems and even financed a demonstration project in regional waste management, encouraged floodplain mapping studies, started to regulate conditions for power plant construction, and has contributed to the development of recreational areas.

River basin compacts have been found to be particularly adaptable to the individual needs of a basin. A compact is developed as a result of agreement among the states in the basin and can be directed either to solve a single problem (e.g., water pollution) or to aid in reaching multi-purpose basin management goals. The powers authorized under the compact may be broad or narrow, reflecting the desires of the states. The compact agency may even have powers not actually exercised by the member states individually, although this point has not been fully decided. Certainly when Federal powers are given to the compact agency by Congress and the United States becomes a signatory party (so that the compact agency becomes a Federal instrumentality), it does receive authority over and above that possessed by an individual state.

Some critics suggest that the development of compacts and their approval both at the state and congressional level requires too much time. Delays in developing compacts often result from disagreements over specific policy matters rather than from any fault of the compact mechanism. The Delaware and Susque- hanna River Basin compacts were negotiated and approved with reasonable speed. At present, states are usually represented equally in a compact agency even though they may have quite disparate interests in the basin because of geography, resources, importance of the basin, or population distribution. It is possible that a form of weighted representation in the compact agency might prove to be a fairer arrangement in some cases. Also, some possibilities of limited modification of the compact without the need for a new ratification by state legislatures or approval by Congress would seem to be reasonable. Congress could also aid in the establishment of a compact by giving advance consent to a certain limited class of compacts.

## Federally-Chartered Corporations

Federal corporations have been used by Congress to carry out both financial and operational missions for the government. The Tennessee Valley Authority is a particular example of a Federal corporation in the river basin management field. In many ways, it has proven to be a successful organizational structure, leading to suggestions that similar corporations might be established to manage other multi-state water activities. There may be particular situations where a limited Federally-owned corporation might provide a useful service but it is questionable whether any corporation of the scope and type of TVA should be created.

Water management corporations at the state level have usefulness in certain situations. An early version was the previously mentioned Wisconsin Valley Improvement Company chartered in 1907 as a private, nonprofit corporation to allocate river water supply and increase utilization of available hydroelectric generating capacity.

Federally-chartered corporations could have a majority of their directors appointed by and responsible to either the Federal government or to the states involved. Congress has the authority to commit the Federal government to a minority role, as has been seen in the case of the Delaware River Basin Commission where it has only one-fifth of the total voting power. The question still must be answered whether a Federally-chartered corporation has the right to exercise state governmental authority or even whether states can give such a corporation state governmental power.

Financing options of Federally-chartered water resources corporations are many. The corporation might be supported by project revenues or user fees, or it might be financed through the issuance of revenue bonds. Congress could pledge the credit of the United States to support securities issued by the corporation. If the corporation is not self-sufficient, it may have to depend on annual grants from either the Federal or state governments. Whether the corporation can be granted the power to tax, or whether either the Congress or the states would even want to grant such power, has not been fully explored and would seem to depend on the exact wording of the appropriate state laws.

Most observers feel that the corporation form, since it is somewhat separate from political control and responsibility, is best suited to water construction, operation, and maintenance projects. This does not mean that it could not be used in water activities involving comprehensive planning or regulation of water uses or users, but it is generally felt that those activities might be better handled by means of some interstate compact form of organization.

## THE POTOMAC RIVER: AN EXAMPLE OF RIVER BASIN
## PLANNING AND PROBLEMS*

The problems that beset a moderate-sized interstate river in the United States today can be better understood by a detailed consideration of the Potomac River. The Potomac is not particularly large, being 383 mi in length from headwaters in western Maryland and northern West Virginia to Point Lookout where it enters into Chesapeake Bay. It has a number of tributaries—the Shenandoah in western Virginia, the Monocacy in central Maryland, and the Anacostia as it reaches the coastal plains near Washington, D.C. (Fig. 10.3).

Unlike some rivers, it is not heavily industrialized but it does flow through a region of large and growing population. Demand for Potomac River water covers the whole range of typical human needs—domestic water supply, waste dilution, recreation and enjoyment of aesthetic beauty, navigation, commercial fishing—and, of course, the inhabitants of the floodplain areas must be protected from flooding. Streamflow is quite variable in the Potomac. At the gaging station near Washington, D.C., flow has varied from an extreme low of less than 800 cfs to a high value of nearly 500,000 cfs. This wide range makes flood control activities and river basin planning extremely difficult. The mean flow in the river is about 11,000 cfs.

The Corps of Engineers has estimated that water use in the Potomac River basin will increase from a mid-1960 level of 1.3 bgd to a value of 4.3 bgd by 2010. Such estimates, if borne out in practice, will result in water shortages in the Washington area by 1985. The problem results not because of total water volume in the river, but rather from the seasonal pattern of flow. Thus, regulation of flow

*The material in this section was adapted in large measure from a comprehensive summary of the problem which appeared in *The Johns Hopkins University Magazine* (Anonymous, 1966).

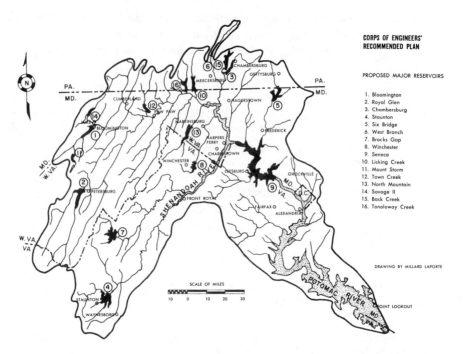

CORPS OF ENGINEERS'
RECOMMENDED PLAN

PROPOSED MAJOR RESERVOIRS

1. Bloomington
2. Royal Glen
3. Chambersburg
4. Staunton
5. Six Bridge
6. West Branch
7. Brocks Gap
8. Winchester
9. Seneca
10. Licking Creek
11. Mount Storm
12. Town Creek
13. North Mountain
14. Savage II
15. Back Creek
16. Tonoloway Creek

DRAWING BY MILLARD LAPORTE

**Figure 10.3** Basin of the Potomac River showing location of proposed Engineer's reservoirs. *Source.* © *The Johns Hopkins Magazine* (1966); Anonymous (1966).

is suggested as the way to supply the ever-increasing demands, by holding the excess supplies of the winter period for release during the summer times of low flow. The problem of water quantity is closely linked to water quality since as more water is removed from the river for use, less water remains for dilution purposes. The river was highly polluted in the 1960s, at least along its lower reaches. Some 2 to 3 millions tons of silt were estimated to be carried by the river each year. In addition, coal mines in the Appalachian ridges resulted in quantities of acid mine waste reaching the river. Most industries in the basin treat their wastes somewhat before they are returned to the river, but still the effect of river use for dilution purposes is noticeable in increasing levels of pollution. Runoff of fertilizers, insecticides, or other trace metals and sewage, of course, provides the largest volume of wastes that go into the river. While most of this is treated waste, present techniques of primary and secondary treatment cannot remove all the organic material or the biochemical oxygen demand from the domestic effluent. A brief example illustrates the problem of using the river for effluent disposal.

In the 1930s, essentially all of the domestic sewage entered the Potomac River untreated. By the mid-1960s, about 97% of the domestic sewage discharged to the river received secondary treatment. A gain like that in treatment of wastes might seem to be sufficient to prevent the development of problems from domestic wastes, but that is not the case. If we assume that secondary treatment removes

80% of the waste load, then such effluent from the Washington, D.C. metropolitan area of 2,000,000 people would have the same effect on the river as the disposal of untreated domestic sewage from a population of 400,000 people. Because of recent increases in population, total volumes of sewage loadings in the 1930s and 1960s were about equal even though there had been a significant increase in waste treatment.

If the population of Washington grows to 5 million by 2010, as some estimate, treatment of the domestic effluent will have to be improved in such a way that 92% of the waste load is removed just to maintain the mid-1960s pollution loading on the stream. But increasing the level of treatment from 80 to 92% is a very expensive operation.

The Potomac River reaches sea level at Washington and is a tidal river and estuary from there to the Chesapeake Bay. This condition results in an accumulation of pollutants in this reach of the river as the tide washes back and forth. At times of low flow, there is little dilution. At the same time, with sewage effluent well endowed with nitrogen and phosphorus, these nutrients support a thick growth of algae. Algae that has died and been carried to the bottom will decompose, often anaerobically, with the release of strong gases. Because there is not sufficient water for dilution in the summer season, either the nitrogen and phosphorus will have to be removed from the effluent, sewage effluent will have to be kept out of the water, or larger volumes will have to be released from upstream storage to increase dilution volumes.

A team of Johns Hopkins University scientists has suggested nine steps that, if undertaken singly or in combination, would result in a lowering of the pollution load in the river at Washington (Anonymous, 1966).

(1) Decrease the BOD of the effluent everywhere through improved waste treatment facilities.
(2) Remove the nitrogen and phosphorus by building new treatment facilities since these plant nutrients are unaffected by the conventional sewage treatment system.
(3) Transport treated effluent downstream by pipeline to the estuary of the Potomac where water volumes are greater.
(4) Transport treated effluent overland by pipeline to the Chesapeake Bay.
(5) Construct barriers or low dams to control water levels and the tidal movements of pollutants in the lower portions of the river.
(6) Construct a major upstream reservoir to even out river flow, provide flood control, confine nutrients and sediments, and provide recreational facilities.
(7) Modify present combined storm and sanitary sewer systems in Washington to eliminate discharge of raw wastes during rainstorms.
(8) Accelerate silt dredging operations and removal of debris.
(9) Increase watershed activities designed to reduce erosion.

Conservationists have made a number of suggestions concerning how the river might be rehabilitated. One plan calls for the removal of 90% of the contaminants in the sewage by 1975, 95% by 1990, and 100% by the year 2000. To achieve 100% purity, an evaporation distillation technique similar to the process now used for sea water desalination is suggested. But the estimated costs of such a treatment facility for Washington is $300 million plus $30 million more for yearly operating costs. The heat necessary to boil off the water would require 1 million gallons of fuel oil each day, the contaminated sludge would be difficult to dispose of, and the heated water at the end of the process would raise the temperature of the river water to over 100°F!

A more feasible plan—piping the Washington effluent 30 mi to an outflow in the Chesapeake Bay—first suggested by Professors Abel Wolman and John Geyer of Johns Hopkins in 1957, might cost only $50 to $75 million. Provided the effluent did not upset the ecological balance of the Bay, the dilution volume in the Bay would be equivalent to a flow of 100,000 cfs, 10 times the average Potomac River flow.

The Corps of Engineers has also put forth its own plan for the future development of the Potomac River Basin. Its nine-volume report provided a plan (with three alternatives) to solve the water supply and pollution problems of the basin. The Corps' basic plan would only satisfy water supply and quality needs to the year 2010. It would also provide needed flood control on the river and establish recreational facilities sufficient for approximately 16 million visitor-days per year. The overall cost of the plan would be $498 million. Following their previous basin plans, the Potomac plan suggested by the Corps called for more than 400 small upstream dams for flood control, for irrigation water, and for sediment retention. Sixteen major storage dams were also planned, one to be at Seneca just a few miles upstream from Washington itself (Fig. 10.3). These major dams, and especially the one at Seneca, have resulted in the bulk of the controversy over the Corps' plan. The Seneca dam is necessary for flow regulation in the tidal river which cannot be handled by upstream dams but it may not be needed for water supply purposes. The Seneca dam would also be very effective in trapping silt from the entire basin area but it would also fill with sediments and lose some of its storage capacity.

Opponents of the Seneca dam have marshalled some significant facts. While reservoirs and dams do lead to a regulation in stream flow, they are not necessarily the best way to protect against flooding in all river basins. Dams cannot be counted on in all situations to prevent flooding. A more permanent solution would be floodplain zoning, although this solution would be difficult to apply in areas already well built up.

The additional water made available to the lower reaches of the Potomac River during the summer period of low flow has been calculated to be rather small and entirely inadequate to handle the pollution load of the river at that time. Dissolved oxygen will still be depleted even though the summer flow might be increased by a factor of three from 1000 to 3000 cfs. The excess flow will also not do much to lower the excess nutrients in the upper Potomac River estuary. It has been estimated it would be necessary to increase the summer flow in the upper estuary to 20,000 cfs if the nutrient level is to be reduced to a level twice as great as in the upper Chesapeake Bay where the nutrient level is viewed as in good balance with the aquatic life.

The proposed Seneca dam would also inundate a long segment of the historic Chesapeake and Ohio Canal, and cover undeveloped woodland now used for recreation. The water surface of the reservoir would, of course, provide a recreation facility for a number of water sports enthusiasts. However, summer drawdown of the reservoir storage to supply low water flow in the river would create a belt of silt around the border of the reservoir and make use of the reservoir more difficult at those times. Finally, the reservoir site is not in an uninhabited area. As shown in Figure 10.3, the reservoir area would extend up the Potomac River from a few miles above Washington, D.C. nearly to Frederick, Maryland on the Monocacy River and almost to Harpers Ferry at the confluence of the Potomac and Shenandoah rivers. Many homesites and farms would have to be abandoned to the reservoir waters. Not only would the social consequences of such a reservoir be great, but the loss in tax revenues would be significant.

The problems facing those who want to plan for the water resources development of the Potomac River basin are thus manifold and the solutions are not easily found. More in-depth study could be called for but a great deal is already known about the river and its problems. More study would delay decisions and, while that would offer temporary respite, it would only make ultimate decisions more necessary and might increase the seriousness of the problems. Yet going ahead with preliminary stages of the Corps of Engineers plan, for example, might commit the planners, and the Nation, to an irrevocable decision made somewhat in haste that would not ultimately prove to be in the best interest of the people of the basin. Since the Seneca dam is a key element in the Corps plan, land development in the area that would be flooded (should it be built) should certainly be discouraged until a decision is reached. High level, unbiased advisory groups need to be established to attempt to articulate problems and solutions. Coordination of the different state agencies that have some control over development of the basin must be encouraged. And public concern for the problem of water management in the basin must be developed. Despite much publicity given to basin water problems in the past few years, residents are still quite unconcerned about the future of their water resources. It is possible that only a serious crisis—such as the sudden lack of fresh water—could replace present apathy with a concerted call for positive action.

The problems of the Potomac River are not unique to that river; they are common to many interstate river basins. Too many Federal and state agencies have some vested interest in the development of the basin. These interests are often in conflict with one another on very sound and logical grounds. State, county, and even local governmental groups frequently disagree strongly since what is best for one group may not be best for another. A strong river basin commission is needed to cut through the maze of suggestions and countersuggestions and to develop and enforce a unified plan. The Interstate Commission on the Potomac River basin, established more than 30 years ago, has achieved some significant results but it does not have the necessary power to develop the overall plan needed to make significant improvements in the basin. An effort in 1957 to extend the power of the Commission was not ratified by the legislatures of all of the states involved. This probably weakened the position of the Commission in future planning for the basin. An effective solution for the Potomac River basin would not only solve an immediate problem on one basin but it might serve as a model for many other basins, which are only now beginning to be faced with the complex and interrelated problems of water development illustrated by the Potomac River basin example.

## PART E:

## MERGING THE PHYSICAL, HUMAN, ECONOMIC, AND POLITICAL ASPECTS OF WATER

Droughts actually have no marked beginning or ending, a malaise that slowly grows in the life of a community and often leaves it just as gradually. Floods are quite different. They have well-defined beginnings and ends, marked by the return of the inhabitants to their homes or what is left to them. In many respects, however, droughts and floods are similar. Both events are related to water—one to too much, the other to too little. Both are periodic and to be expected due to the vagaries of weather and climate. But both are much more than climatic aberrations. For example, both inflict enormous economic losses on individuals, on communities and regions, and on the nation. Both disrupt the normal course of human events and often result in great socioeconomic and cultural changes. Both bring people together in rescue efforts. Both produce unselfish acts of heroism and result in the molding of mass opinion to cooperate and to contribute to the common good in a time of trial.

Droughts usually last much longer in time than do floods though the scars on the people or on the community may be of equal duration. Floods usually bring greater physical damage to property, to personal belongings, and to lives because often it is not possible to move belongings and people out of the way of rapidly-moving flood waters. Thus floods may become more immediately terrifying than droughts which grow and build so slowly that they become a way of life almost before they are recognized to exist.

Droughts and floods are much more than statistical data or records. Too often we spend our time deciding whether the event was due to the rainfall (or lack of it), whether it was the inadvertent actions of humans or an act of God, or whether it was the responsibility of governments or of individuals. By concentrating on the once-in-100-year or once-in-50-year record, the particular depths of precipitation, the river stages, or the volumes of streamflow—by focusing on the statistics—one misses the human, social, cultural, and economic aspects that really are a drought or a flood. For this reason, chapters on both droughts and floods are included in this section of the book rather than earlier with the discussions of precipitation or streamflow.

Inasmuch as droughts and floods are water related, they are important aspects of any water resources study. Much of their importance, however, results from their role in modifying the economic, social, and cultural life of the nation. They are events that strongly show the close interrelatedness of water and human life, and how the physical aspect of water cannot be separated fully from its effect on individuals, communities, and even regions.

## 11

## FLOODS

Floods may be defined as periods when the streamflow cannot be contained within the banks of a river channel. At those times, the river will spread out over the gently sloping land surfaces on one or both sides of the channel (the floodplain), covering them to some depth. The river channel is responsive to the average or normal flow of water in the stream. If, over a long period of time, streamflows are consistently greater or smaller than they were previously, the stream channel will change to accommodate the new water volume in the stream. But stream channels do not respond instantaneously to great increases or decreases in streamflow that occur periodically. As a result, there will often be times when a large stream channel is carrying just a trickle of water and other times when the same channel will be filled, or even overflowing, with water covering portions of the floodplains on either side.

Engineers and hydrologists are concerned with data of past river flows in order to determine flood frequency—the number of times per year, per 50 years, or per 100 years, for example that certain water volumes or river stages might be expected. Such statistics are useful for some planning activities but they are also quite misleading because a once-in-a-100-year storm does not necessarily come only once every hundred years. It may actually come twice in one year or in successive years and then not again for some long time period. And, of course, activities of humans always result in modifications of past statistical relationships. Human interference in the hydrologic relationships within a watershed through channelization, levee building, or changing basin land use has modified the earlier relationships between precipitation, evapotranspiration, and water runoff. Flood stages to be expected once in 100 years may actually occur once every 10 years now as a result of changes in the watershed, although it could still be years before the long-term hydrologic statistics reveal such significant changes.

The floodplain is a significant part of the stream and, in fact, it becomes part of the channel at times of high flow. While no one would knowingly build his

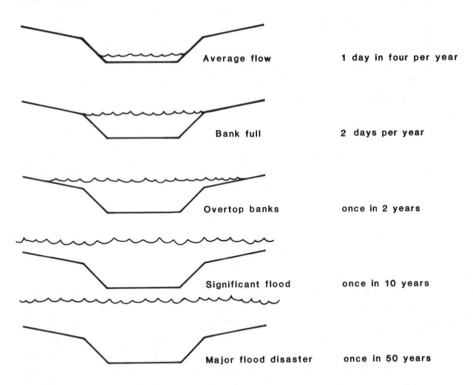

**Figure 11.1** Schematic relation between water level in river channel and frequency of occurrence.

or her house in the middle of a stream channel, yet floodplains which will become part of the stream channel from time to time are increasingly used for dwellings as well as for industrial and commercial establishments. Whole cities, such as New Orleans, are located entirely on floodplains and possibly within a few feet of the average height of the river. Any marked deviation in river stage will bring widespread flooding. Yet people move into such areas with scarcely a thought about the possibility of future flooding.

Leopold (1974) pointed out that flows greater than the bankfull stage are quite common. A large number of our streams will experience bankfull flows two to three times a year (Fig. 11.1). Since floods are directly initiated by too much precipitation or too rapidly melting snow, they are considered to be meteorologic or climatic events and as such they are studied as a probability problem. Floods are treated as random events so that the floods occurring during a certain time interval are considered to be a fair sample of a very large population over time. Ranking the greatest stage of record each year in decreasing order and determining the plotting position from the relation

$$\text{Recurrence interval} = \frac{N+1}{m}$$

**Table 11.1**  Peak Discharges for Each Year of Record, 1950–1969, for
Salem River at Woodstown, New Jersey Along with Rank Order and
Recurrence Interval

| Annual Peak Flows | | | Recurrence Interval (Yrs.) |
|---|---|---|---|
| Year | Discharge (cfs) | Order | $\dfrac{N+1}{m}$ |
| 1950 | 912 | 1 | 21.00 |
| 1951 | 567 | 5 | 4.20 |
| 1952 | 277 | 11 | 1.91 |
| 1953 | 197 | 15 | 1.40 |
| 1954 | 78 | 20 | 1.05 |
| 1955 | 425 | 6 | 3.50 |
| 1956 | 604 | 4 | 5.25 |
| 1957 | 91 | 19 | 1.11 |
| 1958 | 380 | 7 | 3.00 |
| 1959 | 123 | 17 | 1.24 |
| 1960 | 797 | 2 | 10.50 |
| 1961 | 303 | 10 | 2.10 |
| 1962 | 204 | 13 | 1.62 |
| 1963 | 226 | 12 | 1.75 |
| 1964 | 118 | 18 | 1.17 |
| 1965 | 136 | 16 | 1.31 |
| 1966 | 201 | 14 | 1.50 |
| 1967 | 338 | 9 | 2.33 |
| 1968 | 708 | 3 | 7.00 |
| 1969 | 365 | 8 | 2.62 |

where N is the number of years of record and m is the rank of that particular
stage in the total data record, allows one to compute the frequency with which
any particular high stage has occurred in the past. If future records are similar
to past records, such information can be used to predict the likelihood of future
high river stages.

Table 11.1 provides data on peak daily flow (in cubic feet per second) for
each year from 1950 to 1969 on the Salem River at Woodstown, New Jersey;
ranks the flow data in decreasing order; and provides data on the recurrence
interval (in years) as computed from $\dfrac{N+1}{m}$. Figure 11.2 is a plot of the discharge
data vs. recurrence interval. A completely straight line should not be expected
when the discharge data are plotted on a logarithmic scale. The extreme flows
can often deviate somewhat from the straight-line relationship determined on
the basis of the other points. This might be interpreted to mean that within the

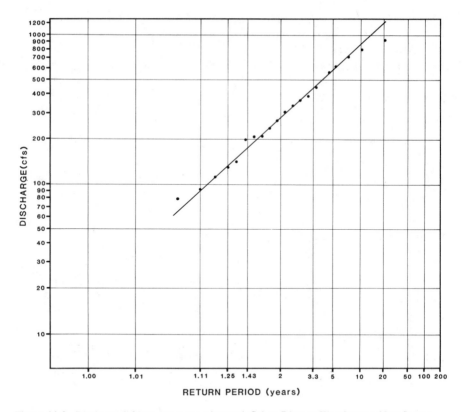

**Figure 11.2**  Discharge (cfs) vs. recurrence interval, Salem River at Woodstown, New Jersey.

particular finite record evaluated, the flood events were not normally distributed and that the expected number of high flows, for example, did not occur in such a short record. It is also possible to draw a smooth curve through the extremely high values and to interpret those data as the actual discharges that would occur once in the particular number of years indicated by the return period. One cannot really say which interpretation is better. Thus, in Figure 11.2, the discharge that will occur once in 20 years may be read as either 900 ft³/second or more than 1150 ft³/second. Both interpretations may be in error due to the continued human modification of the basin and stream channel.

Figure 11.1 suggests that a stream may overlap its banks once every 2 years. This does not mean that all reaches of the stream would overflow, the reason being that banks are not the same height everywhere. As the stream overtops the banks in some areas and flows out over the floodplains, it relieves the flooding pressure along other reaches of the stream so that certain slightly higher areas along the banks may not flood. While these areas may be protected from the more routine floods, their safety should not be overly emphasized; higher discharges may occur that will ultimately overtop even some of these more elevated banks.

# FLOOD CONTROL BY CONSTRUCTION OF LEVEES

To keep streams within their banks during periods of high flows has long been a goal of engineers and local planners. The most accepted technique has been by means of artificial levees built along the banks of the river to raise them above expected flood stages. Much later came ideas incorporating good watershed conservation, the building of reservoirs, straightening channels, or building spillways to divert the water safely to nearby lands that can be flooded with little cost to local residents. But let us consider the use of levees as flood control techniques, using information from the Mississippi River.

New Orleans was selected as the site for a town in 1718 by Bienville LeMayne, a French Canadian, who, with his brother Iberville, sailed from Brest, France for the mouth of the Mississippi River in order to claim the area for France. They located the mouth in March 1699 and later camped on the site of New Orleans.

While some houses were being built at New Orleans in 1718, a great flood inundated the area and actually flowed over the floors of the cabins still under construction. The engineer and builder, de la Tour, directed that a levee be built along the river bank to keep out later floods. This work was not completed until 1727 when the levee extended 5400 ft in length and was 18 ft across at the top. New Orleans, with 700 houses, became the capital of the Province of Louisiana in 1722. With the completion of the first levee in 1727, Governor Perrier then announced that a smaller levee would be constructed 18 mi both up-river and down-river in the following year to protect the whole district from flooding.

In the following 250 years, these early levees, thrown up largely by hand labor and by plows, have become thousands of miles of great mounds of rocks, dirt, and concrete on both sides of the river to hold back seemingly ever-increasing flood peaks. The building of levees has, in part at least, changed the character of the flood flows. Whereas at first they were more numerous but smaller with only local damage, in recent times they have been less frequent but more severe floods, covering wide areas. The great Mississippi River flood of 1927 (to be described in detail later) is one of these less frequent but wide-reaching catastrophic events that seems to have resulted from the "levees only" policy of flood control.

Although a flood in 1735 brought great destruction to the levees which extended from English Turn, 12 mi below New Orleans to nearly 30 miles above the city on both sides of the river, the idea that levees were needed for protection was not questioned. In fact, an ordinance in 1743 indicated that if the inhabitants did not complete their levee building by January 1, 1744, their lands would be forfeited to the Crown. Some forfeiture may have been required since, by 1770, the levees extended only from 20 mi below New Orleans to 30 mi above it (Kemper, 1949). The settlers did not react favorably to the ordinance. As long as the levees were not continuous, flood stages in the river were never very high because the water in flood time had plenty of area over which it could spread once it had topped the banks.

By 1828, there was a continuous levee on the west bank of the Mississippi River from the Red River to below New Orleans and from Baton Rouge to below

New Orleans on the east bank. These levees were from a few inches above the natural levee already built by the river during earlier flood periods to 4 or 5 ft high. A few isolated stretches of levees existed from the Red River northward to the mouth of the Arkansas River on the west bank. The west bank levee had been entirely completed to the Arkansas River by 1844 while some levees were being built near the Yazoo River on the east bank and above the mouth of the Arkansas River. An act of 1849 encouraged the "levees only" policy by granting to the states of Louisiana, Mississippi, Arkansas, and Missouri all of the unsold swamp areas and other land frequently covered with water. Sale of this land would provide funds to reclaim land in these states liable to be flooded. The Federal government encouraged the states to undertake their own flood control activities (which meant levees).

Congress did ask the Corps of Engineers to undertake a survey to determine how best to protect the lower Mississippi River basin from flooding. A board, under the direction of Captain A. H. Humphreys and Colonel S. H. Long (later replaced by Lt. H. L. Abbot), was established in 1850 and their report of 1861 (printed in 1876) was the cornerstone of Mississippi River hydrologic development for the next 65 years. Strongly arguing against anyone who might question the efficacy of levees, the Humphreys and Abbot report tried to determine the maximum flood discharge as well as the levee height needed to contain the flood of 1858 all the way from Cape Girardeau in Missouri to the Gulf. Estimates were made of the amount of water that had been lost from the river through breaks in the levees (or crevasses, as they were called). This loss, as well as the contributions of all of the tributary streams, were added while the losses through Bayous Plaquemine and Lafourche were subtracted in order to achieve the estimated gage heights along the river at various places during the 1858 flood.

Humphreys and Abbot argued that the levees along both sides of the river should be brought up to the height necessary to contain the flows of the flood of 1858 except near the mouth of the St. Francis River and also near Lake Providence (a possible suggestion that some diversion of the flood water might be permitted since these levees could be overtopped). Economically, the cost of this levee work would be $17 million but since approximately 6.4 million acres would be protected, the cost was only around $2.50 per acre for the protection. The report specifically rejected the use of reservoirs, cutoffs, or diversions as useful flood control techniques. Use of levees was supposed to deepen the channel and result in improved navigation as well as to protect nearby lands from flooding.

The question of what to do to deepen the channel for improved navigation had long been debated. Captain James Eads, a self-taught engineer who achieved considerable fame by building the St. Louis cantilever bridge over the Mississippi River, suggested in 1873 that the way to remove bars at the mouth of the river was by parallel dikes or jetties at the mouth of one of the passes. He was so certain that he could open and maintain a channel to the Gulf in this manner that he made an offer to do it at his own expense. No payment was to be made until he had achieved a depth of 20 ft, when he was to be paid $1 million. For every 2 ft in depth he achieved in the channel below that, he was to receive another million up to a total of $5 million for a depth of 28 ft. The remaining charge of $5 million for his efforts was to be paid in annual installments conditioned on the fact that the channel depth would be maintained for 10 years. The argument over the Eads proposal centered on whether the sediment-carrying capacity of the river was just in balance with the river velocity. Eads felt that this was so. He proposed to increase river velocity and so increase its ability to scour its bottom by narrowing the stream channel through the use of parallel dikes. The Corps of Engineers—under its chief, A. W. Humphreys—disputed this contention vigorously.

The government, after much argument, agreed to let Eads try to open a depth of 30 ft in Southwest Pass but attached all sorts of conditions to the work. Eads would not receive pay until certain stipulated depths and widths were achieved and maintained. Eads, a private engineer, won the right to work on a portion of the river over the bitter opposition of the Corps who wanted no outside interference. He had to obtain outside funding, and he faced questions and scrutiny of his efforts at every turn, yet he completed the work successfully.

While the jetties worked beautifully in Southwest Pass to create and maintain a deepened river channel, the technique also gave support to the "levees only" policy of river control. If confinement and increased velocity could deepen the channel at Southwest Pass, then why could not such confinement increase velocity elsewhere, deepen the channel, and make floods less likely?

To resolve these and other problems in the control of the river, the Mississippi River Commission was created in 1879. The first report of the Commission in 1880 gave support to the idea that increasing the velocity of the water increased its ability to carry sediment, although it stated that no fixed relation between sediment load and velocity could be found. It did give strong support to the idea that if the volume of water in any stream could be permanently increased it would increase velocity, as well as the erosive and sediment-carrying capacity of the stream. The whole theory was based on containment of water; any suggestion of outlets or diversion of water to other areas in time of floods was completely rejected. It is interesting to note that Benjamin Harrison (later President of the United States), a civilian member of the Commission, did not sign the first Commission report. He issued a minority report since he did not agree entirely with the idea of complete containment of all river waters. He also felt that levees were of little value in improving the low water navigation in the river.

High river stages in 1882, 1884, and 1890 resulted in breaks in the protective levees. While some remained for a number of years, work was directed toward closure of all crevasses. A great flood came in 1897 which seriously threatened New Orleans. However, once again the city was saved by many crevasses elsewhere that lowered the river stages at New Orleans. The value of crevasses in relieving pressure on the levees elsewhere was not fully noted by those responsible for river development. There was still implicit faith in the levee system. For this reason, the flood of 1912 was a surprise due to the fact that the people had been led to believe that the levee system was complete and danger from floods was a thing of the past. The problem was quickly diagnosed as a defective levee rather than any fault of the levee concept.

The engineers argued that new and stronger levees could be built that would withstand the river stages of the 1912 flood. But the argument was never seriously presented that the 1912 flood was not the worst on record. And who could use the past flood records to foretell the future high stages on the river since, in the past, breaks had always occurred on the river to keep river stages from going too high? As levees became stronger and higher, as crevasses were prevented, there would be no way to relieve the pressure of higher and higher stages. Routine floods might indeed be prevented but what would happen when, at a very high stage, a crevasse developed? Or was faith in the permanence of the levees so complete (in view of a record of continual failures in the past) that no one would ever dare to consider such an eventuality?

In 1917, Congress recognized flood control as separate from navigation. Congress appropriated $45 million for flood control, to be made available when matched—one local dollar for each two Federal dollars. Levee building greatly expanded not only up the Mississippi River toward Rock Island in Illinois but also up the various streams and rivers tributary to the Mississippi River. A flood in 1922 brought fear to New Orleans but again crevasses provided relief. The threat to New Orleans did result in the creation of the Safe River Commission of 100 organized in that city. Its purpose was to achieve flood relief through spillways and diversions below the mouth of the Red River. The Corps of Engineers, while listening to the argument, refused to act. In 1923, $60 million was appropriated by Congress to continue the levee work on the Mississippi River (the earlier $45 million had been exhausted).

And so the stage was set for 1927. No matter how long, how strong, how high the levees along the Mississippi River were built, floods occurred. Time could be measured in the great floods—1858, 1862, 1867, 1882, 1884, 1890, 1897, 1903, 1912, 1922. Yet in 1926, the Chief of Engineers wrote with easy assurance in his annual report, "It may be stated that in a general way the improvement is providing a safe and adequate channel for navigation and is now in condition to prevent the destructive effects of floods."

## THE MISSISSIPPI RIVER FLOOD OF 1927—THE FALLACY OF LEVEES

The year 1927 would be a year of great significance. Lindbergh crossed the Atlantic in the "Spirit of St. Louis," Babe Ruth propelled 60 baseballs out of various ball parks, Al Jolson became part of the first talking motion picture in "The Jazz Singer," and a man named Seagrave drove an automobile at the unheard of speed of over 200 mi/hr. But other records were broken that year. Over 16 million acres of land were flooded in 170 counties in seven states, over $100 million in crop losses occurred, more than 160,000 homes were flooded, over 40,000 buildings were destroyed, some 325,000 people were housed in 154 Red Cross camps while another 300,000 were fed by the Red Cross in private homes (over 600,000 people were made homeless), more than 33,000 individuals were involved in relief efforts, and between 250 and 500 people were killed by a flood—the great Mississippi River flood of 1927. But lists of extremes do not tell the story by any means. The experiences of people, the suffering withstood, the events of tragedy and near tragedy, the spirit of cooperation and compassion, the national unity, these are all part and parcel of floods.

The late winter and spring of 1927 in the Mississippi River Valley were wet but not necessarily unusual. In fact, at Baton Rouge, the cumulative precipitation from January through April was less than normal (Table 11.2). Heavy and continuous rains in March and April at a number of places led to the runoff of a great deal of water in the streams but it would be difficult to look at the precipitation statistics alone and conclude that a catastrophic flood event would occur. The discharge for March, April, and May past the gage at Red River Landing on the lower Mississippi River was no greater in 1927 than it was in 1922 or 1912. Antecedent moisture conditions, the juxtaposition of flood flows in the tributaries with flood flows in the main stream, and the condition of the levees all contribute to whether one particular set of conditions will result in a record flood or just another period of high water with minor flooding.

The flood of 1927 was a very long event. The Ohio and Tennessee rivers reached flood stage early, in December and January, while the Red River was high from December to June. All river storage was essentially used by early in February and the river could not clear itself. Even with normal rainfall, the levees would have had a hard time holding. January was below normal in precipitation in general (rainfall in January 1927, was the lowest ever recorded in New Orleans). But rainfall increased in March and especially in April. Very heavy rains in the White and Arkansas River basins early in April resulted in a crevasse in the levee near Dorena, Missouri, flooding the St. Francis Basin. The river storage capacity was full at the mouth of the Arkansas River as a result of earlier floods from the Ohio, upper Mississippi, and Missouri rivers. The top of the levee was reached on April 20 and the levee at Mound Landing, Mississippi, on the east side of the Mississippi River, broke on April 21. Flood waters were pushed back up the Arkansas River by the high stage in the Mississippi River and another major break occurred on the same day at Pendleton, Arkansas. The Mound Landing break flooded 2.3 million acres and affected 172,000 people in the Yazoo basin while the Pendleton break (and shortly thereafter numerous other breaks) flooded 5 million more acres. The water from the Mound Landing break returned to the river at Vicksburg and this additional water resulted in the Cabin Teele break across the river from Vicksburg which flooded 6 million acres of Louisiana, affecting 277,000 people. By the end of the flood period, there were 42 major breaks in the levees in the three heavily hit states of Arkansas, Mississippi, and Louisiana, and 120 breaks all told. Areas all the way from Cairo, Illinois to the Gulf of Mexico were flooded and in places the flood area extended nearly 100 mi in an east-west direction (Fig. 11.3).

When the Mound Landing break occurred early in the morning of April 21, workers were on the levee in a last ditch effort to keep the river under control. A number of workers were lost as the levee collapsed but no actual count was possible

**Table 11.2** Monthly Precipitation (in inches) at Selected Stations in Lower Mississippi River Valley, 1926–1927, as Compared with Normal

| Station | Nov. 1926 | Dev. from Normal | Dec. 1926 | Dev. from Normal | Jan. 1927 | Dev. from Normal | Feb. 1927 | Dev. from Normal |
|---|---|---|---|---|---|---|---|---|
| New Orleans | 2.92 | −0.87 | 2.18 | −2.28 | 0.61 | −4.02 | 10.15 | +5.68 |
| Baton Rouge | 5.62 | +2.40 | 4.63 | −0.75 | 2.22 | −2.77 | 5.88 | +1.18 |
| Vicksburg | 3.08 | −1.11 | 9.00 | +3.98 | 4.30 | −1.37 | 10.48 | +5.87 |
| Memphis | 4.23 | −0.36 | 9.50 | +5.12 | 3.51 | −1.70 | 2.89 | −1.46 |
| Little Rock | 3.16 | −1.43 | 9.46 | +5.22 | 4.50 | −0.29 | 3.03 | −1.15 |
| Nashville | 4.64 | +0.79 | 13.53 | +9.71 | 3.27 | −1.58 | 4.26 | −0.06 |
| Cairo | 4.49 | +0.47 | 4.15 | +0.82 | 8.16 | +4.34 | 1.36 | −1.97 |
| St. Louis | 2.17 | −0.17 | 1.15 | −1.08 | 3.66 | +1.39 | 0.56 | −2.19 |
| Davenport | 3.88 | +2.08 | 0.65 | −0.86 | 0.77 | −0.78 | 2.36 | +0.86 |
| Total deviation from normal | | +1.80 | | +19.88 | | −6.78 | | +6.76 |

| Station | March 1927 | Dev. from Normal | April 1927 | Dev. from Normal | May 1927 | Dev. from Normal | Cumulative Jan.–April 1927 | Normal |
|---|---|---|---|---|---|---|---|---|
| New Orleans | 7.99 | +2.69 | 14.94 | +10.03 | 3.14 | −0.74 | 33.69 | 19.31 |
| Baton Rouge | 7.21 | +2.36 | 1.13 | −3.56 | 5.73 | +0.93 | 16.44 | 19.23 |
| Vicksburg | 8.57 | +2.32 | 4.74 | −0.42 | 5.21 | +0.95 | 28.09 | 21.69 |
| Memphis | 13.04 | +7.27 | 13.13 | +8.30 | 5.40 | +1.06 | 32.57 | 20.16 |
| Little Rock | 6.89 | +1.95 | 14.81 | +10.30 | 6.82 | +1.72 | 29.23 | 18.42 |
| Nashville | 9.66 | +4.22 | 7.38 | +3.02 | 3.63 | +0.13 | 24.57 | 18.97 |
| Cairo | 8.07 | +4.05 | 9.59 | +6.02 | 10.18 | +6.35 | 27.18 | 14.74 |
| St. Louis | 7.67 | +4.24 | 6.30 | +2.78 | 9.21 | +4.97 | 18.19 | 11.97 |
| Davenport | 3.13 | +0.82 | 4.64 | +1.85 | 5.63 | +1.53 | 10.90 | 8.15 |
| Total deviation from normal | | +29.92 | | +38.32 | | +16.90 | | |

*Source.* U.S. Weather Bureau records.

**Figure 11.3** Flooded areas of the Mississippi valley in 1927 flood.

because of the confusion. The break came suddenly as about 100 ft of the levee was pushed in by the power of the river (Daniel, 1977).

As the water rushed through the crevasse, it seemed to many to roar like a mighty wind. Near the initial break, the water moved forward as a low wall, rapidly spreading out and losing its avalanche-like condition. Later, as more water poured through the break, the water spread out more slowly, creeping over the ground, filling ditches and low places, and then gradually rising first over the steps of buildings and then onto the first floor area. In this respect, these large river basin floods are quite unlike the flash floods resulting from heavy showers in small basins or the breaking of dams. In those, walls of water crash down the river valley destroying all in their path. Here, however, only near the breaks in the levees did walls of water, possibly 4 to 7 ft high, move out from the break to knock down buildings and to gouge out the land. As the water spread outward from the break, over the flat floodplains, speed of flow was replaced by a slow but steady creeping forward of the edge of the water and a gradual deepening of the water. Local inhabitants had considerable time to move to the second floors of buildings or to move to the higher-lying levee areas that were still standing between the river on one side and the flooded plains on the other.

As the water came through the crevasse and rolled down the back side of the levee, large craters (called blue holes because of the water color), sometimes as much as a hundred feet in depth, were scoured out (Daniel, 1977). The moving water would also swirl around buildings, eroding material supporting the foundations, until the buildings would collapse. Often all that would be left of a farmstead would be fields covered with thick layers of coarse sediment, and deep holes where the farmhouse, the barns, and the various outbuildings used to stand.

Greenville, Mississippi was one of the first towns to be hit by the flood waters pouring through the Mound Landing crevasse. Fire whistles warned the people that the levee had gone. People milled around the center of town excitedly. Greenville residents hoped the protective levee around the town would keep the water out but no one was certain. As the wind picked up, scattered groups of people came to the conclusion that the levee around the town would not be sufficient protection and that single-story houses should be abandoned for higher, safer structures. Working hastily and in some fear of the unknown, many residents gathered personal things, blankets, food, and other possessions together to move. More than 100 people congregated in the courthouse even before the water reached the town (Daniel, 1977).

Flood waters quickly overtopped the protective levee in the early morning of April 22. By daybreak, it was slowly filling the gutters in the streets. The water rose around the houses both in town and in the countryside outside—slowly and inexorably. The slow rise of the water generally allowed people to move to higher rooms in their homes, to more substantial buildings in town, or to the portion of the levees still left above water. There, many lived until rescue boats could reach them. The weather was generally cold and rainy during this early period of the flood so that life on the levee, without shelter, was far from happy. But some did not want to leave the levees to go to Red Cross rescue camps because they would be taken away from home environments. However, more than 600,000 did go by boats, wagons, or railroads to the rescue camps or to friends away from the flood zones.

The flooding of the area around Greenville, Mississippi with the Mound Landing crevasse relieved the pressure on the levees below Mound Landing and they generally held. However, the water moving eastward and southward from the Mound Landing break came down the Yazoo River basin to enter into the Mississippi River at Vicksburg. The river just could not hold all that water again so the break at Cabin Teele, Louisiana across from Vicksburg was inevitable. The Cabin Teele break on May 3 came slowly, unlike the Mound Landing break, so the destruction below the break was much less severe. Many of the older homesites were on higher ground and never were covered with water. By the middle of May, the flooding had moved southward to the bayou country south of the Red River.

New Orleans was spared the ravages of flooding during the 1927 flood but only by scraping the centuries-old "levees only" policy. A dynamite blast was set off on April 29 to create a break in the levee at "Caernarvon" on the east bank of the Mississippi between the settlements of Paydras and Braithwaite. The first blast failed to produce more than a trickle of water. Two days later a second blast opened a bigger hole and the levee collapsed, allowing hundreds of thousands of cubic feet of water to pour through (Daniel, 1977). The river gage at Carrollton near New Orleans began to drop and the city was saved from flooding. However, saving of the city came not from the height or strength of the levees but from opening the defense to allow water to flood into less valuable areas. The policy of flood control by levees was being replaced by a series of other measures designed to work with one another to maximize protection.

Catastrophes unite people in the common effort to help. Most of the rescue work was done by heroic local volunteers. The stories of men and women working around the clock to pick up stranded people in skiffs, power boats, rowboats, barges, or almost any conceivable floating craft were legion. Generally, it was first a case of moving people by small boats from house lofts, trees, or isolated mounds to more substantial levees. At these places, larger barges, riverboats, or Coast Guard boats could come in to pick them up to take them to the strategically located Red Cross emergency relief camps.

Though the water often rose slowly—warding off possible death by drowning in the wild frenzy of a flood flow—it was often not possible for people to take much food with them as they climbed into lofts or trees to get above the rising waters. Thus, many of the rescued had not eaten for 4 or 5 days when found in their places of saftey, and many were cold, wet, and in the first stages of diseases. Rescue efforts had to be carried out rapidly if those who had first saved themselves from the rising water were not to die from hunger, cold, or disease.

Relief efforts began almost simultaneously with the first break in the levees at Mound Landing, Mississippi. Herbert Hoover, as Secretary of Commerce but, more importantly, as an experienced veteran of various relief efforts extending from World War I, was put in charge of the overall rescue and relief efforts. The American Red Cross also moved quickly under the leadership of James Fieser,

Vice President in charge of domestic operations. Both groups shared control over the relief operations and were on the scene in one form or another from April 21 to the end. The death toll could easily have grown to thousands had not the rescue and relief agencies acted so promptly and as efficiently as they did at the beginning of the flooding. But while organizations are important to handle the flow of relief supplies—i.e., food, medicines, clothing, and shelter—to bring scattered families back together and to help in the rehabilitation effort, much of the early rescue effort resulted from the unselfish efforts of friends and neighbors, of people helping people, and even of the rum-runners of the river who moved in with their fast power boats to help snatch stranded individuals from collapsing homes and isolated high points. Of the 33,000 workers engaged in the rescue effort, only about 1400 were paid workers. Over 31,000 humanitarians volunteered their services to help others made less fortunate than they by the great flood (Daniel, 1977).

As homeless people started to funnel into higher-lying areas, being brought mostly by boat and train, it was first necessary to establish relief centers. The Red Cross set up 154 centers throughout the flooded area. One of the larger centers was located in the national park area on the hills overlooking the river at Vicksburg, Mississippi. More than 15,000 refugees were housed in a large tent city, fed, given medical attention, entertained, and had their daily needs taken care of in that center. Other centers were perhaps not as ideally located as the Vicksburg center. Some were more isolated, more difficult to reach with supplies and medicines, or less easily supplied with sanitary services. By and large, the camps were efficiently run and presented little in the way of health hazards to the temporary residents.

Life in the camps was accepted in a mixed fashion by the rescued. To the young it was a fairly pleasant time to mix with other young people, quite different from the more isolated conditions of their farm life, and free, at least temporarily, of farming chores. Entertainment was often provided in the form of movies and get-togethers, probably rarely experienced events in the life of most farm families under normal circumstances. For the elderly, camp life was not as good—they missed their own homes and the close relationships with family and friends from whom they were separated. To the middle-aged, life in camp presented a very mixed experience. The overall reaction probably would have been negative. Many were unaccustomed to living in close proximity to thousands of other people in a city-like environment and they longed for the peace and quiet of their more isolated farms. Most had no idea what had happened to their houses and property and a gnawing anxiety preyed upon them as they waited, seemingly without end, for the water to subside. Many had large debts to repay and if they could not make a cash crop for the year they would be in dire need to survive the following winter. Many had lost track of children or parents. Enforced idleness in a relief camp, no matter how attractive it might be made, had more negative than positive aspects, of course.

While certain problems might exist (i.e., poor food, venereal disease, and suggestions of conditions of peonage to keep tenant labor from disappearing), one would have to conclude that the overall performance of the relief camps was extraordinarily good. The existence of the camps provided a wonderful opportunity for home demonstration agents to give instruction on health and child care, on sewing, cooking, preserving, and decorating. This was an opportunity to reach thousands who would never have been reached otherwise and the opportunity was utilized to introduce rural families to developments rapidly being made available to all by our technological society. Many refugees remained in camps for several months, waiting for the waters to leave their lands sufficiently for them to return to what was left of their homes or farms.

When they returned home, they encountered a wide variety of conditions (also some similarities). As the water receded and the mud dried, the odor of decay and death was everywhere. Snakes were a particular problem in the houses that remained. Crayfish, waterbugs, frogs, and cockroaches were abundant and mud was everywhere. Water-soaked and mud-covered furniture was often just shovelled out of the houses to be burned or hauled to dumps.

The statistics of the Tom Newton plantation on the south bank of the Arkansas River, 10 miles northwest of Little Rock, might be considered fairly typical. Daniel (1977) reported that prior to the flood the plantation contained 600 acres of which 550 acres were planted in crops. In addition, it had the usual plantation buildings—a home, barn, smaller outbuildings and 17 tenant houses, mules, hogs, cattle, and several expensive pieces of farm equipment. An after-flood survey reported that 40 acres of land had caved into the river, 100 acres were badly eroded with gullies and pot holes, and 450 acres were covered with sand anywhere from 6 to 18 in. deep. The report concluded that 75 acres of the farm were entirely untillable. Only 1 of the 17 tenant houses remained, the farm equipment was found in deep holes washed out by the water swirling around the equipment and buildings, the barn and all of the farm outbuildings were gone, while 6 mules, 15 cows, and 25 hogs had drowned and 40 tons of hay, 1200 bu of corn, 30 tons of cottonseed, and 75 bu of oats had been washed away.

Rehabilitation efforts continued through the summer and, by fall, a semblance of order had been restored. Local rebuilding would go on for some time but the major effort was over and the rest of the nation had turned to other events. The scars of the flood might last a life-time on the people who had gone through the great flood of 1927 but to most others it became just another statistic in the record book along with Babe Ruth's 60 home runs—records to be broken by later floods and later ball players.

From a water resources point of view, the demise of the "levees only" program for flood protection was certainly the most significant event of the whole catastrophe. General Jadwin, Chief of the Corps of Engineers, on June 30, 1927, still reiterated his stand that levees were the only way to protect the land along the river from flooding. However, he was experiencing a change of heart, for while he realized that only levees would be able to keep high water flows from covering an area, such a policy was self-defeating since levees could not be built high or strong enough. Spillways and diversions were also necessary in places to relieve the pressures on the levees in other more critical places. Reservoirs might be helpful in providing additional storage on the land before the water entered the main channel and some control of the release of water to the river. By fall of 1927, General Jadwin had endorsed a compromise flood control plan involving reservoirs, spillways, diversions, as well as levees. The plan was submitted to Congress on December 8, 1927, with an estimated cost of $296 million. Hearings on the bill revealed, in part, how little was really known about flood control. More than $240 million had already been spent by the Mississippi River Commission on its "levees only" policy but floods had not been controlled nor would they be. A new direction was needed and the plan that emerged from Congress early in 1928 was expected to give that new direction. Congress passed a flood control act on May 15, 1928 and appropriated $325 million for the construction of a flood control system. The expense of the construction would be borne entirely by the Federal government.

The 1928 act did continue major reliance on higher levees but it included relief outlets (no reservoirs, however). Levees were to be raised high enough to carry a flood stage one-quarter greater than the 1927 flood stage and, in addition, there would be a 1-ft freeboard on top of that (Buehler, 1977). This was probably the first time that a plan contemplated a possible flood greater than the record to that date. The work was not completed until 1937 and by that time the total cost had grown to nearly $600 million, about twice the original amount authorized.

The flood of 1927 on the Mississippi River showed that levees had gone about as far as they could to protect low-lying areas. The greater weight of the levees in some cases exceeded the support limit of some of the foundation material and completed levees sank as much as 10 ft. The danger from seepage, boils, and blowouts increased significantly. Previously prepared relief outlets were utilized in the 1937 flood but it was also clear that levees and relief outlets were not sufficient. It would still be necessary to create detention reservoirs on tributary streams to help relieve the pressure on the main river (Buehler, 1977).

The results of the plan of 1928 are still to be fully evaluated. Other floods have occurred but none as serious as the flood of 1927. We have evidently learned something from history but not all we could. The "levees only" plan is dead and the compromise replacement plan is better. Continual modification of any flood control plan is necessary. We will never be able to eliminate all floods but the hope remains that we will adjust our activities to mitigate, at least, the more catastrophic aspects of the floods that we know will always recur.

## RAPID CITY—THE FALLACY OF DAMS

The Rapid City, South Dakota flood of June 9, 1972 was quite different from the great Mississippi River flood of 1927. True, it also involved rushing water, drownings, destruction of lives and property, acts of great heroism, tragedy, suffering, and a united outpouring of help from the rest of the nation. Catastrophes are alike in those respects. Both floods resulted from too much water for the stream channel to contain. Time was the difference in the two events. The

Mississippi River flood was a long event while the Rapid City flood, like its name, came and went within one night (although the scars may last a life-time).

Rapid City, South Dakota, a city of some 44,000 people, lies at one entrance to the beautiful Black Hills with Mt. Rushmore, Crazy Horse Monument, Deadwood, and the legends of many famous western heroes and outlaws. It is visited by thousands each year for rest, recreation, and sightseeing.

Rapid Creek, which flows through the center of the city, is a fairly typical mountain creek draining snowmelt water and rainwater from the slopes of the central portion of the Black Hills. Its origin is 34 mi west of Rapid City at an elevation 4000 ft above the city. The stream has a rapid descent to the city area with only one stop at Pactola Dam located about 20 mi west of the city. The creek is also checked slightly in Canyon Lake, a man-made lake, located on the western edge of the city. The creek flows through the main business district of Rapid City.

The flood of June 9 resulted in the deaths of 236 people and damage of more than $100 million in the city alone. Other creeks in the area also flooded that night so that damage was far more extensive. For example, more than 60% of the town of Keystone, on Battle Creek, was completely destroyed.

The excessive rain that resulted in the flood began during the afternoon of June 9 as a strong low-level flow of moist air from the east was forced upslope on the eastern sides of the Black Hills (Thompson, 1972). Light upper-level winds did not disperse the moisture-ladened air as it rose or cause the thunderstorms resulting from the orographic cooling of the air to move appreciably. Thus, there was a concentration of heavy thundershower rainfall in the Rapid Creek watershed. The local terrain was a factor of significance in the occurrence of the heavy and concentrated rainfall along with the wind direction, the weak upper-level flow, and the high moisture content of the air. The official weather forecast for the day was "partly cloudy with scattered thundershowers, with some possibly reaching severe proportions," a forecast that has been made thousands of times before and in most areas of the country.

The forecast was so routine that it did not raise any questions in the minds of cloud seeders from the South Dakota School of Mines and Technology, Institute of Atmospheric Sciences. Two flights were carried out on that day, one (between 3:05 p.m. and 3:43 p.m.) generally north of Rapid City with four seeding points and the other (between 4:39 p.m. and 5:11 p.m.) generally due south of Rapid City with three seeding points. The approximate direction of the upper-level winds was from southeast to northwest so that the seeding materials would have been carried northwestward into some of the areas of heaviest precipitation. Table salt (NaCl) was used as the seeding agent since it is felt by some that it speeds droplet growth by hygroscopic action, reduces the opportunities for hailstone growth, and does not cause the release of extra heat which would happen when dry ice or silver iodide produce freezing of water droplets. This extra heat might result in added buoyancy to the air in the growing clouds. Salt has a negligible heat of solution so that no cloud growth is to be expected from the use of salt. The experience of those who have seeded with salt is that there may be

### Table 11.3   Possible Relationship Between Cloud Seeding and Flooding*

1. Disastrous flash floods occur in the Rapid City area about once every 9 or 10 years without human intervention.

2. As the population and development of the city increases, the loss of life and destruction of property will incease with each occurrence unless appropriate measures are taken.

3. Water management and flood control practices will have to be developed to preclude such damage, in addition to floodplain zoning and building code changes designed to minimize loss.

4. The June 9 flood was caused by meteorological conditions, beyond the control of man. Indeed, the same weather system had produced similar floods from California to South Dakota.

5. Weather modification activities by the Institute of Atmospheric Sciences (IAS) did not contribute materials to the flood. Had none been conducted, damage would have been the same.

6. The statewide effort in rain augmentation conducted by the South Dakota Weather Control Commission was in no way responsible.

7. Regulation, initiated by the Weather Control Commission on a statewide basis, of all weather modification activities is needed to preclude even the appearance of hazardous activity. Operating programs must have available information sources that will permit a safe shutdown of operations and a clearly stated procedure for so doing in the event of a flood threat, regardless of whether or not seeding could contribute to the hazard.

8. State and IAS weather modification operations should be resumed and continued.

9. The city, county, and state police, the National Guard, and personnel of Ellsworth Air Force Base acted quickly, efficiently, and properly in doing all that could be done with available resources (in warning and evacuating people in the flood's path).

10. State radio and the several official agencies communicated well and as effectively as they could have.

11. An improved procedure for disseminating storm and disaster warnings to official agencies and to the public is urgently needed.

*Conclusions of a Report to Governor Richard Kneip, commissioned by the South Dakota Weather Control Commission, prepared by Pierre St.-Amand, Ray Jay Davis, and Robert D. Elliott (as quoted from *Rapid City Journal*, June 29, 1972 by Reed, 1973, and used with permission of the *Rapid City Journal*).

a small increase of rainfall from clouds already producing light rain or some fall of rain from clouds not already giving rain.

Because of the juxtaposition of the afternoon cloud seeding and the torrential downpours that evening, there has been a careful investigation of any possible relationship between the seeding and the flooding. Most informed investigators have concluded that there was no connection between the very heavy and sustained rainfall and the seeding activity. Table 11.3 sums up the conclusions of a special outside commission brought together by the South Dakota Weather Control Commission.

One possibly unique aspect of the Rapid City flood experience was that a number of South Dakota residents had been questioned specifically on their opinions about weather modification. This had occurred in January and February of 1972 in preparation for the first season of a state weather management program to increase precipitation and to decrease hail. Following the June 9 flood incident, considerable information was given out in the newspapers and on radio and television about the seeding in the area as well as the conclusions of the panel of outside experts that there was no connection between the seeding and the flooding. Still, to determine public reactions to the event as well as to the campaign to present a fair and unbiased picture of cloud seeding, the same respondents were interviewed again in September to determine their feelings about weather modification at that time (Farhar, 1974).

There is, of course, some question whether asking people directly about a particular issue might bias their responses. Questions dealing with the possibility of flooding due to cloud seeding might suggest a relationship that may not actually exist. The results of the survey do not necessarily bear out these suggestions.

While 71% of all respondents knew that cloud seeding had occurred before the time of the flooding, 85% of the respondents in the vicinity of Rapid City were aware of it as against only 68% in the rest of the state. Local residents were much more aware of the occurrence of cloud seeding, as they were about all aspects of the storm and the resulting tragedy. However, when asked about possible links between cloud seeding and flooding, respondents from the Rapid City area were no more in agreement than other residents in the state that any causal relationship existed. Four percent of the respondents both locally and statewide felt that the seeding was the sole or primary cause of the flooding while 57% of the local respondents and 55% of the more distant respondents said that seeding was probably not, or definitely not, a cause of the flooding. Of the remainder of the respondents who voiced an opinion, 28% of those in the Rapid City area felt that the seeding might have been a contributing or a minor cause. Twenty-six percent of the respondents in the rest of the state felt this way. Thus, 30% of the respondents statewide felt there was some connection between seeding and flooding while 55% felt that there was no connection. As would be expected, those who felt there was a connection between seeding and flooding were considerably less favorable toward cloud-seeding technology than those who did not believe in any seeding-flooding connection. This relation was found both before and after June 9.

No organized opposition to cloud seeding developed in South Dakota. This might be due to the fact that local groups have control over whether to participate in the cloud-seeding program and that the weather modifiers of the Institute of Atmospheric Sciences are held in high regard by most of the residents of the state. They are viewed as competent and reputable. As an interesting illustration of this matter of opposition, the Rapid City flood has been used by opponents of weather modification in the Pacific Northwest area although South Dakota residents remain generally favorably inclined toward weather modification.

Since the demise of the "levees only" policy on the Mississippi River, the use of dams with reservoir storage has been mentioned as a useful flood control technique. To operate a dam-reservoir complex for flood control, it is necessary to have the reservoir near its minimum storage capacity at the start of the flood season. As the chances for floods decrease, the level of water in the reservoir may be allowed to rise.

The watershed above Rapid City is controlled, in part, by the Pactola Dam located about 20 mi upstream. About 75% of the Rapid Creek watershed above Rapid City drains into Pactola Dam. The gates of the dam were closed at the beginning of the heavy precipitation so that runoff from the watershed above the dam was held in the reservoir and did not contribute to the flood in Rapid City. The runoff that produced the flood, therefore, came from the 51-mi$^2$ drainage area above Canyon Lake on the west side of town or from the 91-mi$^2$ drainage area above the Rapid City gage somewhat to the east of the lake (Winchester, 1972). That discharges from small basin areas could produce such destruction is witness to the intensity of local thunderstorms and, of course, provides one argument against full reliance on dams for flood control. Rain falling in portions of the basin between the dam and the area to be protected can still produce flash floods over which the dam will have no control whatsoever.

A second argument against full reliance on dams can be seen from what happened at the small dam on Canyon Lake. Canyon Lake covers about 40 acres with a depth of 3 to 15 ft. The 20-ft-high earth dam to create the lake was built by the WPA in 1938. While the spillways of the dam were opened at 8:30 p.m. in order to relieve the pressure from the rapidly rising water, it is felt that debris, houses, trees, even automobiles or parts of trailer homes, brought into the lake by the rampaging water helped plug the spillways and lessen their ability to pass water. The water rose to overtop the dam. Thus, an appreciable wall of water was poised ready to sweep through the town when the dam gave way at 10:40 p.m. Release of this extra volume of water coincided quite closely with the arrival of the natural flood crest rushing down Rapid Creek and the two flows swept destructively through town.

The failure of the dam was certainly not the cause of the flood and resulting destruction that was already occurring before the dam failed. However, the additional water coming as a wall 5 to 10 ft high and adding to the flood already cascading down Rapid Creek could have been the "straw that broke the camel's back," so to speak. The added velocity and strength of the water was sufficient to crush homes, move automobiles and trailers, and overturn trees with a speed that made escape for many virtually impossible. Dams can be useful in flood control but, if the reservoirs behind them are full when the flood strikes, there is no additional storage capacity, or if the dams fail, an additional volume of water is released which will just compound the severity of the flooding. We have had several examples recently of the failure of dams (Teton Dam, Idaho; Johnstown, Pennsylvania) which reinforce the seriousness of these problems.

Rainfall began in the Black Hills during the afternoon of June 9. The 6-hr rainfall expectancy, based on a 100-year return period, varies from 3 to 3.5 in. in the Black Hills area, yet more than four times that amount fell in many areas of the Black Hills during the next 6 hrs. Nemo, on Box Elder Creek, measured

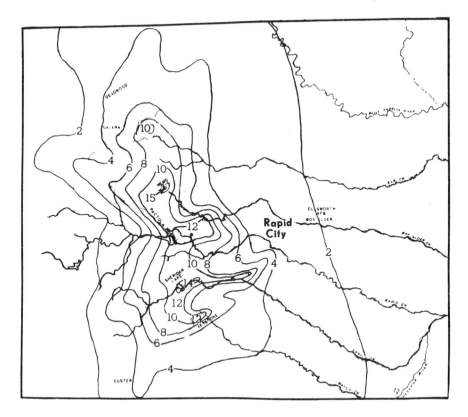

**Figure 11.4**  Preliminary isohyetal map for storm of June 9-10, 1972, based on regular climatological network supplemented by about 200 unofficial precipitation reports. Large figures indicate inches of rainfall.
*Source.* Thompson (1972), with permission of the American Meteorological Society.


15 in. of rainfall while a gage near Sheridan Lake, southwest of Rapid City and located on the divide between Spring Creek and Rapid Creek, measured 14.5 in. in a 5-hr period (Thompson, 1972). Figure 11.4 shows the distribution of rainfall for the storm of June 9-10 based on regular climatic network data supplemented by about 200 unofficial precipitation observations. Streams draining the area rose from normal levels to a condition 8 to 10 ft over bankfull in less than 3 hrs. Dozens of bridges were destroyed by the floods, 1200 homes and 100 business buildings were completely destroyed, and 5000 automobiles were destroyed or damaged. But the sudden surging terror of that night and the loss of some 236 lives were without doubt the real legacy of the whole tragedy.

By 6 p.m. on June 9, heavy rains had brought local flooding across the highway to the northwest of Rapid City. Rain began in Rapid City itself at almost the same time. But already creeks in the Black Hills, 10 to 20 mi west of Rapid City, were over their banks. Battle Creek, at the foot of Mt. Rushmore, rose 8 ft in a matter of minutes, inflicting heavy damage on the town of Keystone and bringing death and destruction to campers along its banks.

The rain in Rapid City was keeping most people at home. There were spot announcements from time to time on TV and radio about cloudbursts to the west in the Black Hills and generally unconfirmed reports about damage. The spillways to Canyon Lake were opened at 8:30 p.m. and some 30,000 ft$^3$ of water per second were flooding down the creek into the lake. By 10 p.m., when the mayor and city engineer left the Canyon Lake dam, firemen and policemen were warning residents around the lake and along the creek below the dam to leave for higher ground. Unfortunately, few heeded the warnings that were given 45 minutes before the break came. But already the creek was over its banks and water was swirling through the low-lying areas along Rapid Creek (Winchester, 1972).

The 10 p.m. news on the local radio station devoted a small segment of time to the heavy rain and flooding but reported that they had no reports of serious injuries. Scores were already dead by then. At 10:30 p.m., the mayor received a call indicating that a 4-ft-high wall of water was surging down Rapid Creek toward the city, and he radioed a warning message from a police car to TV and radio stations for all residents living adjacent to the creek to leave immediately. But it was already too late.

The dam on Canyon Lake broke about 10 minutes later along with the arrival of the flood crest coming down Rapid Creek. With a terrifying roar, a wall of water anywhere from 5 to 10 ft high (who was going to measure at this point) surged into Rapid City. The next 2 hrs were one continuous nightmare of death and destruction as houses and stores were smashed by the raging water, automobiles and trailer homes were swirled around and smashed into telephone poles and buildings or carried like huge surfboards on top of the surging waves (Winchester, 1972). The main electrical transmission lines were destroyed, plunging the city into a chaotic blackness. With natural gas escaping from dozens of broken mains and the sparks from the still-live power lines, a number of fires started.

The waters began to recede shortly after midnight and the search began in earnest to rescue those trapped in cars, in trees, in broken homes or hanging on to almost any conceivable object. For many, the search ended sadly as missing loved ones turned up days or weeks later washed far away and drowned. But help came quickly. Even by daybreak June 10, Salvation Army groups, National Guardsmen, local volunteers, Vietnam veterans, Red Cross workers, and others were on the scene to try to comfort those whose lives had been tragically changed in just a few hours the night before.

The lessons of Rapid City are many. The roles of dams in flood control work must be carefully evaluated and understood by those living downstream. Complacency has no place when living along the banks of small creeks that might be subject to flash flooding. Warnings must be heeded and heeded quickly even though many times nothing will come of it. As a people, we must become better aware of the terrible destructive potential of nature unleashed—it is too easy to become apathetic.

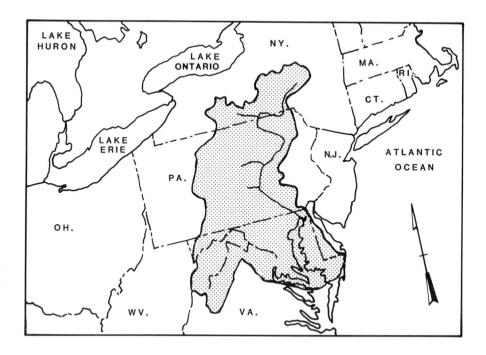

**Figure 11.5** Map showing general location of Potomac and Susquehanna basins and Chesapeake Bay region affected by Agnes.

The local theater group in Rapid City was preparing for a new play entitled "You Know I Can't Hear You When the Water's Running"—more prophetic than they could ever have guessed and possibly a good summary of the events of the night of June 9 when so few could hear the warnings.

## AGNES IN PENNSYLVANIA, JUNE 1972—THE FALLACY OF PREVIOUS RECORDS

Agnes, a hurricane, a tropical storm, an extra-tropical cyclone, call it what you want, has been labelled by the Corps of Engineers as the most costly natural disaster, in terms of property damage, in the history of the world. Only about 125 people lost their lives in the course of the whole storm, considerably fewer than in one night in Rapid City. But the property damage totalled over $4 billion of which $3.5 billion occurred in the combined Susquehanna River, Potomac River, and Chesapeake Bay areas (Fig. 11.5). The previous record flood flows in the Susquehanna River Basin (of main interest in the present discussion) were

**Table 11.4**   Flood Losses (in Millions of Dollars) in the
Susquehanna and Potomac River Basins and Adjacent
Areas of Chesapeake Bay Attributed to Hurricane
Agnes

| | |
|---|---|
| Total Damage Within the Baltimore District | $3469 |
| Specific Damages | |
| Residences | 918 |
| Business | 838 |
| Agriculture | 349 |
| Transportation | 226 |
| Schools, churches, hospitals | 99 |
| Utilities | 85 |
| Marine | 60 |
| Debris removal | 47 |
| Government (Federal, non-Federal | 36 |
| Miscellaneous | 809 |

*Source.* Based on U.S. Corps of Engineers estimates.

established in March 1936 as a result of both heavy rains and rapid winter snowmelt. Even Agnes-related flows were not higher in some portions of the area. The
1936 flood property damage was of the order of $55 million which, if multiplied
by a factor of five to account for inflation, produced losses of only $275 million
in 1972 prices, far less than the losses from Agnes. Table 11.4 lists the approximate
losses within the Susquehanna and Potomac River basins and the adjacent areas
of Chesapeake Bay as reported by the Corps of Engineers.

Agnes was first identified as a depression (a low pressure area) near Cozumel
(close to the Yucatan coast) on June 15 (U.S. Dept. of Commerce, 1972). On the
following day, as the central pressure deepened and the wind speed increased,
it became tropical storm Agnes. It was a rather large circulation system with a
poorly defined eye or center. By June 17, Agnes began moving northward at about
10 mph and, by the morning of June 18, hurricane-force winds (75 mph or greater)
were found near the center, which was located some 250 miles west of the Florida
Keys (see Fig. 11.6). During the next day, the 19th, Agnes moved due northward
and was found about 200 mi west of Ft. Myers. Maximum winds over land were
between 25 and 45 mph. Jacksonville, on the opposite shore, with winds from the
south and east off the Atlantic, had the highest winds from Agnes in Florida,
56 mph, on June 19. Agnes was a small hurricane as hurricanes go and neither the
central eye nor the wall of clouds around the eye had fully developed. On June 19,
tides were 3 to 5 ft above normal on the west coast of Florida and 6 to 7 ft above
normal in the panhandle area near Cedar Key and Apalachicola.

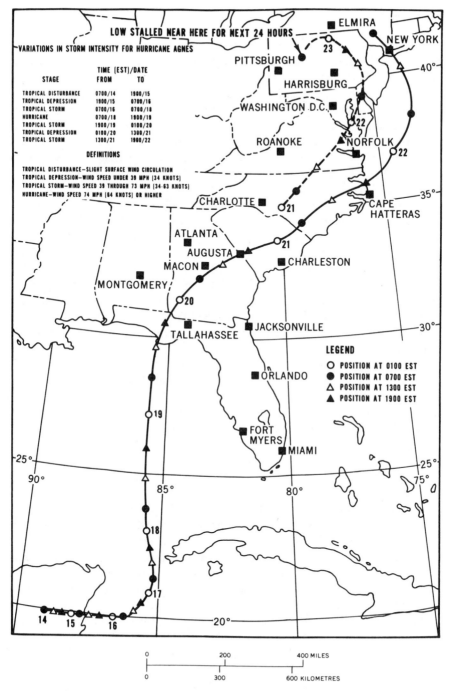

**Figure 11.6** Path of Agnes as a tropical disturbance, tropical depression, tropical storm, hurricane and extra-tropical storm, June 14–23, 1972.
*Source.* Bailey, Patterson and Paulhus (1975).

353

Agnes came ashore as a tropical storm near Panama City on the afternoon of June 19. Winds were of the order of 40 to 50 mph. Movement over land weakened these winds so that by the time it passed over Georgia during the 20th, Agnes was no more than a depression. The major effect of Agnes now was heavy rain.

Rains in Florida had been typical for a hurricane, generally ranging around 7 in. but totalling 12.69 at Big Pine Key. In South Carolina, heavy rains were encouraged by the upslope motion as southeasterly winds, moving around the center of the low, were forced upward by the slopes of the Blue Ridge Mountains. Five to 10 in. of rain, in a 48-hour period, fell over the Georgia–South Carolina area affected by Agnes as the storm moved slowly northward. While some flash flooding did occur in mountain and piedmont streams, the rain was generally considered a real blessing since the Southeast was in the midst of a critical dry spell.

As Agnes moved closer to the Atlantic Ocean on June 21, it intensified in strength and late that night became an extra-tropical storm. It moved out to sea again early on June 22 where it became a large low pressure area with pressures close to 1000 mbs stretching all the way from upper New York state to North Carolina. The warm Atlantic Ocean replenished the moist Gulf air that had brought heavy rains to the Southeast and renewed the storm for its upcoming assault on the Northeast.

Agnes moved across western Long Island, near New York City, late on the 22nd and then, strangely, curved westward into southern New York state and southwestward into central Pennsylvania on June 23. Heavy rains continued falling from Agnes as the warm, moisture-ladened air from the Atlantic was forced up the slopes of the Appalachian–Allegheny ranges. On the 23rd, Agnes was generally absorbed by a broad low pressure center in central Pennsylvania which moved slowly toward the Buffalo area and then over southern Ontario on June 25. Moving up the St. Lawrence River Valley, it once again strengthened upon reaching the ocean and it was still influencing shipping in the North Atlantic as late as July 7.

Agnes did spawn a number of tornadoes in Florida and Georgia, as all good hurricanes do, but heavy rains seemed more her specialty. While the rains in the Southeast brought welcomed relief to a dry area, those in the mid-Atlantic states came right after a previous period of heavy June rainfall. The ground was already well soaked. Five to 10 in. of rainfall through Virginia and Maryland brought flash flooding to many of the streams along with new record high river stages at a number of gaging stations. Severe flooding occurred in the James and Appomattox River basins in Virginia and in the Potomac River Basin. Sixteen inches fell near Chantilly, Virginia, while 13.65 in. were recorded at Dulles Airport near Washington, D.C. Rains in Delaware totalled only 4 to 6 in. so that flooding was not severe. But in Westminster and Woodstock, Maryland, 11.55 and 11.35 in., respectively, fell on June 21, among the very highest one-day records for the state. New Jersey rains were much like Delaware's and while some local flooding occurred, it was nowhere as severe as in Pennsylvania or New York.

Torrential rains in Pennsylvania began falling on already wet ground on Wednesday, June 21, and continued through the next day. Maximum 24-hour totals were greater than 7 in. in a band from above Williamsport southward through Sunbury, Harrisburg, and York to the Maryland border. Harrisburg had 12.55 in. in 24 hrs while a gage in western Schuylkill County showed 14.5 in. in 24 hrs. Total storm rainfall ranged between 8 and 12 in. through the central portion of the state while 18.8 in. fell in western Schuylkill County.

Heavy rains fell in the highlands of southern New York beginning in the evening of June 20, and while individual 24-hour totals were not as heavy as at places further south, the 3 days of fairly continuous rain did bring storm totals to 10 to 13 in. in Allegheny and Steuben counties (Robinson, 1976). Thus, a large area of south-central New York state and central Pennsylvania, centered fairly well over the Susquehanna River Valley, received a 3-day total rainfall of 9 to 13 in. on ground that was already saturated. Runoff to the streams had to be heavy and rapid since there was no place to store any more water.

Small streams in Pennsylvania started flooding on Wednesday evening, June 21, and the major rivers followed suit on June 22 and June 23. Susquehanna River flood levels exceeded all previous records (set in March 1936) by 3 to 6 ft. Crests were generally 12 to 18 ft above flood stage. On June 24, the crest of the Susquehanna was 18 ft above flood stage and 7 ft above the record flood of 1936. At Harrisburg, the flood crest was 16 ft above flood stage, sufficient to put the first floor of the governor's mansion underwater. The Juniata River at Lewiston, north of Harrisburg, crested at 42.1 ft, 19 ft above flood stage on June 23. This was still slightly below the 1936 record flood stage. Severe flooding also occurred on streams in eastern and western Pennsylvania with the Schuylkill River reaching record levels in some places (Pottstown, 8 ft above the previous record of 1902). Sections of Philadelphia were flooded. At Pittsburgh, the Ohio River reached its highest level since 1942 and levels on the Ohio River were 9 to 12 ft above flood stage.

More than two-thirds of the property damage caused by Agnes occurred in Pennsylvania and, in spite of the warnings (Agnes had been around for a week and had been talked about on radio, TV, and in the newspapers), 50 deaths also occurred there. The whole state was declared a disaster area. Roughly 250,000 people were forced to evacuate their homes. Public water and sewage facilities were flooded out, water had to be rationed in a number of communities and firemen, unable to reach blazes because of the floodwaters, had to stand by helplessly and watch buildings go up in smoke. Table 11.5 indicates the real impact of Agnes on property in the various states over which it passed. Since New Jersey suffered a relatively small amount of damage, the figures for Pennsylvania–New Jersey can be considered to be almost entirely due to damage in Pennsylvania.

One of the most heavily damaged regions in the entire Susquehanna River Valley area was the Wyoming Valley of north-central Pennsylvania, in Luzerne and Columbia Counties, where the communities of Pittston, West Pittston, Swoyersville, Forty Fort, Kingston, Edwardsville, Wilkes-Barre, Plymouth, Nanticoke, Shickshinny, and Bloomsburg (Romanelli and Griffith, 1972) are located. This is a coal-producing area where the major industries are coal mining followed by some agriculture and heavy and light industry and commercial activity. Precipitation

**Table 11.5** Impact of Hurricane Agnes on Property in East Coast States

| | Virginia | Maryland | West Virginia | D.C. | New York | Pennsylvania–New Jersey | Total |
|---|---|---|---|---|---|---|---|
| Total families suffering loss | 6,438 | 3,477 | 856 | 506 | 39,553 | 71,144 | 121,974 |
| Dwellings destroyed | 95 | 103 | 107 | 0 | 628 | 2,219 | 3,152 |
| Dwellings, major damage | 1,336 | 866 | 259 | 0 | 4,912 | 33,480 | 40,853 |
| Dwellings, minor damage | 3,008 | 1,564 | 216 | 350 | 27,560 | 30,700 | 63,398 |
| Mobile homes destroyed | 125 | 49 | 118 | 0 | 132 | 1,266 | 1,690 |
| Mobile homes, major damage | 435 | 44 | 86 | 0 | 313 | 2,010 | 2,888 |
| Farm buildings destroyed | 11 | 17 | 0 | 0 | 93 | 433 | 554 |
| Farm buildings, major damage | 27 | 44 | 50 | 0 | 355 | 1,240 | 1,671 |
| Small businesses, destroyed or major damage | 177 | 82 | 17 | 0 | 1,336 | 2,946 | 4,558 |

*Source.* Based on figures (preliminary) compiled by the U.S. Corps of Engineers.

within this area due to Agnes varied from about 5 in. in the northeast to 13 in. in the southwest.

Protective structures at Wilkes-Barre consisted of nearly 5 mi of earthen levees and 160 ft of concrete floodwall on the Wilkes-Barre side (east) of the Susquehanna River. There were also drainage structures, eight pumping stations, and an impounding basin for Solomon Creek with 3200 ft of additional earthen levees. All of the protective structures had been installed to withstand the record flood discharges found in the March 1936 flood.

Wilkes-Barre received only about 6 in. of precipitation during the June 20-25 period which caused some locally severe runoff conditions and flooding in low-lying areas. Stream and river flows, however, were well below previous records until the flood waters from the nearly 10,000 mi$^2$ of upstream drainage area reached the city. The peak flow of 345,000 cfs on June 24 was well above the 1936 record peak of 232,000 cfs. The protective devices worked well until the new record stage was set on June 23 when the levees were overtopped. Additional sandbagging could not stem the much higher flows still to come. Sufficient warnings were given to prevent major loss of life when the levees were overtopped, but property damage was heavy.

During June 22, water was climbing in the river at Wilkes-Barre at a rate of about 1 ft/hr. Starting at a stage of 8.3 ft at 7 a.m. on June 22, the water level of the Susquehanna River rose to 17.2 ft just 9 hrs later and in 6 more hours (10 p.m.), it stood at 24.8 ft (a rise of 16.5 ft in 15 hrs). By 1 a.m. on June 23, the river level had risen another 4 ft while at 7:00 in the morning, the level stood at 35.4 ft, fully 27 ft higher than it had just 24 hrs earlier.

The rate of rise now slowed appreciably for, in the next 24 hrs, the stage rose only to 40 ft and 12 hrs after that (now 7 p.m., Saturday, June 24), the river crested at 40.6 ft. The river stage remained over 40 ft for more than 15 hrs on Saturday, June 24, before receding rather rapidly during the 25th and 26th.

On June 22, informed watchers on the river bank were pessimistic about the chances of the city being spared a serious flood but most of the residents went to bed with seemingly little concern. Many had been through the 1936 flood or later smaller floods but there was little expectation that a high crest of 37.5 ft would be exceeded and nearly everyone felt that any crest below that would cause only minor flooding. During the night of June 22-23, sandbagging efforts were redoubled and early in the morning of June 23, Mercy Hospital in South Wilkes-Barre and Nesbitt Memorial Hospital in Kingston across the river were ordered to evacuate their patients. There was considerable movement toward evacuation during the morning of June 23 and at 11:16 a.m. warning sirens indicated to all that the floodplains and low-lying areas of the city were to be evacuated at once. Most of those who did not heed that warning had to be rescued by boat or helicopter later. By evening of that day, 29% of the city was under water.

As the Susquehanna River waters came over the levees, only one sandbagger lost his life, unlike the situation at Mound Landing in Mississippi, and the levee on which he was working did not give way until Saturday morning, June 24. South Wilkes-Barre, a fairly new subdivision, was hardest hit. Water depths ranged from 8 to 30 ft, covering some homes to the rooftops. Somewhat to the north, the college buildings of Wilkes College suffered damage estimated at nearly $18 million. East of the college, Public Square Park was covered with 8 ft of water and later several feet of mud. Stores in this area were severely damaged. During June 23 and 24, the thousands of people who had not heeded the warnings to evacuate became the objects of wide-ranging rescue efforts by local volunteers and military personnel who were moved into the area rapidly. Water depths in the central business district of Wilkes-Barre averaged about 9 ft, causing extensive damage to stores and businesses.

Extremely heavy damage was also inflicted on other nearby communities. Pittston is located on fairly high ground and so escaped the most severe damage. However, West Pittston is located between the river and an old meander scar of the river. Much of it is on land that was an island in the former river channel. With the flooding of June 24, the river once more followed both of its river channels in West Pittston and the central portion of the city became an island. About 25% of the town was flooded.

Opposite North Wilkes-Barre and just north of Kingston on the west bank of the Susquehanna are the towns of Swoyersville and Forty Fort. The towns are built on fairly low ground but were protected by 3 mi of earthen levees and half a mile of capped steel sheet pile wall as well as some diversion and drainage structures constructed by the Corps of Engineers. These flood protection works should have provided protection against a flood discharge equal to the previous maximum flood of record, March 1936. The protective works were started in 1953 and finished in 1957 so they had been effectively in place for 15 years doing what they had been designed to do.

The flood discharge of 1972 was 345,000 cfs as opposed to the previous record of 232,000 cfs for which the protective works were designed. Clearly, they could not contain the flood of 1972 and they were overtopped. Nearby Wyoming Valley Airport on the north end of Forty Fort was covered to a depth of about 16 ft while a break in the levee system at Forty Fort resulted in great erosion at the Forty Fort cemetery. More than 2000 bodies were swept out of the cemetery.

Kingston and Edwardsville are located on the west bank of the Susquehanna across from Wilkes-Barre on land that is about 20 to 30 ft above the river. The towns had been protected by levees and drainage structures, pumping stations, and a large concrete pressure culvert some 6600 ft long with an impounding basin to carry the flow of Toby Creek. These improvements also were designed to protect against the flood of record, March 1936. When the capacity of the protective devices was surpassed on June 23, the levees were overtopped and the entire community of Kingston was covered by water from 15 to 20 ft deep. A major segment of Edwardsville, somewhat further from the river, was flooded. In Kingston, essentially every structure was flooded and all inhabitants were evacuated; residential buildings and businesses suffered severely and numerous homeowners had to find replacement housing; many businesses just closed; streets and roads were severely damaged along with the public utilities.

In Plymouth, more than 30% of the community was partly or totally destroyed by the flood water. While levees were in place to protect against a March 1936-type storm, actually one small segment of the levee had caved in 2 months earlier as a result of the rupture of a sewer main beneath the levee. Contracts had been let by the Corps of Engineers to repair the break and the work had started but had not gotten far by the time of the flood. As the flood approached, Civil Defense workers gave the word for residents to evacuate the portion of town protected by that segment of the levee while volunteers and National Guardsmen worked diligently to fill the break. Tons of sand, shale, rock, and dirt were brought in for the job but the water rose too rapidly. After working all night June 22, it was clear that the levee would still be 6 ft below the remaining levee and would not hold the rapidly rising river. To protect those working on the levee, the word was given to pull out at 5:50 a.m. on June 23. A wave of water rushed over and through the previous break at about 6:30 a.m., flooding that portion of town. Ultimately water nearly 10 ft deep poured through the break in the levee, which was about 50 ft long.

Shickshinny is located on a creek tributary to the Susquehanna in a small valley between two mountain ridges. The creek usually is quite shallow in summer and even dries up at times but not in June 1972. The heavy rains of the early portion of Agnes broke the Shickshinny Dam on June 22 sending a volume of water roaring down the creek and cutting a path several hundred yards wide through all obstacles. The surging water from the broken dam was more than 20 ft deep and it took everything with it, homes, cars, roads, and bridges. Shickshinny was a disaster area even before the Susquehanna had reached its flood stage. When that happened the following day, rising waters forced many who had survived the onslaught of the water from the broken dam to seek higher ground. The western section of town was completely cut off from the eastern section, water was in short supply, and every business establishment in Shickshinny suffered very heavy damage. Some looting was reported.

The military, the National Guard, the Red Cross, the Salvation Army, and various church groups, all moved rapidly to help the unfortunate victims, to provide food and shelter, to locate lost loved ones, and to begin the Herculean task of clean-up and rebuilding. Power lines had to be restored, water and sewerage systems put back into working order, roadways and bridges opened, and buildings too severely damaged to be repaired had to be levelled. Tons of debris of all types along with the inevitable mud had to be cleaned up and moved to dumps. Even this presented a real problem for the load on the existing dumping grounds was too great and new dumps had to be opened. But reconstruction work started as soon as the waters began to recede and united action by those who suffered damage and those thousands who volunteered to help brought a semblance of order to the devastated area in a short time. The scars of possibly the worst national disaster in U.S. history will not soon be removed.

## CONCLUSIONS

Flood control activities will not necessarily eliminate floods. They may help in reducing flood peaks or in providing a channel able to contain significant discharges of water above the mean but they will not eliminate all danger from flooding. The present chapter has discussed the characteristics of flood flows, the interpretation of discharge records, and the nature of the stream channel and floodplain system in an effort to increase our understanding of the problems of flooding. But the real emphasis of the chapter has been to point out the complex interactions involved in any episode of flooding and to consider certain fallacies concerning flood control work. Floods are extreme events and, as such, they involve statistics of great volumes of water, tremendous dollar values of losses, and staggering death totals. They also involve socioeconomic programs, political decisions, public responses, changes in attitudes, and they leave scars that remain for long years. They well illustrate the close relation of the human, cultural, economic, and physical aspects of water resources.

Three specific flood events—the 1927 Mississippi River flood, the Rapid City flood of June 9, 1972, and the floods from Hurricane Agnes in Pennsylvania in June—have been briefly recounted. While they provide interesting historical detail, one theme running through each of the accounts is the fallacy of relying on existing flood control techniques for ultimate protection against the ever-present dangers of future floods. Levees are quite useful in keeping moderately high river flows within the river banks but there is a limit to which levees can be built. The more streams are confined within definite banks and prevented from flowing out over the floodplain, the higher the flood peaks will be. Ultimately, the levee will be overtopped and the damage will be extreme. There must be ways to relieve pressure on the levee system by diversions, spillways, or even by reservoirs and conservation measures. That was the prime lesson of the Mississippi River flood of 1927.

Dams and reservoirs can be useful for flood control purposes but most engineers and planners realize that they are not sufficient by themselves. The Rapid City flood illustrated two significant limitations of the dam-reservoir system in flood control work: (1) flood waters may originate from basin areas between the dam and the point of flooding so that the dam offers little protection; (2) the dam may break, releasing even greater volumes of water to do damage. Both of these events occurred at Rapid City and the result was a tragic night of terror.

Use of past records to determine the type, height, and strength of flood control works would seem to be a justified technique except that past records may have little validity in the present or future. As land-use changes occur, past hydrologic relationships may no longer hold. With the removal of forests and the elimination of much of the floodplain for storing high river flows, increasing river stages will be experienced for a given amount of precipitation. New record stages will be established even with the same rainfall amounts; flood control devices will be overtopped or destroyed. Building to contain past record flows results in a sense of security that just cannot exist as a result of land-use changes. That was one lesson learned from Agnes in 1972.

Clearly, there are fallacies in all existing flood control structures or activities but this does not mean we should cease to build them, for carefully undertaken plans that involve levees, reservoirs, conservation measures, and diversions will provide significant relief from many potential floods. Zoning or otherwise restricting building in flood-prone areas will also be a valuable aspect of any flood control plan, for while it may not prevent floods, it will decrease the opportunity for property damage and loss of life. We must seriously evaluate the human development of all floodplain areas, for the destructive potential from floods is too great to be neglected.

## 12

## DROUGHT

Unlike floods, earthquakes, or hurricanes—during which something exciting and violent occurs rapidly and then is finished—drought is more often like a cancer on the land that seems to have no recognized beginning. It may grow slowly, possibly enlarging and spreading to cover vast areas until finally it is diagnosed as existing in our midst, draining the vitality of the nation. Nothing cataclysmic or immediate or violent happens in a drought. One day is generally just like the next. The sun may shine more than usual, temperatures may be somewhat higher than normal, and precipitation may be slightly less frequent, but life goes on pretty much as always. Often, in fairly dry regions, weeks or even months may pass before there is any recognition that things are different from normal and then there is always the thought that conditions will shortly change and return to normal. And so usually nothing is done.

Any detailed description of the course of events in a real drought reads like a science-fiction thriller—not to those who are facing the dreary and unchanging day-to-day conditions but to those who study and interpret what has happened after the drought has run its course. Droughts often result in significant scientific advancements, cause major economic and demographic shifts, increase national unity and purpose, and may produce international cooperation. Yet an event with no recognized beginning or ending, an event that develops only slowly, an event that only gradually modifies life is hardly looked upon at the time as an epic struggle, a heroic battle, or even a story worthy of many newspaper headlines.

Droughts never exist as only isolated climatic aberrations. If such were the case, they could be dealt with more simply, with less disruption to the individuals involved or to the nation. But droughts are always associated with widespread economic tribulations, with disruptions of family life and social order, with problems of insect, fungus, or virus attacks on both plants and animals (and even humans). Droughts seldom lead to revolution for those who have suffered their consequences are usually too weakened to attempt revolts. Yet, the final conquering of a drought and its associated problems frequently results from revolutionary changes in the life of a large segment of a people or a nation.

The earth has seldom known a year without drought some place. There are times when several large areas are undergoing the rigors of drought simultaneously. Drought is commonplace and familiar yet it has scarcely received the detailed attention it deserves. Only seldom do studies of drought attempt to look at its many interrelated aspects. Yet drought is not just a meteorologic phenomenon; it does not just affect agricultural production or steamflow or water supply managers. Drought affects all aspects of life. Perhaps only when viewed as a complex sociological, economic, political, geographic, hydrologic, agricultural, and meteorologic phenomenon will droughts assume the significant role they rightly deserve.

## DEFINING DROUGHTS

Droughts differ significantly from floods in another respect—the ability to define them. It is a relatively simple matter to define what is meant by a flood since there are sharp temporal limits to observe and we can determine just when the river exceeds its banks and begins to cover the floodplain. Not so with a drought which can be defined in nearly as many ways as there are workers in the field.

Clearly, drought results from a lack of rainfall over a significant period of time but it cannot be defined on the basis of rainfall alone. Preferably, drought should be defined on the basis of an inadequate amount of soil moisture over a period of time inasmuch as this considers both rainfall factors as well as plant cover and soil type, and also takes into account the climatic need for water and the removal of water by evapotranspiration.

Meteorologists might want to consider droughts as the result of a persistent large-scale fluctuation in the atmospheric circulation resulting in increased subsidence over a reasonable area. Such circulation patterns would favor little or no rainfall.

Palmer (1965) suggested the need to consider a somewhat different definition of drought. He defines it as a significant deviation from the normal conditions existing in an area. Those definitions that use rainfall or soil moisture conditions to define drought consider that a permanent drought exists in a desert area. Palmer, however, would only consider a drought in a desert area if the precipitation in a particular period of time was significantly less than normal.

Finally, there are economic definitions of drought. It can be argued that drought is not significant unless crop yields or other economic production suffers sufficiently to result in significant financial losses. Similar to the Palmer definition, economic drought would generally not exist in a desert unless the already low level of economic activity were significantly affected by the deviation of conditions from normal.

A recent World Meteorological Organization (WMO, 1975) report included a detailed survey of the various definitions of drought. The many definitions are classified into subheadings as follows: (a) definitions based on rainfall alone; (b) definitions based on rainfall and mean temperature; (c) definitions based on soil moisture and crop parameters; (d) definitions based on climatic indices and estimates of evapotranspiration.

All told, the WMO report provides 14 different references for definitions in category (a), 13 for definitions in category (b), 15 for definitions in category (c), and 17 for definitions in category (d). Their review of the problem of definition of drought makes it clear why it is neither useful or practicable to attempt a rigorous definition. It is probably sufficient to say that drought involves water shortage over some period of time. However, it is a relative rather than an absolute condition so that statements including actual amounts of rainfall for prescribed time periods have little value. Water shortage adversely influencing the established economy of an area—an economic definition—may be as good a definition as is possible under the circumstances (Subrahmanyam, 1967), and it serves the purpose of linking the physical and economic aspects of water. Even in defining drought, one has difficulty in separating socioeconomic aspects of water from physical aspects.

## THE CANADIAN PRAIRIES—DROUGHT AS A SCIENCE-FICTION THRILLER

In 1857, Captain John Palliser hastily explored portions of the flat prairie in the triangular area having Saskatoon, Saskatchewan as its apex, the international boundary with the United States as its base, and the boundaries with the provinces of Manitoba and Alberta as its corners. He described this area as one that would never support a viable agriculture and thus it should never be settled because of its dry, inhospitable conditions. Yet in 1927, this same triangular area produced some $90,000,000 worth of wheat (Gray, 1967). True, there were millions of acres within the Palliser Triangle (as it is often called) that never should have been farmed, but there were also many millions of acres that are nearly ideal farmland if treated properly. Palliser, as many before and after him have also done, made a broad generalization based on only one quick view of an area, without detailed study of soils, climate, drainage, and farming techniques. By 1901, an estimated 20,000 people lived in the Palliser Triangle area outside the city areas and, in 1915, yields of 30 to 40 bu of wheat to the acre were common in the Triangle. The Prairie Provinces of Canada produced some 360 million bushels of wheat that year from just less than 14 million acres (a bumper year)—a yield nearly twice as great as ever before.

The tremendous yields of 1915, the desperate need for wheat during World War I, and governmental encouragement for rapid extension of wheat growing into areas that had never before been farmed, all proved to be pressures too hard to resist. In the next 4 years, 5.5 million new acres of wheat land were added in the Palliser Triangle alone. Interestingly, precipitation in 1915 differed little in amount from 1914 when crop failures were commonplace but, in 1915, the limited rain fell at the right time and in the right amounts. Total amounts rather than distribution were used by those arguing for further development of the prairies and the record average yields of 1915 (not topped for another 40 years) were convincing proof that these areas could indeed become the "breadbasket" of Canada. Crop failures throughout the region in 1917, 1918, 1919, and 1920 did raise serious questions concerning the possible role such an area could play in the agricultural plans of the country but, by then, the die had been cast, the country was being developed, farms were established, communities and their attendant services, roads, and schools were spread throughout the area. It would have caused tremendous problems to remove all of the people, many of whom were convinced that the area had a sound future in agriculture. Even so, the period 1921 to 1926 saw a heavy emigration from the Triangle area as well as from other sections of the Prairie Provinces. More than 10,000 abandoned farms were counted in Alberta in 1926 while the population in the area north of Medicine Hat dropped 30% within 5 years. In Saskatchewan, some of the poorest land did go back into weeds and finally to grass, but not nearly enough. The rains returned in the 1920s and from 1922 to 1928 good crops were experienced by most of the farmers in the area. But just as importantly, wheat prices were high. Large yields with low prices would have resulted in low income per acre for the farmers just as would high prices and poor crop yields. High prices along with excellent yields, however, boosted farm incomes and the spirits of the farmers themselves. Cash farm income in Saskatchewan averaged nearly $300 million a year during the 1920s. Problems of drought, of blowing dust, of insect pests seemed very far away and the warnings of Captain Palliser were largely unheeded.

Yet by August 1, 1931, when the Canadian Red Cross launched an appeal for food and clothing for 125,000 destitute farm families suffering their third year of crop failures, the word was out to all of Canada that the scourge of drought was ever-present in the Palliser Triangle. One might question why 3 years of crop failures were needed before the word spread across Canada, but then drought is insidious. Each year the farmers said that next year would be different, the rains would come again as they had in the 1920s, good yields with high prices would return, and there would be no further problems. Nothing really happened from 1929 to 1931. Some rains came but not enough or at the right times to save the crops. Farm income fell, farmers went into debt, a few left the land and moved elsewhere, but always there was next year with its promised return to normal conditions. And the people of Canada had other problems to trouble them as relief roles mounted, wages were cut and the depression became a way of life. Still, the response to the call of the Red Cross was heartwarming. Some 247 carloads of food and clothing were donated for the impoverished farm families and, more importantly, East and West were more firmly united than they had ever been as human suffering gave purpose and direction to the lives of many throughout the nation.

As in most droughts, conditions were not uniform throughout the affected area. In some areas or on some soils, good yields were still possible. Thus, the pattern of suffering and privation was spotty. A tour through the area might

**Table 12.1**  Cash Return/Acre of Wheat, Crop-Reporting Districts, Saskatchewan, 1930–1937

| District | 1930 | 1931 | 1932 | 1933 | 1934 | 1935 | 1936 | 1937 |
|----------|------|------|------|------|------|------|------|------|
| South Eastern | $5.83 | $1.29 | $3.63 | $3.29 | $2.01 | $1.40 | $3.96 | $2.31 |
| Regina-Weyburn | 4.56 | .11 | 3.32 | 5.13 | 2.01 | 2.95 | 7.39 | nil |
| South Central | 3.25 | .61 | 2.37 | 1.18 | 1.28 | 5.98 | 2.82 | nil |
| South Western | 5.74 | 1.60 | 4.96 | 1.32 | 1.60 | 4.20 | nil | nil |
| Central | 4.28 | 2.66 | 3.53 | 1.93 | 3.98 | 7.38 | 7.93 | nil |
| West Central | 8.65 | 4.52 | 5.39 | 1.23 | 4.33 | 5.00 | 3.34 | nil |

*Source.* Gray (1967), with permission.

reveal vast areas laid waste but also other areas where life went on almost normally although yields might be reduced. Table 12.1 illustrates this shifting pattern of drought during the 1930s for the various crop-reporting districts of Saskatchewan. Government estimates suggested that a cash return of $5.30/acre was necessary for a farmer to survive without relief even without paying taxes.

Of all the days during this continuing battle against drought, possibly May 12, 1934 was the most dramatic (Gray, 1967). On that day, huge clouds of dust rose from the American and Canadian prairies, reaching upward to over 10,000 ft. Dust covered everything from the Rocky Mountains to the east coast. At the same time, word came that the western prairies of Canada faced the greatest grasshopper invasion in history. Within a day, half a million dollars was appropriated to establish stations to mix poison bait to combat the grasshoppers.

Drought was bad enough, but now blowing dust and grasshoppers were added to the problems of the farmers to lower even further the already limited yields. Before it was all over, new scourges were heaped upon these disasters. Say's grain bug growing in the Russian thistle attacked nearby stands of wheat and barley. Gophers multiplied in the abandoned farm fields and moved out of them by the thousands to destroy whatever was growing nearby. Sawflies, caterpillars, cutworms, some never before seen in Canada, added their destruction to the list. A bacterial wilt destroyed the potato crop in Alberta and wheat rust added its destruction to that already visited upon the wheat.

Worst of all were the grasshoppers. They moved like huge armies, in numbers beyond all possibility of estimate, and they ate nearly everything in front of them. Showing a particular fondness for things handled by humans (attracted by the salt or the moisture in the perspiration, no doubt), they would eat the handles of pitchforks and hoes, the armpits out of shirts right on the farmers' backs and clothes left to dry in the wind. Any vegetation in their path was destroyed—trees, bushes, crops, gardens, weeds. The grasshoppers contributed to the spread of the deserts already being created by the drought. The grasshoppers also speeded the departure of many farm families who abandoned their fields to the vagaries of wind and insects. Not only would these abandoned lands blow, but they would also serve as areas for the growth and development of more grasshoppers and other insects and diseases to affect nearby farmed land. And who was responsible for the care and control of these abandoned farms? No real responsibility could be placed, but until those lands were brought under control, little could be done to control the interrelated complex of factors that really constitute a drought.

Blowing dust had always been common in the western provinces though never before like that May storm of 1934. But then, conditions had changed in the previous few decades in the prairies. Many more acres of prairie sod had been turned over as cropland expanded. Protective surface covers had been destroyed. Summer fallow had been developed as the only practical and efficient way to insure a crop in these semiarid lands. Possibly most important, however, was that many farmers had moved into the area from the more moist eastern portions of the country. They were quite often good, hardworking farmers who brought with them the ways of farmers from more moist areas. Farming techniques in areas with 40 in. of rainfall were quite different from farming techniques in areas with 15 in. of precipitation. Old habits had to be unlearned and new habits developed and this required time—time often not available to the new migrants to this western rangeland.

Summer fallow had proven to be of value for the conservation of moisture. By raising only one crop in 2 years, or two every 3 years, farmers were able to conserve the moisture of the alternate year to help insure germination and growth of crops when they were planted. During the fallow period, nothing was to be grown on the land but rather it should be maintained bare—covered by what the good farmers described as a fine "dust" mulch. Summer fallowing received considerable study as well as publicity during the Dust Bowl years of the 1930s in the United States and in Canada. It was thought that good summer fallowing practice required the maintenance of a fine, dry, dust mulch on the surface; that the surface should be reworked after rain to prevent the development of surface caking or clods and to reestablish the mulch. In theory, moisture was drawn to the surface by capillary action to be lost by evaporation. Cultivation of the surface would break up the capillaries through which this water movement occurred and so reduce the evaporative loss of water. Summer fallowing would conserve water and allow the storage of some, at least, of the meagre summer rains for later crop use.

Summer fallowing did conserve moisture but, at the same time, it greatly increased the possibility of wind erosion and the blowing of great quantities of dust so common in the mid-1930s. Studies showed that weed growth, much more than capillary action, removed moisture from the soil. Instead of producing a fine dust mulch at the surface, farmers were advised not to work their summer fallow fields. Plowing, discing, and harrowing the soil should be done only to the extent necessary to control weeds. Plowing of dry soil should never be attempted because of the possibility of wind erosion.

During the 1930s, the thinking about a dry, dust summer fallow mulch changed markedly with the suggestion of a plowless fallow. But fields that had the stubble burned off to kill the weeds still were affected by wind erosion. Stubble fields left unburned did not suffer wind erosion; but weeds, of course, would grow and vital moisture would be lost. There was need for equipment that would eliminate weed growth yet leave the stubble on the field. A blade-type cultivator that sliced through the soil just below the surface was developed. The idea of a dust mulch did not die easily, however, Farmers, accustomed to neat, cultivated fields, did not like to see rough, unplowed fields covered with stubble and dead weeds through the summer fallow period. New words such as "stubble mulch" and "trash cover" were coined to explain the new mulching technique. The success of the technique in reducing both moisture loss and wind erosion ultimately led to its widespread adoption by farmers throughout semiarid areas where farming with summer fallowing was necessary. While plastic, paper, or hay mulches are also effective in reducing water loss and wind erosion, such techniques can hardly be applied over thousands of acres of farmlands. Stubble mulch, however, can. The results of this human interference with the natural operation of the water budget has been a real conservation of water without attendant problems of wind erosion in semiarid areas.

The change in the approach to summer fallowing helped in the continuing battle to control blowing dust but no single solution was sufficient. Blowing dust from abandoned fields would cut the young growing vegetation on a neighboring field or would drift over the farm gardens or fields of newly emerged wheat. Thus, the problem existed not only on those fields that were being farmed, but also on those that were abandoned. The owners of these farms had moved away and could not be located. In many cases, they no longer owned the farms anyway since ownership had been lost due to nonpayment of taxes or the land had been repossessed by banks or other money-lending groups. The government itself was the owner, ultimately, of much of the abandoned land.

As one step to try to eliminate the problem of abandoned land, Alberta made it an offense punishable by a stiff fine for an individual to allow soil from his farm to blow onto a neighbor's land. As might be expected, no one was ever charged with such an offense since nearly all were guilty to one degree or another. It was not until concerted efforts were made by the government and farmers working together to apply good farming techniques not only to the occupied farms but also to the abandoned farms that the problem of blowing dust began to be controlled. But an interesting part of the solution involved crested wheat grass as well.

Crested wheat grass was first imported from Russia to the United States in 1898. Tested in various experiment stations in Montana and the Dakotas, it was found that it did well in areas of low rainfall but, for some reason, little more was done with this information until 1915 when workers at the University of Saskatchewan learned of the tests and imported some seeds into Canada. Initial testing in Canada showed some promise, for the grass was particularly tolerant of the severe Canadian winters, but other favorable aspects were not readily apparent. Almost by chance, it was discovered that crested wheat grass liked rough treatment and did not do well when well protected and watered.

Crested wheat grass seed did not require much moisture to germinate and then it wanted to be left alone. It required no help from humans to survive in its struggle against weeds and drought. Under hot, dry conditions, the grass turned brown and looked dead. Without water, it would stay this way through the whole summer. With the fall rains, it returned to life and overwintered well. By the following year, it had prospered so well that it had no difficulty in crowding out most of the annual weeds in competition with it. Roots of the crested wheat grass would penetrate 6 ft or more into the soil looking for water. Later work with the grass showed that it would provide a better crop when mixed with other grasses and that its greatest value was as a pasture grass rather than as a hay crop. Addition of Russian rye grass, brome, or alfalfa would provide more adequate summer pasturage when the crested wheat grass turned brown.

The story of crested wheat grass, only part of the overall story of the fight against drought and its attendant conditions, is interesting in showing how adversity produces outstanding scientific achievements. While grown in dry areas in Russia and the United States in fairly limited amounts, it was not until the Canadian experiments proved the overwhelming value of the grass that reintroduction into both Russia and the United States occurred to any great degree.

The insect and bacterial pests which time and again afflicted the drought-weakened crops produced their own stories of heroic achievements. Early work by Norman Criddle showed that grasshoppers were attracted by the moisture in fresh horse manure and they could be easily killed by poison mixed with the manure. Dry manure held no attraction for them and in this condition the bait was relatively ineffective. Later work developed other ways of providing the poison in an attractive form to the grasshoppers and widespread cooperation by the government, researchers, farmers, and other field workers was necessary to put the program into action. In 1934, for example, Saskatchewan stockpiled 90,000,000 lbs of sawdust, 35,000,000 lbs of bran, and 180,000 gal of sodium arsenite in numerous mixing centers (Gray, 1967). Farmers spread this poison bait around their fields in an all-out attack on the grasshoppers. The campaign was a success even though 10% of the crop was destroyed and $20 million in damage was done. It was estimated that 40% of the crop in the Palliser Triangle was saved by these Herculean efforts to destroy grasshoppers. While the crop saved amounted to little because of the dry conditions, this seemed of less importance. A war had been entered into against the grasshopper and marked success had been attained.

Grasshoppers were less of a problem in 1936 and 1937 as the poison bait campaign was making definite progress. But no one expected 1938. The crop got away to a good start when clouds of grasshoppers came from the south once more. With the peak of the grasshopper invasion, rust developed in the wheat crop in epidemic proportions. Thunderstorms were next and with them came hailstorms of extreme severity buffeting the remaining crop. As if these attacks were not enough, wireworms, which had been growing in numbers, chose this time to attack with renewed vigor. They destroyed more than $8 million worth of potential wheat production. Potential crop yields were cut in half in less than a month in the Palliser Triangle as a result of these staggering devastations. The farmers of Saskatchewan lost more than $50 million worth of crop production in that year.

Another pest, the sawfly, produced a new problem and another opportunity for sound agricultural advancement (Gray, 1967). Unlike grasshoppers which would invade and destroy a whole field with ease, the sawfly confined its attacks primarily to the edges of the wheat field. The sawfly needs a hollow-stemmed plant in which to lay its eggs. The larvae hatch from the eggs then feed within the hollow stem of the grain until it nears ripeness. The larvae then move to the bottom of the stems, girdle them from inside, and retreat into the roots to overwinter. The girdled stem, of course, collapses before the grain ripens.

Because the sawfly does not go far into a field in search of a hollow-stemmed plant, the efforts to control blowing dust by strip-cropping came into serious jeopardy with the rise of the sawfly. Strip-cropping produced more edges of the wheat for the sawfly to infest. Many farmers were considering abandoning strip farming when the answer came from the experiment stations. Brome and oats are also attractive habitats for sawflys to lay their eggs but both of these plants are poison to them. Brome grass needed more moisture than oats so it was seeded along the roadsides in the ditches where more moisture would accumulate while both brome and oats were seeded in strips around the wheat fields. The menace of the sawflies was finally overcome as this program of seeding became more widespread. And incidentally, as roadsides and ditches were cleaned for the seeding of brome, the area developed some of the best kept roadsides in Canada.

Gophers became a real scourge, multiplying rapidly in the abandoned fields and foraging for anything they could find in nearby cultivated areas. The Prairie Provinces paid a bounty of 1¢ for each gopher tail turned in. Hunting gophers not only became a source of candy money for the children, but gopher meat meant survival for more than one farm family during the drought years. Gopher stew, gopher pie, smoked gopher, and pickled gopher all had their supporters. In 1934, the youngsters around Indian Head collected $765.14 for killing 76,514 gophers while some 10 times that number were killed that year in each of the Prairie Provinces (Gray, 1967). In spite of these efforts, the gopher population was not brought under control until the problem of abandoned farmland was solved.

The story of the drought in the Palliser Triangle in the 1930s is more than a story of lack of precipitation. It is also a story of summer fallow and "dust mulches" vs. "trash covers." It is a story of grasshoppers and sawdust and poison bait, of sawflies and brome grass, of gopher stew and gopher pie, of machinery to cut weeds without disturbing the stubble cover, of experiment stations working with the farmers to understand their problems and to suggest solutions that the farmers were willing and able to try. It was a story of international exchange of ideas and information about crested wheat grass and about farm machinery and farming methods and of national unity and purpose to defeat a cancer that affected the life of the whole nation. And, of course, it was a story of people—the farmers who stayed on the land as well as those who lost everything including hope and left, the experimenters and researchers who worked tirelessly to perfect new techniques, new machinery, new insecticides to blunt each new hazard as it developed. These are the ingredients of a heroic campaign or of a science-fiction thriller that turned out not to be fiction.

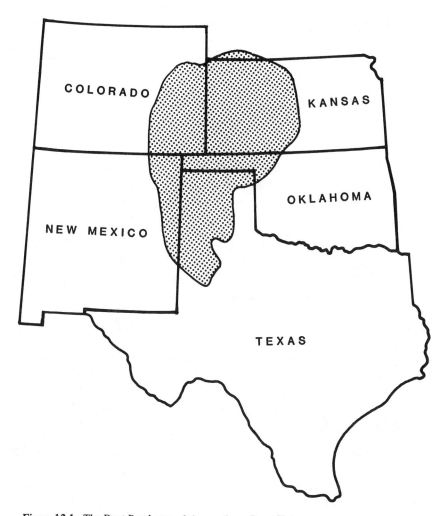

**Figure 12.1**  The Dust Bowl area of the southern Great Plains.

## THE AMERICAN DUST BOWL–AN UNLEARNED STORY

Stretching south from northwestern Kansas and northeastern Colorado, across the panhandles of Oklahoma and Texas into eastern New Mexico, is an area that gained worldwide publicity during the 1930s–the Dirty Thirties, as they were known by many–as the heartland of the Dust Bowl of the southern Great Plains (Fig. 12.1). The story is too well known to repeat in any detail. Factual accounts of the dryness, the grasshoppers, the intense heat, the withered crops, and the everlasting dust merge imperceptibly into fiction. The stories of the Oakies and the Arkies loading all they possessed into the family car to leave a seemingly God-forsaken land for one of promise in California became the subject of novels and factual tales describing the uprooting of a whole group of people and the changing of a way of life. More clearly fiction (but possibly not to some who experienced those days) was the story of the pilot whose engine failed him in the middle of a dust storm and who, after he bailed out, took several hours to dig himself down to earth (Johnson, 1947). While the message of the Dust Bowl is engraved on many minds, it can still be labelled an unlearned story for it is still quite possible for the story to be repeated again as it already had been before the Dirty Thirties.

Any pioneer or settler could tell you that the most impressive thing about the High Plains of the western United States was the wind. The wind never seemed to stop blowing. There was nothing on the flat, treeless prairies to stop it or even to slow it down. It blew almost constantly. Yet, the new farmers coming to the semiarid southwest in the late 1800s saw something else—a broad, flat, prairie grassland—a farmland free for the asking on which there was money ready to be made. As the Indian menace subsided, as communications improved, as demand for cattle and crops to feed an industrializing nation grew, so too did the flood of immigrants into the southwestern prairie tableland.

Some good years were experienced by the newcomers, money was made and the land appeared to be one of hope and promise. Then came 1890. The rains nearly ceased and the sun shone down hot from a cloudless sky. Crops failed over much of the Southwest in 1890 and again in 1891. The drought continued into 1892 but this time it was accompanied by hordes of grasshoppers. They ate everything but always the thought was that next year would be better. However, 1893 was a year of financial panic in the East, businesses failed, hunger marches were organized, riots developed, and there was threat of revolution. The drought continued in the prairie tableland through 1894 and 1895. Families could really not survive on the 160 acres given to them by the government even under the normal semiarid climate of the region but, during a severe drought, there was just no reason to try to hang on. Slowly at first, and then by the hundreds, the settlers loaded their wagons and headed east to return to families or friends and a securer way of life in a more humid climate. A few, with still a sense of humor, printed across their wagons, "In Kansas we trusted, in Kansas we busted" (Johnson, 1947).

The rains came back in 1896 and, by 1899, those farmers who, for one reason or another, had remained beyond 1895 were boasting of their great foresight in staying on in this rich farmland. New settlers were already pouring back into the area and railroads and real estate operators were touting the wonders of the open and available farm utopia. Even though warned by J. E. Payne of the Cheyenne Wells Dry Land Station through many newspaper stories of the hardships, the droughts, and the fact that two out of every five harvests would be failures, the new settlers came on unmindful, eager to share in the riches that were being realized (Johnson, 1947). Payne listed four items of advice to the newcomers to help increase chances of successful settlement in the prairies:

(a) Have a milk cow, if possible, or even more than one, take good care of them and sell cheese, or butter.
(b) Keep as many chickens, turkeys, and geese as possible.
(c) Plant fodder crops for the cattle—amber cane, yellow milo maize, and corn; then it might be satisfactory to plant a small area for fall wheat.
(d) Plant a farm garden.

The advice, which generally went unheeded, recognized the need for new settlers to be self-sufficient first, to be able to withstand the anticipated drought periods without all their fortunes tied to one crop. But money was to be made in ranching and in wheat farming even on the small quarter sections as long as the rains held and scarcely anyone listened to the advice.

Some crops failed in the Texas Panhandle in 1908, a few more areas felt the withering effect of drought in 1909 while, by 1910, continued widespread drought broke the growing boom for farmland. But then in 1911, while drought conditions were patchy, a new hazard was added to the picture—dust. Dust had always been a condition of springtime. With the wind and the plowing, some dust would just naturally be in the air. It was part of a way of life just like mosquitoes in some areas or chiggers. In northwestern Kansas, however, in the spring of 1911, the dust rolled across the open plains, turning the noon sun into a red sunset. It cut the young wheat and even pulled some out by the roots. The Kansas Agricultural Experiment Station issued a bulletin to the farmers suggesting that if they would plow perpendicularly to the prevailing wind and leave a clod mulch rather than a dust mulch during the summer, they would have little trouble with blowing dust.

But the bulletin was scarcely heeded. And 1912 was another bad drought year for most of the southwestern plains. Late winter of 1913 brought continuous strong winds to northwestern Kansas and dust once more began to blow in an ominous way. Dust storms lasting 3 and 4 days at a time and marked by black clouds rolling across the prairies were seen in various places. Some drifts of dust and sand were as much as 25 ft high. In many areas in northwestern Kansas, the topsoil had blown away right down to the plowsole while roads were choked with dust. The worst areas of blowing were abandoned by the farmers who camped around the edges waiting for the dust to subside.

By June, the wind had died down, the dust was settled, and the farmers were back plowing and planting kafir corn because of its rapid growth and good rooting system. The fall of 1913 was moderately wet and the wheat crop got off to a fine start. More rain fell during the following year. The start of World War I brought such a demand for wheat that the blowing dust of 1911 and 1913 was soon forgotten—another lesson unlearned.

Under the pressure of wartime with its need for motor vehicles, tanks, and other troop carriers, and with the worldwide demand for wheat to feed nations whose ways of life had been disrupted by the swirling battlefields, two great new forces were added to the life of the Great Plains. First was the development of the lightweight, all-purpose tractor within the price range of farmers, and second was the demand, encouraged by the government, for more wheat land to be planted, for more of the prairie sod to be broken in the interest of increased agricultural production (Bonnifield, 1979). In 1913, three 30-hp kerosene-powered tractors had pulled a single gangplow with 50 bottoms, sufficient to turn over a lane nearly 60 ft wide at every pass across the field (plowing at a record rate of an acre every 4 minutes and 15 seconds). This equipment was clearly not available to the average farmer but the development of the internal combustion engine did put small gas tractors within the price range of the average farmer.

The tractor made it possible to turn over the prairies in a manner never before possible. Good land and bad, light soils and heavy, well drained and poorly drained, it made little difference. The nation needed more wheat so it was the patriotic thing to do and, besides, there was money to be made. By 1918, the Department of Agriculture made some suggestions about the need to be careful in the selection of new lands to plow, pointing out that certain areas were marginal and should not be brought under cultivation. However, much of the damage had already been done and few were ready to listen even had the message been more forcefully delivered. The Colorado Agricultural Experiment Station again called for the adoption of a diversified agriculture rather than widespread monoculture of wheat but few followed such sound advice. And the government wanted more wheat. Forty-five million acres of wheat were harvested in 1917 in the United States but the government suggested an increase of 7.5 million more acres for 1918. There was no time for diversified agriculture. Between 1914 and 1919 wheat production increased by 27 million acres and nearly half of this increase came from land that had never before been plowed.

Several other agricultural statistics were not only interesting but ominous in their own way. Farms became much larger. During the decade from 1910 to 1920, in the southwestern plains area, farms increased in size from an average of 465 acres to over 770 acres. The number of small farmers had decreased. Although a vast amount of land was being taken out of grazing and plowed for wheat, the number of cattle on grass in 1920 was still 50% greater than 10 years earlier. Some grazing areas were overstocked to the detriment of the forage crops growing on them. Farming was also becoming more expensive as the need for more power equipment increased. Mortgage debt of farmers rose markedly during the decade. And the number of tenant farmers more than doubled during the decade. Tenants are not as interested in good conservation practices on land they do not own so this development exacerbated an already perilous land situation.

Wheat, which had been well over $2.00/bu during the war years, dropped to 70¢/bu 9 months after the stock market crash of 1929. This price was fixed because of large purchases by the Federal Farm Board in order to take wheat out of the market. The Farm Board also asked for appreciable reduction in wheat acreage, but no one listened. Even though prices were low, the way to make more money was to plant more wheat and indeed a greater number of acres were brought under cultivation. Blowing dust was more commonplace now than earlier but the dust always subsided and dust was becoming a way of life. It was expected and, in a sense, it was satisfying because it showed that the soil was being plowed, the farmers were working, and everything was normal.

Some farmers made money in 1930 but many did not because of low prices. Yields were good, so the way to make more money was to plant more. The year 1931 was a banner wheat year for the southern Great Plains. Rain and snow both came in goodly amounts and dust was less of a problem. Yields of 50 bu to the acre were found on many farms. But yields and prices often do not go hand in hand. On June 10, No. 1 hard wheat brought 50¢/bu at Amarillo; by August 7, it was down to 20¢/bu. Not only was there a depression worldwide but there was just too much wheat. No one wanted it. Many farmers decided to hold their wheat rather than to sell it, still hoping for higher prices; but the world economic situation was not going to allow an increase in prices. Farmers finally had to sell for cash to pay some bills even though they were losing money. Bushels of wheat were equivalent to cash money and even though each bushel brought very little, cash was still necessary to pay off interest or to buy food and clothing. No matter what the price, it was clear that more bushels for 1932 would bring in cash to meet the ever-present bills and mortgage payments. The winter of 1931-32 saw even more land planted to wheat.

But 1932 was not a good wheat year in many respects. A late hard freeze, hail storms, cutworms, and the beginning of a drought all reduced yields appreciably. Even though the price of wheat came up somewhat, the lowered yields made it a poor year. Even though some of the big "wheat kings" failed, it did not stop the continued plowing of the land for more acres of wheat. The farmers were trapped. They needed more wheat for more cash to pay more bills so they needed more acreage.

Dry conditions continued in 1933 over much of the Southern Plains and hardly any snow fell during the winter of 1933-34. Vast acreages of wheat were planted, but it did not look good as 1934 started. Winds picked up in February and March and dust began to move ominously. Then came April 14. To those who were there, it seemed as if all of Kansas just rose in one big swirl of dust, a cloud that choked and smothered, and cut, and covered every living thing (Johnson, 1947).

A tremendous cloud of dust rolled southward out of Kansas and into Oklahoma and Texas, a cloud stretching east-west from horizon to horizon. Visibility in the cloud could be measured in feet rather than in miles and the darkness of night existed at noonday. The blackest part of the dust cloud passed in about an hour but the eerie darkness held on for 3 more hours. Dust piled up everywhere. Crops, farm gardens, and roadways were covered; small prairie animals and birds were suffocated in vast numbers. It was incomprehensible to those who had not seen it, but should it have been? The warnings were out long before—in 1911 and 1913 and nearly every other year, for that matter—for blowing dust was a way of life.

May rains usually were significant in the Southern Plains because they would settle the dust and bring good crops. They had failed in 1933 but they would not fail a second year in a row. However, they did. The sun continued to shine down hot from a cloudless sky and dust continued to blow back and forth—drifting, cutting, covering as is the way of dust.

On May 10 and in the days immediately following, the very end of the world seemed at hand. Across the whole prairie, from Texas into Canada, the wind raised up a tremendous cloud of dust reaching altitudes in places of more than 15,000 ft. Not as black as the cloud in mid-April, but rather more light-brown. The sun was obscured and a darkness covered the land. Individuals out in the dust were blinded by it and some wandered for hours trying to find their way to shelter. With visibility reduced, road accidents were greatly increased.

The wind blew from the west and the cloud of dust rolled eastward. By May 12, it was clearly seen over New York, Washington, and in eastern Canada. It shut out the sun for 5 hours and made a tremendous impression upon individuals far from its source of origin. Ships at sea reported seeing the cloud of dust. Unbelievable, but true, accounts of the vast acreages of top soil that had blown away began to appear in newspapers. When parts of Kansas were being deposited in New York and Massachusetts, it was enough to worry everyone. No longer was it a local problem; it became a national one.

In many ways, 1934 was hardly unique. Earlier years had seen dust. The Oklahoma Panhandle had had 40 destructive wind and dust storms in 1933 alone. But they were localized and did not affect a whole state, let alone the whole country. They would pass and be forgotten.

In the week following May 10, 1934, heavy rains fell in much of Kansas and Texas. The wheat that still remained was saved. Surprisingly, yields were quite good. Twenty-three Texas counties harvested 18 million bushels, about two-thirds of their normal crop, while yields from the Oklahoma Panhandle were 14 times greater than in 1933. Kansas harvested grain from a million more acres than in 1933. One could almost forget the lessons of April 14 and May 10.

The rest of the story, in most respects, is similar to that in the Palliser Triangle. The drought of the middle Thirties intensified. Grasshoppers and other insect pests multiplied and added their destruction to that of the drought. Cattle, already grazing in too large numbers on the remaining prairie grasslands, could no longer be fed; hundreds of thousands were shipped to market to bring what little they could or were shot in the fields to save shipping expenses. Farmers abandoned their lands and the abandoned fields blew out of all control, adding to the problems of those who remained. The problem of cutting the acreage planted to wheat finally began to take care of itself.

Arguments for improving the overall farm situation abounded. Farmers called for a fair price for wheat, feeling that nature would return to normal on its own. Government wanted more land out of wheat and returned to grassland, not only to provide more fodder for the cattle but to stabilize the soil and to reduce the blowing. Farmers wanted subsidies to help tide them over until the "next year" when things would be better; government called for controlled plantings of shelter-belts of trees, good conservation measures, contour plowing, and basin listers.

Meanwhile, the weather bureau in Amarillo in the Texas Panhandle classified the frequent blowing dust storms into three types: (a) northwesterly black dusters which lasted 10 hours or more covering a quarter of a million square miles; (b) lighter-colored dusters from the southwest, often less spectacular, but more intense, larger, and longer lasting; and (c) severe local storms, which were the most frequent of all but generally covering areas of less than 10,000 mi$^2$. Dark brown or black dust storms came from Kansas, red ones usually from Oklahoma, and tan or dirty yellow storms were from Texas and New Mexico. You could tell the direction by the color.

A concerted attack by the government and by the farmers themselves, now banded together against a common foe, began in the spring of 1936. Farmers were to be paid 20¢ an acre to plow their own fields on the contour and to provide little listers or dams in the furrows to trap and hold rain. By April, more than 20,000 farmers signed up to contour-list some 4 million acres. The plowing went on through wheat fields as well as through unfarmed fields. The number was up to 5-1/2 million acres by May, with nearly 40,000 farmers in the program (Johnson, 1947). Wiser heads now recognized that the problems of the Great Plains had been long in developing and would require a long time and a concerted effort by many to bring under control. But they could and would be controlled. Basically, the problems were:

(a)  too much land under cultivation with no attention to soils and their characteristics;
(b)  grasslands overgrazed;
(c)  cash-crop farming, especially of wheat, rather than a more balanced, diversified subsistence agriculture;
(d)  wrong cultivation methods practiced.

Farmers had to accept responsibility for much of the problem since they had planted too much wheat, had not diversified their agriculture, were too optimistic about weather and crop prices, and had not learned by experience. Some of the blame, however, could be attributed to factors beyond their immediate control. Many more items contributed to the problems of the Thirties: for example, the government land policy of 160 acres to each homesteader, too small for subsistence in semiarid areas; government encouragement of wheat expansion during the first World War; lack of real knowledge of the soil characteristics of the Southern Plains; the whole credit structure for farmers; state laws on leasing, taxation, tax delinquency; the role of tenant farmers; lack of adequate zonation of land for agriculture, for grazing, and for fallowing.

Can the story be repeated or do we now know enough to prevent another dust bowl? Drought will return to the Southern Plains once again, we are certain— only the time is unsure. Wheat prices will fluctuate up and down. There will be economic recessions and booms. And there will certainly be increasing demand for wheat to satisfy the every-growing world population. Agricultural production will be expanded again and again and land taken out of cultivation in the Thirties may once again feel the plow.

Already there are statements from high officials that agricultural production is needed to provide exports to balance our oil imports—that agriculture will provide us with the needed favorable international balance of payments. All the ingredients are certainly present for a return to the Dirty Thirties. Possibly conservation measures have been so well learned that the blowing dust will be avoided, but the boom and bust agriculture with drought and insects and low prices and

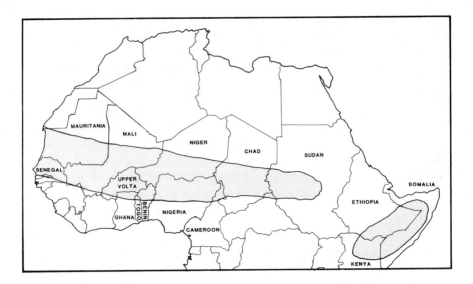

**Figure 12.2** The Sahel region of Africa.

hope for tomorrow will return unless we have finally learned a lesson that count-less former generations have seemed incapable of learning.

## THE SAHEL–SOCIOECONOMIC CHANGE OR CLIMATIC CHANGE?

Stretching across North Africa just south of the Sahara Desert in a belt some 200 to 300 km wide is the Sahel, taking its name from the Arabic word meaning borderland (Fig. 12.2). Basically, it is a climatic term since the Sahel covers the border steppe area between the desert wastes of the Sahara to the north and the seasonally-moist savanna area to the south. Geographically, it covers much of the six nations of Senegal, Mauritania, Mali, Upper Volta, Niger, and Chad. The rainfall in this belt increases markedly from north to south. Along the northern portion of the Sahel, the annual rainfall averages 100 to 350 mm (4 to 14 in.) while along the southern portion it ranges from 350 to 600 mm (14 to 24 in.), seemingly quite adequate for serious agricultural development but more limited than it might appear. Much of the moisture comes in a few heavy showers so that 80 to 90% of the rainfall is lost rapidly as runoff or evaporation (Brabyn, 1975).

The Sahel has always been subject to drought. Drought is a way of life for the nomadic herdsmen who inhabit the northern reaches of the Sahel and quite common even for the more sedentary agricultural population that dwells in the southern portion of the area. Just a small change in the rainfall from year to year or in the distribution of the limited rain that falls brings marked changes in crop yields or in the forage available to the cattle of the nomadic herdsmen.

But the vagaries in rainfall and the continuing struggle against its meagreness is well known to the inhabitants of the Sahel. In fact, over hundreds of years of occupancy of this border area, they have developed a particularly effective adjustment to the ever-present dry conditions, the harshness of the climate, and the constant lack of adequate food or fodder. Amidst many changes that do exist in the area in terms of precipitation, diseases, epidemics, and nomadic wanderings, two constant factors upon which the life patterns of the inhabitants of the area can be established are the permanence of the semiarid conditions and the marginal adjustment with starvation and famine (Bugnicourt, 1975).

Thus, the recent threat of mass famine in the area, the great exodus of thousands upon thousands from the northern portion of the Sahel toward the south, and the great loss of animal life that seems to be a direct consequence of the drought of the late 1960s and early 1970s, come as a surprise to many observers. No one argues that the Sahara has pushed farther south in the past decade or that the climates of the past few years have been drier than usual. But are they abnormally so, sufficient to disrupt a way of life that for centuries has been carefully adjusted to long periods of drought? This is a question that climatologists alone cannot answer for the reason that it requires study of socioeconomic and ethnocultural changes as well.

Climatologists have studied the precipitation problem in the Sahel without much agreement. Three possibilities seem to exist. The drought of the past few years is either:

(a) a low point in a normal cyclic variation in precipitation that will swing toward more moist conditions in another few years;

(b) an ominous forewarning of a global shift in circulation conditions that will increase the extent of the subtropical dry belts worldwide, leading to vast global shifts in population and agriculture;

(c) a result of cultural changes by the people of the region that have tipped the always-delicate balance between man and environment in such a way that it is no longer possible for the area to sustain the population it is being forced to support—that the current problem is man-induced.

Strong arguments can be made for each of these points and only time will be able to help us determine whether either of the first two possibilities is germane. The suggestion that it is just part of a normal cyclic variation in rainfall has proved to be true many times in the past. There have been drier periods before but without the great privations of the present, and the area has returned to more moist conditions. Worldwide changes in the climate have also occurred in the past, if one considers the time period from the glacials and interglacials of the Pleistocene to the present. Changes even greater than those of the present have been found in the area and will certainly be found again.

It is the third suggestion that seems attractive to many who view the situation more from a humanistic point of view, without strict reference to weather records and millimeters of precipitation. If one accepts the fact that drought is a way of life in the area, that the inhabitants had long ago adjusted to the ever-present dryness and near starvation conditions throughout the area, one might argue that the drought of the past few years should not have produced the great migrations, the suffering, and the starvation that have indeed been found. Some other factors have seemingly accentuated the problem, changing a short period of below-normal precipitation into a massive struggle for survival of whole nations.

What socioeconomic or ethnocultural changes have occurred in the area in the past few decades that might contribute to the present situation? One of the unique adjustments of humans to the land in the Sahel had been the long-standing balance between nomadic herdsmen to the north and sedentary farmers to the south (Bugnicourt, 1975). The herdsmen bartered their meat for millet from the farmers. Each needed the other and while the adjustment was not always harmonious, an accord had been reached within the capabilities of the land. But changes have now been introduced into this system. A money economy is being superimposed onto the existing barter system, changing the nature of the relationship of meat for millet. The farmers are developing their own herds of cattle while the demand for meat in the larger cities of the area has put the herdsman more at the mercy of the meat buyers from distant areas. To obtain the money needed in the newly emerging socioeconomic environment, the herdsman has been encouraged to build up his herds in the expectation that the meat buyers will want his additional cattle. Herds have always meant wealth and power to the herdsman so there has always been pressure to let the herds develop as rapidly as possible.

In previous years, this effort to increase herds had always been balanced by the high death rate from thirst and disease. Thus, herds had been maintained in general balance with the carrying capacity of the environment. The adjustment was a continuing affair. Recent actions by central governments in the Sahel, however, have led to the digging of many deep wells, and the introduction of pumping equipment so that water from these wells is more readily available. There has also been a great reduction in disease and death of cattle through vaccination and other modern medical practices. No longer is there a natural adjustment in herd size to the potentialities of the environment. Except in drought times, thirst is not as much of a threat to life as previously and many more cattle now survive the epidemics that used to sweep the area.

These advances in technology would seem to be worthwhile but they have clearly upset the precarious balance between the inhabitants, their herds, and the environment. The environment does not supply sufficient food for the larger herds and more cattle are grazing on each piece of land. Clearly, the vegetation is being stripped from the area and with it the shade, the protection of the soil from blowing, and often the very possibility of rejuvenation of a browse crop for later use.

Around the new deep wells, the herds that congregate have completely denuded the land for kilometers in all directions. The watering holes, rather than being oases in the middle of drier areas, are now dry deserts surrounded by packed bare soil. The herds must go further and further from the watering holes to find food and then return to the wells for water. Thirst is no longer a frequent cause of death but starvation has become common (Bugnicourt, 1975).

Superimposed on these problems of the nomadic herdsmen of the north are changes in the lives of the more sedentary farmers to the south. Encouraged by more stable governments, supported by better education and a money-based economy, they have been encouraged to extend their agricultural production not only in area but also in type of crops. Agricultural self-sufficiency is the goal. Groundnuts and cotton have been added to the millet and farm gardens. The availability of more water for irrigation has resulted in the extension of agriculture further northward into some of the marginal land previously used by the nomadic herdsmen for the support of their cattle. And with improved medicine and sanitary conditions has come an increase in population. Possibly some 6 million people are now trying to subsist in an area where before no more than 3 million were able to eke out a bare existence.

This would seem to be the real crux of the present situation. Technical and socioeconomic changes, introduced from outside and with the best of intentions, have upset the already delicate equilibrium between the inhabitants and their fragile environment (Bugnicourt, 1975). As a result, a drought that should have resulted in some death to the weaker or older cattle, as well as certain privations for an already-stressed people, has produced mass starvation and death, has possibly forced a change in a whole culture, and has required international help and cooperation on a massive scale.

Even with a return to moister conditions, if that occurs, the old way of life, the old balance with nature, has been destroyed. During the early stages of the drought, cattle were replaced in large numbers with sheep and goats because they were able to survive the dry conditions better. But sheep eat plants down to the ground level, making regrowth more difficult, while goats will eat roots and all including tree bark and even shrubs poisonous to other animals. They have completely stripped much of the area, possibly introducing some irreversible changes on the land. The expansion of the Sahara desert southward, whether by changing climate or overgrazing, may be difficult to turn back. Increased precipitation may not be the answer. The chances are that we will never have the opportunity to see for additional technological and socioeconomic developments will probably establish a new balance with nature. Having once disturbed nature, humans may never let the balance be reformed but instead will seek a new and hopefully better balance for the good of the inhabitants. The cynic may wonder if the results of the past few decades are not sufficient answer to encourage all cessation of outside interference. The emigrants from the northern portion of the Sahel, while wanting to return to their homelands and their nomadic way of life, recognize that that is an impossibility. To many, there is no return. As part of a developing group of nations, they are being buffeted by outside influences of change and "progress" so that old ways, old customs, old adjustments to the environment are no longer possible. Whatever the reason for the recent drought, conditions far removed from the Sahel itself have brought changes to the area that will result in an entirely new adjustment with the land, for better or for worse. Drought in the Sahel is not only climatic but also technologically and socioeconomically induced.

## CONCLUSIONS

Unlike floods, droughts cannot be fought with big engineering works. Nor can we really determine 50-year or 100-year frequencies of droughts for we are not concerned with a particular amount of water at a particular place, as we are with the river stage at some gaging station. There is very little in the way of available technology to eliminate droughts even though there are a number of things we can do to make the effects of droughts less severe.

Descriptions of droughts in the Palliser Triangle and the Great Plains of the United States have shown how the development of new farming techniques, the use of "trash mulch" and basin listers and other technological developments involving such things as cover crops, efforts to mitigate insect attacks, fallowing, and contour farming can help to control many of the important side effects of any drought. Of course, as we saw in the development of the Sahel drought, technological developments that are not adjusted to the culture of the people can also lead to serious problems and might even increase the severity of already dry conditions.

Droughts are not always the result of overuse or misuse of water or even of the running out of adequate supplies of surface or groundwater. Droughts result when meteorological conditions do not supply adequate precipitation to satisfy plant water needs for an extended period of time. But that is only the physical side of drought. There is not enough water supplied in the form of rain or snow for the needs of the people. The result of this lack of the physical property—water— has many economic, social, and human consequences. People suffer, farms are abandoned, migrations of large groups of people follow, economic chaos may occur.

Water, its distribution and use, its management, even its conditions of excess or deficit, should not be viewed from only a single point of view. One cannot fully understand or appreciate the role that water plays in our world from either physical or human-social contexts alone. The story of floods and droughts illustrates the broad interdisciplinary nature of water and suggests why it must be studied from an interdisciplinary viewpoint.

**PART F:**

**THE FUTURE**

The previous chapters have described the hydrologic conditions under which water occurs, the current availability of as well as the demands for water, and the potentialities for obtaining additional supplies of water. Any realistic appraisal of the water resources situation must also consider future demands as well as future strategies for meeting those demands. This section will attempt to provide an outlook for the future based on an understanding of the water resources problems and limitations already discussed.

## 13

FUTURE DIRECTIONS IN WATER
RESOURCES MANAGEMENT

## THE ROLE OF WATER RESOURCES MANAGEMENT

Water resources management has been defined as management that is directed toward providing water of the right quality, in the right quantity, at the right place, at the right time, and at the right price to meet our various demands. We recognize that while the chances of actually running out of water either in North America or in the world are small indeed, there may be periodic shortages of water from place to place due to local changes in water supply and/or demand. Management is needed to insure that the temporary shortages are met from available supplies existing either at the same place or at some other place. The techniques of management are manifold and may include storage to detain surplus water available at one time of the year for use later, transportation facilities to move water from one place to another, manipulation of the pricing structure for water to reduce demand, use of changes in legal systems to make better use of the supplies available, introduction of techniques to make more water available through watershed management, cloud seeding, desalination of saline or brackish water, or area-wide educational programs to teach conservation or reuse of water. The techniques available have been discussed.

Clearly, there is a management problem rather than a water problem. But it is a problem with techniques available to provide solutions. The future is far from bleak inasmuch as with careful planning and forethought we should be able to balance available supplies with demand and so meet foreseeable problems. The solutions do not promise to be inexpensive or easy to apply. Changing water laws, for example, may take years of costly court battles and will not be accomplished without the rights of some being jeopardized for the good of others. Building large-scale storage and transport facilities to move water from areas of excess supply to other areas of excess demand will have a very high price tag and may take several decades depending on the magnitude of the program. Desalination plants are expensive and demand large amounts of energy, another resource in short supply, so that adopting a technique to help relieve one problem may, in fact, exacerbate another pressing problem. Thus, while the tools and techniques of management are available, they cannot always be applied easily, inexpensively, or without creating new problems in some other area. The concern about running out of water may well be replaced by a concern that management will be unable to achieve the changes necessary to keep us always adequately supplied with water at a price we are willing to pay.

## THE SECOND NATIONAL WATER ASSESSMENT

The United States Water Resources Council has just completed its Second National Water Assessment entitled *The Nation's Water Resources 1975-2000.* It includes figures on projected water supplies and demands and it also provides a very concise summary of critical problems for the next several decades based on an analysis of a comprehensive water supply adequacy model applied to 106 water resources subregions across the United States. Ten critical water resources problems were identified, six of which are discussed here in detail:

(1) *Inadequate surface-water supply*

By the year 2000, the Water Resources Council anticipates that the problem of inadequate surface-water supply will be severe in 17 of the 106 subregions, all located in the Great Plains or southwestern part of the country (Fig. 13.1). Because of the nationwide increase in annual demands for fresh water, there will also be periods during low-flow months in which additional subregions, some even in the more humid East, will have inadequate supplies.

The analysis considered the general inflexibility of river-based ecosystems to adjust to fluctuating streamflows. When off-stream uses of water place an additional severe demand on the water in the stream, adequate protection for fish and wildlife dependent on the streamflow may no longer be possible. The Council suggests that an instream flow of 1035 bgd is entirely adequate for fish and wildlife needs, while our actual instream flow is 1233 bgd. Thus, nationwide, we are in good shape although certain streams are in a relatively unfavorable condition. For example, the Lower Colorado Region has an average daily flow of 1550 mgd while some 6864 mgd might be needed for fish and wildlife protection. Thus, in some areas, competition for water is a way of life and trade-offs must be made which will limit future uses and river development. Since surface and subsurface water supplies are interrelated, we cannot view the surface problem without a concern for related subsurface water problems.

(2) *Overpumping of groundwater supplies*

A second problem viewed by the Water Resources Council concerned the localized overpumping of groundwater reserves. Groundwater volumes for the nation as a whole equal about 50 years of surface runoff (well in excess of the total capacity of all of the lakes and reservoirs of the nation including the Great Lakes). Yet, in certain areas, especially in the Great Plains from southern Nebraska to

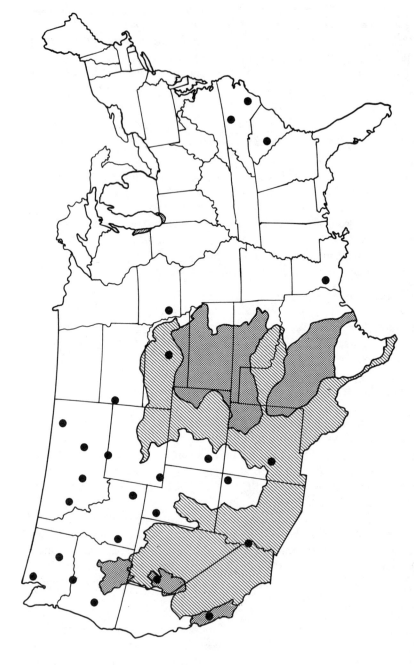

**Figure 13.1** Areas affected by inadequate surface-water supply and related problems. Hatched areas show streamflow 70% depleted in average or dry years, dotted areas streamflow 70% depleted in dry years only. Black dots indicate places with inadequate supply to support conflict between offstream and instream uses. *Source.* U.S. Water Resources Council (1978).

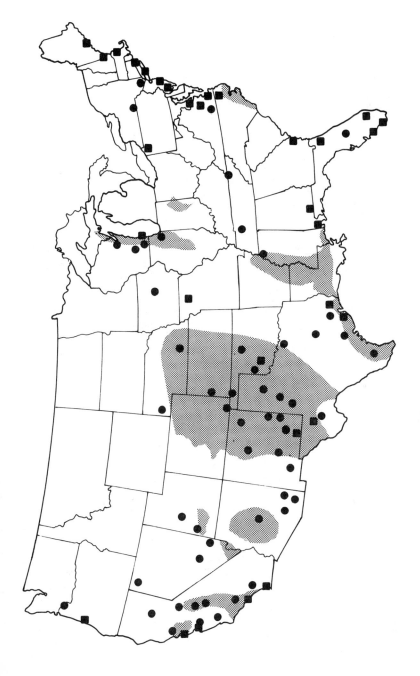

**Figure 13.2** Areas affected by significant groundwater overdraft. Black dots indicate areas of declining groundwater levels while black squares show places where saline water intrusion into fresh water aquifers is occurring. *Source.* U.S. Water Resources Council (1978).

387

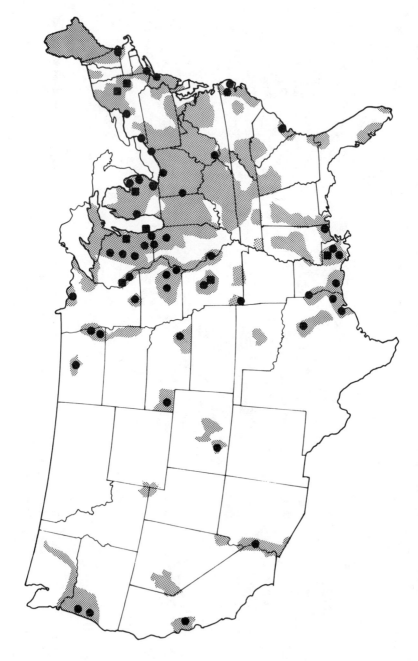

**Figure 13.3** Areas affected by surface-water pollution problems from point sources (municipal and industrial waste) as indentified by Federal and state/regional study teams. Black dots indicate areas of pollution from coliform bacteria from municipal wastes or feedlot drainage. Black squares show areas of pollution from PCB, PBB, PVC, and related industrial chemicals. *Source.* U.S. Water Resources Council (1978).

western Texas, in parts of central Arizona and California, and along the southern portion of the Mississippi River valley, significant groundwater overpumping is occurring even now and will continue into the future (Fig. 13.2). Eight of the 106 water resources subregions have critical overpumping problems at the present time while 30 more have moderate problems and 22 others have minor overpumping problems. Thus, more than half of the water resources subregions are experiencing some sort of overpumping problems at the present which may become more serious in the future as demand for fresh water increases.

Clearly, before critical problems develop in too many areas, some remedial action is necessary. This may include developing additional sources of water, beginning artificial recharge to keep salt or polluted water from entering productive aquifers, reducing water use through water management techniques, or relocation of water-demanding activities. Since irrigation is the single greatest use of groundwater, the results of overpumping will be most clearly seen in our agricultural production. In this case, relocation of water-demanding activities may not be possible because of the need for large arable acreages. A less desirable alternative may have to be a curtailment of irrigation agriculture.

(3) *Pollution of surface water*

We have, for many years, used our surface water bodies as places to dump waste products. As long as the wastes were not too concentrated and did not involve strongly toxic or nonbiodegradable materials, few problems developed for nature has a tremendous capacity to purify water. However, in recent years, with excessive population pressures in certain areas, with the introduction of industrial wastes of a highly complex nature, and with other demands for fresh water increasing rapidly, it is no longer possible to use surface waters for pollution abatement. But large portions of the nation still suffer from polluted surface waters. Pollution is received into the streams and lakes from both point and nonpoint or dispersed sources. Point sources of pollution—generally from factories and municipal treatment systems—can be controlled in time although the control may involve costly treatment programs (Fig. 13.3). However, nonpoint source pollution, which comes largely from runoff from agricultural, forested, or urban areas or from mining operations, is more difficult to control (Fig. 13.4). The Environmental Protection Agency has estimated that one-third of the oxygen-demanding load on our surface water, two-thirds of the phosphorous and three-quarters of the nitrogen being discharged to our streams, comes from nonpoint agricultural sources. Storm runoff from urban areas can also lead to rapid deterioration of those water bodies. Significant amounts of sediment, nutrients, organic material, and trace elements are carried by the storm water runoff into surface water bodies. As urban areas increase in size in the next few decades, it can be expected that nonpoint source surface water pollution will also increase even though pollution from point sources may decrease Eutrophication, the growth of algae in receiving waters due to the presence of an excess supply of nutrients, goes on in all water bodies. It has been accelerated in recent years by human activities. Large areas of the Ohio, Potomac, Mississippi, and Tennessee rivers and portions of Wyoming, Colorado, Utah, Arizona, and New Mexico along with

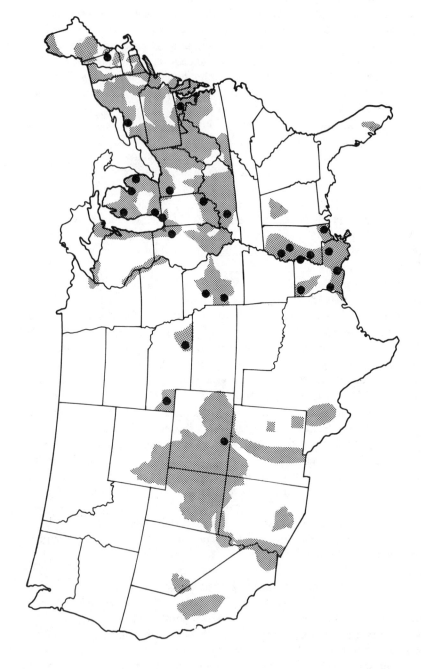

**Figure 13.4** Areas affected by surface-water pollution problems from nonpoint sources (dispersed) as identified by Federal and state/regional study teams. Black dots show areas where pollution problems are from herbicides, pesticides, and other agricultural chemicals. *Source.* U.S. Water Resources Council (1978).

**Table 13.1**   Water Quality Problem Areas Reported by States*
(Number Reporting Problems/Total)

| | Middle Atlantic, Northeast | South | Great Lakes | Central | Southwest | West | Islands | Total |
|---|---|---|---|---|---|---|---|---|
| Harmful substances | 6/13 | 6/9 | 5/6 | 4/8 | 4/4 | 2/6 | 3/6 | 30/52 |
| Physical modification | 7/13 | 3/9 | 3/6 | 8/8 | 3/4 | 6/6 | 5/6 | 35/52 |
| Eutrophication potential | 11/13 | 6/9 | 6/6 | 8/8 | 2/4 | 6/6 | 4/6 | 43/52 |
| Salinity, acidity, alkalinity | 3/13 | 6/9 | 2/6 | 6/8 | 4/4 | 4/6 | 2/6 | 27/52 |
| Oxygen depletion | 11/13 | 9/9 | 6/6 | 6/8 | 4/4 | 6/6 | 4/6 | 46/52 |
| Health hazards | 11/13 | 8/9 | 5/6 | 8/8 | 3/4 | 5/6 | 5/6 | 45/52 |

*Middle Atlantic, Northeast:*

| | | |
|---|---|---|
| Connecticut | New York | **Central:** |
| Delaware | Pennsylvania | |
| District of Columbia | Rhode Island | Colorado |
| Maine | Vermont | Iowa |
| Maryland | Virginia | Kansas |
| New Hampshire | West Virginia | Montana |
| New Jersey | | |

Central:

Colorado          Nebraska
Iowa              North Dakota
Kansas            South Dakota
Montana           Wyoming

Southwest:

Arizona           Oklahoma
New Mexico        Texas

*South:*

Alabama           Louisiana
Arkansas          North Carolina      West:
Florida           South Carolina      California      Oregon
Georgia           Tennessee           Idaho           Utah
Kentucky                              Nevada          Washington

*Great Lakes:*                        *Islands:*

Illinois          Minnesota           American Samoa  Puerto Rico
Indiana           Ohio                Guam            Trust Territories
Michigan          Wisconsin           Hawaii          Virgin Islands

*Localized or statewide problems discussed by the states in their reports.
*Source.* Office of Water Planning and Standards (1975).

much of Maine, Wisconsin, Florida, and Louisiana are problem areas where eutrophication is occurring at present. The process results in excessive demand for oxygen to break down organic matter that has built up in the water body. Eutrophication results in fish kills and the rapid aging of lake areas.

In a 1975 report to Congress, the Office of Water Planning and Standards listed

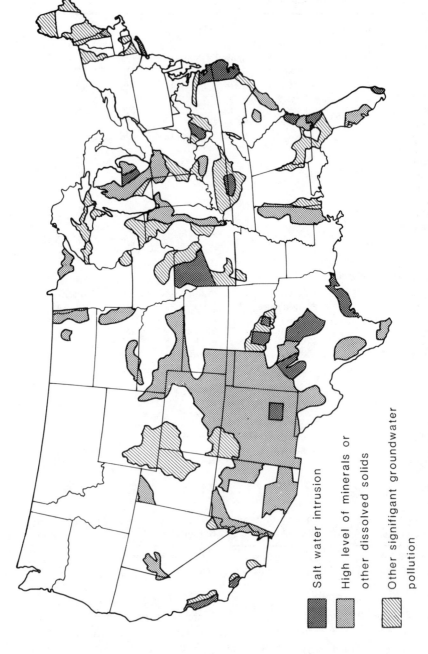

**Figure 13.5** Areas affected by groundwater pollution problems as identified by Federal and state/regional study teams. *Source.* U.S. Water Resources Council (1978).

Salt water intrusion

High level of minerals or other dissolved solids

Other signifigant groundwater pollution

water quality problems by regions of the country (Table 13.1). Summing up the problems for all 50 states, health hazards were found to exist in the water of 45 and oxygen depletion in the water of 46 states. These are our two most widespread problems. Eutrophication potential appeared as a problem in 43 of the 50 states. The list is far from reassuring to those who want to believe that our various water sources are fairly good.

(4) *Groundwater pollution*

While less subject to pollution than surface water, groundwater supplies can easily be contaminated by drainage from septic tanks, infiltration of waste products from animal feedlots, from solid waste landfills and from subsurface disposal (deep-well injection) of industrial waste products. Saline water inflow in areas of excessive pumping of groundwater aquifers can also result in pollution of groundwater areas.

Once pollution of a groundwater aquifer has occurred, recovery is very slow. The water supply over a large area may be removed from productive use and it may take years before the aquifer can be pumped again for domestic purposes. Since some 40% of the population of the United States obtains its drinking water from groundwater, the increasing pollution of this source is a significant problem to the nation. Figure 13.5 indicates the areas of this country where groundwater pollution is a problem at the present time. As in the case of many other problems identified by the Water Resources Council, the southwestern part of the country is an area with significant groundwater pollution problems.

(5) *Drinking water quality*

The significant advances that have been made in improving the quality of drinking water over the past half century have perhaps resulted in a false sense of security about how safe our water really is. With increased ability to detect waterborne disease, somewhat over 4000 cases of waterborne illnesses per year have been reported. The actual number, when the more difficult to detect cases of chemical poisoning are included, may actually be increased by a factor of 10. Only in recent years have we begun to consider the particular effects of combinations of chemicals that might be carcinogenic but whose symptoms might not show up for several decades.

Areas where drinking water quality problems exist now, or may in the near future, are shown in Figure 13.6. The southwest and the upper midwest along with the New York and mid-Atlantic regions are the areas of prime concern at the present.

(6) *Erosion and sedimentation*

Natural erosion and sedimentation are significant problems in their own right but, when exacerbated by human activity, they can become critical to the welfare of the country. The Water Resources Council reported a 1975 average cropland loss of 8.6 tons of topsoil per acre while forest and pastureland lost about 1 ton per acre. Under accelerated erosion, some cropland areas lost as much as 25 tons/ acre or 3/16 in./acre. Overgrazing, poor cultural practices, off-road recreational vehicles, and surface mining all increase the opportunity for accelerated erosion. Storm runoff from construction sites results in a particularly serious problem of erosion and sediment deposition into water bodies.

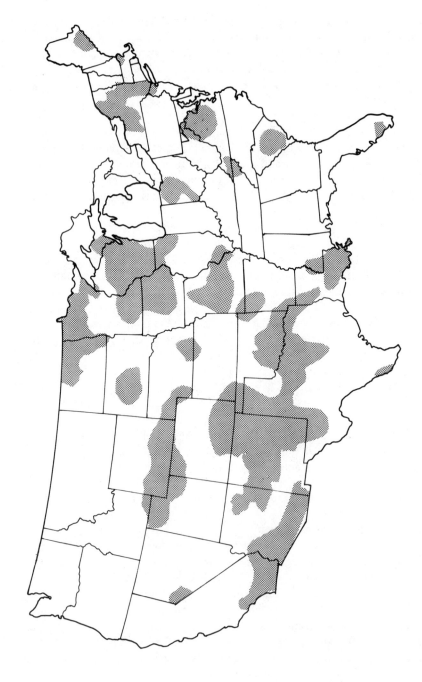

**Figure 13.6** Areas affected by quality of drinking water problems as identified by Federal and state/regional study teams. *Source.* U.S. Water Resources Council (1978).

Sedimentation can modify stream channels and increase the likelihood of flooding. It must be counteracted by expensive dredging operations. Dredging causes its own problems as the dredged material needs to be disposed of as spoil banks covering large areas of valuable wetlands and floodplains. Sediments are not only transporters of pollutants, such as pesticides and phosphates, but they also cover fish and wildlife habitats, and, especially, nesting and spawning areas.

Other critical problems identified by the Water Resources Council include flooding, which resulted in the deaths of 107 and property damage of some $3.4 billion in 1975, wetland drainage which eliminates a valuable resource for fish and waterfowl use, one of the more productive of our ecosystems, drainage of wet soils which can bring into production many acres of potential agricultural land, and the degradation of estuaries and coastal waters through the discharge of domestic and industrial wastes.

## CLIMATIC CHANGE AND WATER RESOURCES

Much has been written recently about the possible effects of climatic change. We have been going through a period of very equitable climate in the United States so that any change from that pattern gives cause for concern and the opportunity for journal articles and magazine supplement features. Such fluctuations or changes, however, should not be interpreted as something peculiar from a climate viewpoint.

Let us assume, for the present, that we are undergoing a significant fluctuation of climate. The question that must be asked is what such a fluctuation will do to our water resources situation in the United States or in the world. While the question is logical, answers are not necessarily simple or straightforward for a number of reasons. First, our ability to forecast climate is anything but exact. In spite of extensive study and the existence of nearly 100 years of instrumental records in a number of places, backed by longer periods of record based on tree rings, journals and diaries, measurements of ocean sediments and glacier cores, we still have little understanding of past climatic fluctuations that we might use to forecast the next 25 to 30 years. A water policy for the future must look at least 30 years ahead because of the long lead time needed for planning and constructing massive storage, treatment, and transportation facilities.

If one were to ask a number of climatologists to predict whether the climate would be generally warmer or colder, wetter or drier in the United States by the year 2000, only two decades away, all possible answers would be given. We recognize that carbon dioxide build-up in the atmosphere will lead to increased absorption of long-wave solar radiation and a heating of the atmosphere. We also recognize that increased dust of certain sizes will lead to more scattering of short-wave solar radiation and possible cooling of the earth. Changes in ozone, cloud cover, and albedo in the earth-atmosphere system will produce other known changes. Lack of knowledge of the interrelation of all the elements of climate makes it difficult to predict specifically what will happen to factors such as precipitation or evapotranspiration without good information on what will happen to each of the other climatic elements. Thus, we are not now in a position to predict the climates of future years with any assurance.

In the matter of shorter range forecasts, Wallis (1977) reported on a study of the 30-day climate forecasts now issued by the National Weather Service as a possible aid to water managers of two small water supply and irrigation reservoirs in New York state. He showed that for the period 1954–1973, the 30-day forecasts of precipitation were correct only 30% of the time. If normal had been forecast each month of the period, it would have been correct 36% of the time. Clearly, if forecasts for larger areas (good-sized catchments) are desired, the monthly forecasts become even more general and of less value to water resource planners. Wallis concluded that past use of long-range forecasts have not been particularly helpful in water resources work. There is little reason why we should expect better results in the immediate future.

A second point must also be made. Assuming that we could predict future climates with some degree of precision, we are then faced with the difficult task of making the transfer from a climate forecast to a water supply value. Our ability to achieve a rational transfer function that will give us a realistic approximation of some desired water supply value is still quite uncertain. Many models exist to convert climatic data into estimates of streamflow or evapotranspiration. Some, such as the Thornthwaite model described earlier, can provide estimates of annual streamflow from temperature and precipitation data with a correlation coefficient of about 0.85 or 0.90 between measured and computed streamflow for most basins. These models are less accurate for shorter time periods, such as for months or for weeks. Individual storm events show very little agreement at all between predicted and actual runoff. A World Meteorological Organization study carried out from 1968 to 1974 evaluated 10 different models on 6 watersheds for a 2-year forecast period with very discouraging results.

If we cannot transfer from a current weather situation to a water supply value with much certainty, how will we be able to take a new climatic regime (postulated under the assumption of a climatic change) and achieve any useful water supply information? The answer is that at present we are probably not able to help water resource managers too much in terms of information about the future climate or water supply situation.

There still are certain things that can be done to prepare for future contingencies. We can be certain that climatic fluctuations will occur and we can evaluate various courses of action to follow under the range of environmental changes that might occur. We must try to achieve a better understanding of the relationship between climate on the one hand and hydrologic response on the other, to evaluate the areas where more knowledge is necessary, and to determine what type of data are needed in order to make rational decisions. Models of hydroclimatic relationships are desirable in order to identify responses to a whole range of different climate-input situations. While much of the previous work along these lines has only been concerned with the relations between climate and hydrology, clearly any evaluation of the adequacy of water resources under a particular set of climatic conditions must also include economic, social, legal, and engineering inputs. But these influences are difficult to understand in their own right and, when mixed with climate and hydrologic questions, the resulting relationships are far from distinct.

For example, under certain circumstances, economic unknowns may have a greater impact on water resource design than even hydrologic questions. The question of forecasting future water demand, for example, is beset with problems quite apart from considerations of population growth and industrialization. A water demand does not really exist in the usual sense of the word until a supply has been obtained. Use then develops and it may expand up to but not beyond the limits of supply. Future forecasts of demand are always constrained, to some degree, by available supply (Lofting and Davis, 1977).

These same authors suggest that there are at least three areas of economic doubt that make forecasts of future water demand questionable. One is the implication of Public Law 29-500 which bans the discharge of all pollutants by 1985 and regulates all irrigation in order to minimize water use and return discharge. Since one present use of surface water is to dilute pollution, this law, if enforced, could result in significant changes in future water demand. A second question involves the possible significant increase in irrigation agriculture in order to meet world food needs. The large-scale increase in irrigation agriculture would suggest the need for interbasin transfers of water on a massive scale. A third question involves the threat of future climatic change, whether it materializes or not. This threat may well change what farmers plant and how they program their farm operations. But such a threat might suggest a different set of ideas on the part of economists and planners concerning what they believe farmers will do. Such unknowns lead to significant economic uncertainty.

Possible climatic changes leading to a worsening of our water supply situation should, of course, increase the pressure to develop more flexible water laws that will have more widespread applicability. Clearly, the historic concepts of riparian rights and prior appropriation are useful concepts and have their place. But water use has become too complex and there are too many competing demands for water for these rather simplistic and fairly rigid legal frameworks to solve. The distinction between the humid East and arid West is becoming less clear for the reason that even in humid areas, we no longer have adequate supplies of water and water cannot be reserved for riparians alone and allowed to flow unused to the sea while others in need wait unsatisfied. The time is ripe for an overhaul of our water laws and the establishment of a nationwide system that will recognize the prior rights of users and the rights of landowners along streams without forgetting that later applicants for water must also be heard and given an opportunity to show beneficial use of water. Modernizing our water laws will not be easy or quick but it is certainly an undertaking that should begin as soon as possible by our best legal minds.

## WATER USE IN THE PRODUCTION OF SYNTHETIC FUELS

The specter of falling petroleum reserves and rising OPEC prices has made it imperative to develop alternate energy sources. One source that has recently received considerable attention is the production of synthetic fuels by converting carbonaceous material from one form to another. Because of an abundant supply of coal and oil shale in the United States, synfuel production from these carbon sources has received the most study. All of the energy or heating value in the coal or oil shale cannot be transferred to the synfuel. That portion of the energy not converted is transferred to the environment, usually as heat, and is disposed of by evaporating water. Water is also used extensively in the mining and preparation of the raw material for processing as well as in the disposal of the unwanted by-products of production.

Considerable quantities of water can be used in the production of synfuels. This would not necessarily be significant if it were not for the fact that the greatest concentration of easily mined coal and almost all of the oil shale in the United States are located in arid or semiarid areas of the country where water supplies are already quite limited and often fully allocated to other demands. The area of greatest possible production of synthetic fuels runs in an arc from Montana and the Dakotas south and southwestward through Wyoming and Colorado to New Mexico and Utah. Much of the area is found in the upper Colorado River basin whose waters are already in very great demand for agricultural and domestic purposes. Other portions of the area are found in the Missouri and Rio Grande River basins that have little excess water for new industrial demands.

Since little real information is available on water needs in synfuel production, published estimates vary appreciably. Water for cooling is usually the major demand. The higher the efficiency of the conversion process the less will be the amount of heat to be removed by cooling water. Thus, coal liquefaction tends to be somewhat less demanding of water than coal gasification when based on the thermal output of the product. Probstein and Gold (1978) suggested that conversion efficiency may not be the only quantity to use in characterizing water consumption. They point out that one coal gasification process may actually consume less water than another because the moisture in the coal is collected and utilized even though such reuse may add to the cost. Those authors describe a hypothetical mine-plant unit designed for coal gasification. Assuming that the plant is designed not to waste water, although it may not minimize water consumption either, they estimate that it will use some 18 gal of water per million BTU of heating value in the gas. A standard plant might produce 250 million cubic feet of gas per day requiring about 4.5 million gallons of water per day. Other authors more commonly quote values of water consumption of from 9 to 40 mgd.

For synfuel development to be a real possibility in water-short areas, process water must be reused, cooling water use must be minimized, and water sources not in demand for other uses (such as brackish groundwater or municipal effluents) must be utilized. There seems to be no absolute water requirement since plant design as well as social, political, and environmental factors can modify water use within fairly wide limits.

Probstein and Gold (1978) studied the conversion of coal to clean gaseous, liquid, and solid fuels as well as the conversion of oil shale. The authors considered

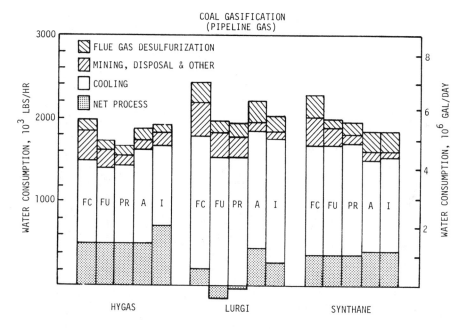

**Figure 13.7** Net water consumption for standard size coal gasification plants. (FC, Four Corners region; FU, Fort Union region; PR, Powder River region; A, Appalachian Basin; I, Illinois Basin.)
*Source.* Probstein and Gold (1978), with permission of the MIT Press.

five general coal mining regions in their analysis. In western United States, they included the Powder River and Fort Union regions in Montana, Wyoming, and the Dakotas, as well as the Four Corners region where New Mexico, Arizona, Utah, and Colorado meet. In eastern and central United States, they provide information on coal conversion in both Illinois and the Appalachian area. Since western coals are mainly low sulfur bituminous and lignite and eastern coals are generally high sulfur bituminous, water requirements per heating value will vary (Figs. 13.7 and 13.8). Oil shales from the Green River Formation of Colorado and Utah are the only shales considered.

The quantity of water consumed depends on four factors for a given-sized coal conversion plant. These involve the product itself, the amount of heat that must be disposed of by wet cooling, the location of the plant, and the specific conversion process. For oil shale conversion, water consumption depends mainly on the method of disposing of the spent shale, the shale retorting process, and the degree to which wet cooling is needed. About 1400 BTUs are transferred per pound of water evaporated in a wet evaporative cooling tower. Where water supplies are not limiting, usually 25 to 60% of the unrecovered heat is dissipated by evaporating water. This drops to 10 to 30% of the unrecovered heat in water-short areas (Probstein and Gold, 1978).

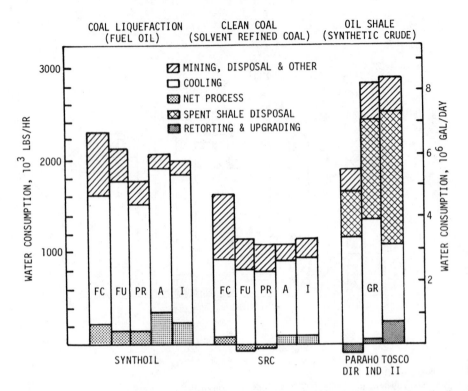

**Figure 13.8** Net water consumption for standard size coal liquefaction, clean coal, and oil shale plants. (FC, Four Corners region; FU, Fort Union region; PR, Powder River region; A, Appalachian Basin; I, Illinois Basin; GR, Green River formation.)
*Source.* Probstein and Gold (1978), with permission of the MIT Press.

The data provide an estimate of the magnitude of the water problem that may develop in water-short areas in the process of developing a large-scale synfuel program. Oil shale production requires much greater amounts of water than are required for solvent refined coal. Coal liquefaction requires nearly twice the water required to produce solvent refined coal. Thus, to use oil shale or coal to produce oil will require considerably more water than will the production of coal alone. Water is in such short supply in most of the areas where synfuels can be produced that it may limit the development of large-scale synthetic fuel production. Certainly this new requirement for water must be carefully considered in the list of priorities in future water resources planning for these water-short areas.

## FINANCIAL CONSIDERATIONS IN WATER DEVELOPMENT PROJECTS

Many acts of Congress authorizing water development projects call for some cost-sharing as part of the financial approval for the project. However, each of the agencies concerned with such water projects is provided with different cost-sharing guidelines. We do not have a consistent cost-sharing policy across the different construction and funding agencies.

For example, the Federal government usually bears all of the costs of maintaining waterways although non-Federal interests may provide land or easements plus public terminals and port facilities. Usually, project construction costs for irrigation on Federal reclamation projects must be repaid without interest over a 50-year period. However, revenue from hydroelectric power and from other water users may be applied to the repayment of the construction costs of irrigation projects in those cases where the irrigators are not able to repay the cost. At present, about 60% of the repayment of such construction costs come from power revenues while irrigators are repaying only 10 to 15% of the construction costs. There is some feeling that too much of the cost of multipurpose projects has been assigned to nonreimbursable aspects, thus lowering the amounts that should be repaid by non-Federal sources. As an example, in 21 project units in the Missouri River Basin, the proportion of costs designated for repayment without interest ranged from 0 to 72%, and in only 2 of the 21 units were irrigators supposed to pay more than 40% of the construction costs.

Major reservoirs built for flood control purposes are constructed and maintained entirely at Federal expense. Local projects including levees, flood walls, and channel improvements are also carried out at Federal expense but land easements and operation and maintenance expenses are to be borne by non-Federal sources.

Sale of power generated by Corps of Engineers and Bureau of Reclamation projects is usually able to pay for all project costs allocated to power including interest. Usually, revenues in excess of those needed to repay the costs allocated to power are used to repay some of the costs allocated for irrigation or other users.

Fifty percent of the "separable" costs of providing recreational facilities are supposed to be borne by Federal sources, the other half by state or local sources. All "joint" costs allocated to recreation, however, are to be paid for by Federal funds.

The list could go on but it is clear that the Federal government has made it very attractive for local groups to request water resources projects. It is quite possible that if local citizens were asked to bear even a reasonable portion of waterways improvements or of other water projects that many of them would not be undertaken in the first place. It does seem highly unfair that the citizens of the whole country are asked to foot the bill for a stream project in a particular locality and, of course, it does raise the specter of wasteful projects in an effort to spread government largesse to all portions of the country. Why would any area refuse a Federal project since it would be almost entirely funded by the rest of the nation? Certainly, to bring some reason and sense of fiscal responsibility to the construction of water projects, a new method of allocating costs for water projects must be found and individual citizen groups who profit most must be required to shoulder the bulk of the costs.

The National Water Commission (1973) lists six deficiencies in the present cost-sharing system which should be corrected.

(1) Cost-sharing policies should be consistent among alternative means for accomplishing the same purpose . . .
(2) Cost-sharing policies should be consistent among Federal agencies for the same water purpose . . .
(3) Cost-sharing policies need not require a uniform percentage of cost-sharing for all water developments . . .
(4) Cost-sharing policies should require uniform terms for the repayment of non-Federal cost shares . . .
(5) Cost-sharing policies should promote equity among project beneficiaries and taxpayers . . .
(6) Cost-sharing policies should not lead to expansion of the Federal role in water resources. (pp. 494–495)

The report outlines appropriate recommendations to achieve these ends.

## BASIC DATA NEEDS

Data are vital in any analysis of supply and demand and in the planning of alternative courses of action. There has been a fairly long-standing effort to collect information on the water supply aspect of water resources. Precipitation and water runoff data have been obtained for a number of years at stations strategically located across the country. In the future, there must be more information on water quality and on the relation between water and other aspects of the natural environment as well as data on the impact of water on social, cultural, economic, and legal problems.

With the collection of more and better data, the means to store, retrieve, analyze, and disseminate those data are vitally needed. Users of the data must be aware of their existence. Such a vast amount of data has been collected from so many different places and in so many different formats that it is almost impossible for any user to be fully aware of all of the data that are available. Often data collection activities are repeated or new studies are undertaken to collect information that is already documented. This problem need not exist if an effort is made to perfect systems to archive and retrieve the data.

Water plays a central role in all aspects of our human environment. With increased understanding of the environmental impact of all of our actions, it is clear that whatever we do to our water resources will have a significant influence on the biological and ecological aspects of our water resources as well as on economic and legal considerations. Evaluation of proposed water projects requires data on so many different aspects of water that we never have exactly the right type of data in sufficient quantity and quality to answer all the questions that planners and managers might ask. We already have examples where data deficiencies have resulted in poorly designed structures. Less obvious, but quite important from an economic point of view, is the amount of overdesign that has been included with many of our water resources structures just to provide a sufficient safety factor in view of the lack of always adequate data.

We should not look at the data picture with the idea of deciding what one or two additional pieces of data are most relevant and important for future observation purposes. Rather, the need is to establish a comprehensive and integrated data collection program to meet the demands of managers and planners. Water data are no longer required only by water specialists (such as hydrologists and engineers) but are becoming important to the economist, lawyer, political scientist, and sociologist. There will be an increasing need for water quality data to be associated with water quantity information, information on trends, and information on the response of one sector of the hydrologic system to changes in another sector. The need is for continuous "real time" data for immediate use as well as for data tabulations and data storage. The data must be available to permit rapid analysis to achieve answers to operating questions and computer models must be available to permit sensitivity analyses of various changes in input conditions.

Two agencies in the Federal government are mainly responsible for the collection of the basic water supply data—the National Oceanic and Atmospheric Administration (NOAA) and the U.S. Geological Survey (USGS). The former agency is in the Department of Commerce and is more responsible for precipitation, evaporation, and river stages. The latter agency is in the Department of the Interior and is more concerned with river flow, the occurrence and distribution of groundwater, and the quality of water. While there is a great deal of overlap in the water work of both agencies, and some cooperation between them, it is clear that improved cooperation and coordination would be possible if the two agencies were to be brought together under a single organization with a single budget.

## SCIENTIFIC AND TECHNOLOGICAL ASPECTS OF WATER POLICY

The Office of Science and Technology Policy prepared a report for the President's Committee for Water Resources Policy Study (Office of Science and Technology Policy, 1978; also Ackermann, Allee, Amorocho, Haimes, Hall, Meserve, Patrick and Smith, 1978). The following paragraphs sum up some of the more salient features of that report. (They serve as a useful summary of many of the aspects of water resources that have been detailed in previous pages.)

*Climate and water supply:* While we are unable to forecast climate with any useful precision, we know that significant fluctuations can be expected from time to time that will result in severe water shortages as well as excesses. Expected future water shortages may well be made more severe by climatic changes. Care must be exercised not to base our plans and policies for water utilization on the amount of water that is available in good or favorable years. We should be well aware of the normal fluctuations from good to poor years and consider possible year-to-year fluctuations in future planning documents. We need research into the "resiliency" (the ability of our water resources systems to adjust) of our water resource systems with respect to limiting climatic conditions. We need to know the possible large-scale consequences of droughts of different severities.

*Floods and droughts:* These are acts of nature and can be expected to recur from time to time. Water resources management is directed, in large measure, toward making the impact of floods and droughts less severe. Better land-use management is one way in which we can improve our adjustment to both floods and droughts. In terms of drought management, better control of land development—for example, by not permitting large irrigated agricultural developments in areas subject to periodic shortages of water—would help eliminate some of the opportunity for serious droughts to develop. Also, developing plans for reduction in water use (low-cost solution) rather than for increases in water supply (high-cost solution) might provide a satisfactory solution to drought problems or other problems associated with water shortages.

In terms of floods, land management in the floodplain areas, along with integrated floodplain development with storage and flood control structures, would result in the most satisfactory adjustment to the flood hazard. Remote sensing techniques and other programs aimed at quantitative precipitation forecasting, especially in mountainous areas or in areas with few precipitation gages, must be developed rapidly to improve our ability to forecast flood conditions.

*Groundwater-surface water use:* Groundwater is the source of water for nearly half of the people in the United States. It is, however, poorly managed, overpumped, and subject to contamination. The close relation between surface water and groundwater must be recognized in state water laws. Rights to both sources of water must be protected and supplies regulated so that wasteful misuse is not encouraged. There need not and should not be a separate treatment of groundwater and surface water. At the same time water policy should be based on current understanding of the relation between use and recharge, drawdown, and infiltration into aquifer systems. At times, we seem to develop water policy in a vacuum or without consideration of the basic knowledge of hydrologic relationships that we already have.

*Water conservation in irrigation:* Conservation of water would be useful under all conditions of shortage but, since irrigated agriculture results in the greatest consumptive use of water, it is clearly the place to start in any effort to conserve. It has been suggested that less than half of the water stored in reservoirs or diverted into canals for irrigation purposes reaches the crops. Department of Agriculture figures suggest that off-farm losses, mostly evaporation and seepage in canals, amounts to 22% and on-farm losses, ditch seepage and evaporation, amounts to 37%. The Soil Conservation Service estimates that present withdrawals of some 195 million acre-feet could be reduced to 147 million acre-feet with the introduction of a high efficiency irrigation program. A program to increase farm irrigation efficiency from 50 to 80% might not cost more than $40 per acre of irrigated farmland and might even result in increased crop yields.

Present irrigation schemes are based on the idea of using low-cost water so they are predicated on applying large volumes of water to the soil. As water costs increase and water becomes less plentiful, research is needed to improve the application of water to plants, to determine the effect of timing of irrigation, and the actual water needs of various crops. The irrigation program cannot, of course, be considered separately from the fertilizer program or the other aspects of an integrated farm operation. There should also be an increased research effort on the use of saline water on crops and on the expanded utilization of salt-tolerant varieties.

*Water quality:* The significance of nonpoint sources of pollution will increase in the future as urban centers grow and agricultural and mining activities expand. Research is needed on the effect of combinations of pollutants on water quality to insure that no combinations can develop that will lead to toxic conditions in the receiving waters or the build up of levels that can be harmful to either humans or fish and wildlife. At the same time, better systems for draining agricultural land are needed to control this source of nonpoint pollution. This might also include the development of fertilizers and agricultural chemicals to minimize inert but undesirable ions. Considerable work needs to be done on the control of salinity in agricultural soils.

*Erosion and sedimentation:* Erosion and sedimentation problems continue to be serious. While natural problems, they have been exacerbated by human activities. Since they occur on such a vast scale, it is only practical to control erosion rates within limits consistent with the maintenance of equilibrium in the drainage basin. Improved agricultural practices should be developed and utilized to control erosion from cropped land. Good conservation practices should be strictly enforced on construction sites where erosion and runoff are excessive.

*Water for energy:* Water is essential in energy production but water supplies are not always readily available where energy resources are most abundant. The need for energy may limit the possible uses of water and it may be necessary to move fuels for processing to areas where water exists. Processing of mineral resources into the fuel products that we need as a country may well provide significant opportunities for environmental degradation to develop. Already problems of thermal pollution, the discharge of toxic trace elements, and mine drainage have resulted in serious pollution problems.

The energy policy of the nation must only be developed with a clear understanding of the value of water. The real value of water for alternative uses must be included in the formulation of energy strategies (Probstein and Gold, 1978). Since water for irrigation is worth between $5 and $40 per acre-foot and water for energy is much more valuable, it is likely that the development of an energy policy for this country will result in some significant shifts in the uses of water. The socioeconomic impacts need to be studied. Saline water can possibly be used for certain energy purposes and these uses need to be clearly recognized.

Unilateral efforts by individual states or nations to solve water problems will only yield limited success. Water at the surface is related to underground water and water in one basin is related to the water in another basin through the action of the hydrologic cycle. Research work by individual states is likely to duplicate that in other areas with an unnecessary expenditure of effort. Individual states and nations may also lack the scientific and technical know-how to solve the very complex water problems of the present and the future. Thus, pooling of interest and knowledge to solve common problems offers more likelihood of common success. The development of the concept of the international river and river basin should lead on to the concept of the international water resources system. Just as weather knows no boundaries, so also water knows no boundaries and can only be fully studied on a continental basis.

Water resources management is shifting. No longer is the development of new supplies the major function of management. Waste water treatment is becoming an important aspect of management along with the wise allocation of existing supplies among competing uses. New uses of water are developing but old uses can often be met with less waste of water so that the available supply can be made to go further. The continuation of this trend might suggest a reduction in, or even a partial breakup of, the old Federal water resource development program.

There has been increasing opposition from the President and from other governmental agencies to Federal water development projects. The environmental movement, supported in large part by the increasing number of urban dwellers, is generally opposed to large Federal water development projects. And many responsible leaders as well as the public in general feel that the job of Federal water development has largely been accomplished.

Concern for waste water treatment (the Federal government has taken over much of the nation's effort in this field through recent congressional action) and the need for urban water management, including urban floodplain management, are problems of significance for the future even if water development problems decline.

Water demand does not have to be inflexible. Efforts to modify demands to make them equal supply are viable alternatives to efforts to increase supply. The encouragement of water-conservation technology should be a goal of any water policy that is developed. The possibility of meeting present and future water needs by adjusting our levels of water use may be among the most realistic of our solutions to water resources problems.

# REFERENCES

Ackermann, W. C., D. J. Allee, J. Amorocho, Y. Y. Haimes, W. A. Hall, R. A. Meserve, R. Patrick and P. M. Smith, 1978: Scientific and technological considerations in water resources policy. *EOS, Transactions of the American Geophysical Union*, 59(6): 516–527.

Alderfer, R. B. and W. C. Bramble, 1942: The effect of plant succession on infiltration of rainfall into Gilpin soil in central Pennsylvania. *Penn State Forest School Research Paper 5*, University Park, PA.

Alpatiev, A. M., 1954: *Water Balance of Cultivated Crops*. Leningrad: Hydrometeorological Publishing House.

American Waterway Operators, Inc. 1979: *Inland Waterborne Commerce Statistics, 1979*, Washington, DC.

Anonymous, 1949: Invisible drought. *Newsweek*, 34(6): 43–44.

————, 1956: Roaded catchments for farm water supplies. *Journal of Agriculture, Western Australia*, 5: 667–679.

————, 1966: The Potomac. *The Johns Hopkins Magazine*, 17(8): 12–33.

————, 1974: The dam(n) builders. *Newsweek*, 84(10): 24.

————, 1979: Advanced wastewater treatment Nature's way. *Environmental Science and Technology*, 12(9): 1013–1014.

Armstrong, E. L., 1976: Irrigation and hydropower. *EOS, Transactions of the American Geophysical Union*, 57(11): 791–797.

Arthur D. Little, Inc., 1973: *Report on Channel Modification* (Vol. I). Prepared for the Council on Environmental Quality.

Aughey, S., 1880: *Sketches of the Physical Geography and Geology of Nebraska*, Omaha, NE, 326 p.

Bader, Henri, 1978: "A critical look at the iceberg utilization project," in Husseiny, A. A. (ed.), *Iceberg Utilization*. New York: Pergamon Press, 34–44.

Baier, W., 1967: Recent advancements in the use of standard climatic data for estimating soil moisture. *Annals of Arid Zones*, 6(1): 1–21.

_____ ..... G. W. Robertson, 1967: Estimating supplemental irrigation water requirements for climatological data. *Canadian Agricultural Engineering*, 9: 46–50.

Bailey, J. F., J. L. Patterson and J. L. H. Paulhus, 1975: Hurricane Agnes rainfall and floods, June–July 1972. Report prepared jointly by USGS and NOAA. *Geological Survey Professional Paper 924*. Washington, DC: Government Printing Office.

Bardo, G. and M. Cassell, 1971: *Tabulations of 128 Waterborne Disease Outbreaks Known to Have Occurred in the United States 1961–1970*. Cincinnati, OH: U.S. Environmental Protection Agency.

Bates, C. G. and A. J. Henry, 1928: Forest and streamflow experiment at Wagon Wheel Gap, Colorado. *Monthly Weather Review*, Supplement 30. Washington, DC: Government Printing Office.

Baumann, D. D. and D. M. Dworkin, 1978: Water resources for our cities. *Resource Papers for College Geography*, No. 78-2, Association of American Geographers, Washington, DC.

Beckinsale, R. P., 1969: "Human response to river regimes," in Chorley, R. J., (ed.), *Water, Earth and Man*. London: Methuen & Co., Ltd., 489–509.

Belt, G. H. and J. G. King, 1977: Augmenting summer streamflow by use of a silicone antitranspirant. *Water Resources Research*, 13(2): 267–272.

Benton, G. S. and R. T. Blackburn, 1950: A comparison of precipitation from maritime and continental air. *Bulletin of the American Meteorological Society*, 31(7): 254–256.

_____, _____ and V. O. Snead, 1950: The role of the atmosphere in the hydrologic cycle. *Transactions of the American Geophysical Union*, 31(1): 61–73.

Bergevin, G. and R. N. Seemel, 1976: The LaGrande hydroelectric complex. *Civil Engineering-ASCE*, 46(9): 60–61.

Biswas, Asit K. 1970: *History of Hydrology*. Amsterdam: North Holland Publishing Co., 336 p.

Blackwelder, Brent, 1976: Benefit claims of the water development agencies: the need for continuing reform. Washington, DC: Environmental Policy Institute. (processed)

Blake, N. M., 1956: *Water for the Cities: A History of the Urban Water Supply Problems in the United States*, Maxwell School Series III. Syracuse, NY: Syracuse University Press.

Bonnifield, Paul, 1979: *The Dust Bowl: Men, Dirt, and Depression*. Albuquerque: University of New Mexico Press.

Bowen, I. S., 1926: The ratio of heat loss by conduction and by evaporation from any water surface. *Physical Review*, 27: 779–789.

Brabyn, H., 1975: The drama of 6000 kilometers of Africa's Sahel. *The UNESCO Courier*, 28(4): 5–9.

Brakke, T. W., S. B. Verma and N. J. Rosenberg, 1978: Local and regional components of sensible heat advection. *Journal of Applied Meteorology*, 17(7): 955–963.

Branson, F. A., 1956: Range forage production changes on a water spreader in SE Montana. *Journal Range Management*, 9: 187–191.

Buckman, H. O. and N. C. Brady, 1960: *The Nature and Properties of Soils.* New York: The Macmillan Co., 567 p.

Budyko, M. I., 1956: *The Heat Balance of the Earth's Surface.* Leningrad: Gidrometeoizdat (trans. N. A. Stepanova, Office of Climatology, U.S. Weather Bureau, 1958), 255 p.

Buehler, Bob, 1977: U.S. floods and their management. *EOS, Transactions of the American Geophysical Union*, 58(1): 4–15.

Bugnicourt, J., 1975: Sahel. *The UNESCO Courier*, 28(4): 11–31.

Bureau of Reclamation, 1963: *Lining for Irrigation Canals.* Washington, DC: U.S. Department of Interior, 149 p.

_____, 1977: *Carriage Facilities–Canals.* Denver, CO: U.S. Department of Interior, 17 p.

Burrill, M. F., 1956: "Geographic distribution of manufacturing," in Graham, J. B. and M. F. Burrill (eds.), *Water for Industry.* Publication 45, American Association for the Advancement of Science, Washington, DC, 23–34.

California Department of Water Resources, 1966: *Implementation of the California Water Plan.* Bulletin No. 160–66, Department of Water Resources, Sacramento, 143 p.

_____, 1976: *Water Conservation in California.* Bulletin No. 198, Department of Water Resources, Sacramento, 95 p.

Carder, D. J. and G. W. Spencer, 1971: *Water Conservation Handbook.* Soil Conservation Service, Dept. of Agriculture, South Perth, Western Australia.

Carreker, J. R. and C. Cobb, Jr., 1963: *Irrigation in the Piedmont.* Georgia Agricultural Experiment Station Technical Bulletin, New Series, 29, 65 p.

Carson, M. A., 1969: "Soil moisture," in Chorley, R. J. (ed.), *Water Earth and Man.* London: Methuen and Co., Ltd., 185–195.

Changnon, S. A., Jr., 1968: The LaPorte weather anomaly—fact or fiction? *Bulletin of the American Meteorological Society*, 49(1): 4–11.

_____, R. J. Davis, B. C. Farhar, J. E. Hass, J. L. Ivens, M. Jones, A. Klein, D. Mann, G. M. Morgan, Jr., S. T. Sonka, E. R. Swanson, C. R. Taylor and P. J. VanBlokland, 1977: *Hail Suppression, Impacts and Issues.* Urbana, IL: Illinois State Water Survey, 427 p.

Chittenden, H. M., 1909: Forests and reservoirs in their relation to streamflow with particular reference to navigable rivers. *Transactions, ASCE*, 62: 249–251.

Chorley, R. J. (ed.), 1969: *Water, Earth and Man.* London: Methuen and Co., Ltd., 588 p.

Christidis, B. G. and G. J. Harrison, 1955: *Cotton Growing Problems.* New York: McGraw-Hill Book Co.

Clark, Chapin D., 1972: "Water law doctrine—some basic considerations," in *Laws for a Better Environment*, Seminar at the Water Resources Research Institute, Oregon State University, Oct. 14, 1971.

Clements, F. E. and R. W. Chaney, 1937: Environment and life in the Great Plains. *Carnegie Institution, Supp. Publ. 24.* Washington, DC.

Cluff, C. Brent, 1978: "Use of floating solar collectors in processing iceberg water," in Husseiny, A. A. (ed.), *Iceberg Utilization*. New York: Pergamon Press.

Cooke, G. W., 1972: *Fertilizing for Maximum Yield*. London: Crosby Lockwood and Son, Ltd.

Cooper, C. F. and W. C. Jolly, 1969: *Ecological Effects of Weather Modification: A Problem Analysis*. Report on Contract 14-06-D-6576, University of Michigan, School of Natural Resources. Washington, DC: U.S. Department of Interior, Bureau of Reclamation, Office of Atmospheric Water Research.

Costin, A. B., L. W. Gay, D. J. Wimbush and D. Kerr, 1961: Studies in catchment hydrology in the Australian Alps III. Preliminary snow investigations. *CSIRO, Division of Plant Industries, Tech. Paper, 15*.

Coughlan, R., 1959: A $40 billion bill for fun. *Life Magazine*, 47(28): 69-74.

Culp, R. and G. Culp, 1971: *Advanced Wastewater Treatment*. New York: Van Nostrand.

Dagg, M., 1970: A study of the water use of tea in East Africa using a hydraulic lysimeter. *Agricultural Meteorology*, 7(4): 303-320.

Dalton, J., 1802: Experimental essays on the constitution of mixed gases; On the force of steam or vapors from water and other liquids in different temperatures, both in a Torricellian vacuum and in air; On evaporation; and On the expansion of gases by heat. *Manchester Literary and Philosophical Society Memoirs*, 5: 535-602.

D'Angelo, A., 1964: "Report on universal metering," to Hon. Robert F. Wagner, Mayor of New York City, Oct. 7.

Daniel, Pete, 1977: *Deep'n As It Come: The 1927 Mississippi River Flood*. New York: Oxford University Press, 162 p.

Davenport, D. C., R. M. Hagan and P. E. Martin, 1969: Antitranspirants research and its possible application in hydrology. *Water Resources Research*, 5: 735-743.

Davis, P. N., 1971: "The law's response to conflicting demands for water: the United States and the Soviet Union," in Fox, I. K. (ed.), *Water Resources Law and Policy in the Soviet Union*. Published for the Water Resources Center, University of Wisconsin, Madison: The University of Wisconsin Press, 53-74.

Deming, H. G., 1975: *Water, the Fountain of Opportunity*. New York: Oxford University Press, 342 p.

Dewsnup, R. L., D. W. Jensen (eds.) and R. W. Swenson (assoc. ed.), 1973: *A Summary-Digest of State Water Laws*, National Water Commission. Washington, DC: Government Printing Office.

Doorenbos, J. and A. H. Kassam, 1979: Yield response to water. *FAO Irrigation and Drainage Paper 33*. Rome: Food and Agriculture Organization of the United Nations.

Eidmann, F. E., 1959: Die interception in buchen-und fichtenbeständen. *C. R. Ass. Int. Hydrologie Sci. Hannover Symp.*, 1: 5-25.

Eliassen, R. and R. H. Cummings, 1948: Analysis of waterborne outbreaks 1938-1945. *Journal of American Water Works Association*, 40(5): 509-528.

Ellis, H. H., J. P. DeBraal, and K. Koepke (undated): Unpublished research, U.S. Dept. of Agriculture, Law School, Madison, WI, 1965-1970.

Farhar, Barbara C., 1974: The impact of the Rapid City flood on public opinion about weather modification. *Bulletin of the American Meteorological Society*, 55(7): 759–764.

Federal Power Commission, 1968: *World Power Data*. Washington, DC: Federal Power Commission.

Ferejohn, John A., 1974: *Pork Barrel Politics: Rivers and Harbors Legislation, 1947–1968*. Stanford, CA: Stanford University Press, 288 p.

Fernow, B. E., 1902: "Relations of forests to water supplies," in *U.S. Forest Service Bulletin 7*, Dept. of Agriculture. Washington, DC: Government Printing Office.

Feth, J. H., 1973: Water facts and figures for planners and managers. *Geological Survey Circular 601-1*, Washington, DC.

Fink, D. W., K. R. Cooley and G. W. Frasier, 1973: Wax treated soils for harvesting water. *Journal of Range Management*, 26: 396–398.

Forsgate, J. A., P. H. Hosegood and J. S. G. McCulloch, 1965: Design and installation of semi-enclosed hydraulic lysimeters. *Agricultural Meteorology*, 2(1): 43–52.

Fox, I. K. (ed.), 1971: *Water Resources Law and Policy in the Soviet Union*, Published for the Water Resources Center, University of Wisconsin. Madison, WI: The University of Wisconsin Press, 256 p.

_____ and D. Herfindahl, 1964: Attainment of efficiency in satisfying demands for water resources. *American Economic Review: Papers and Proceedings of the Annual Meeting of the American Economic Association*, 54: 198–206.

Freeman, K., 1952: *Ancilla to the Pre-Socratic Philosophers*. Oxford: Basil Blackwell, 162 p.

Freese, Nichols and Endress (Consulting Engineers), 1969: *Feasibility Report on the Canyon Lakes Project*, Fort Worth, TX.

French, N. and I. Hussain, 1964: *Water Spreading Manual*. Range Management Record, No. 1, West Pakistan Range Improvement Scheme, Lahore, Pakistan.

Gardner, W. R., 1968: "Availability and measurement of soil water," in Kozlowski, T. T. (ed.), *Water Deficits and Plant Growth* (Vol. 1). New York: Academic Press, 107–135.

_____, D. Hillel and Y. Benyamini, 1970: Post-irrigation movement of soil water: 2. Simultaneous redistribution and evaporation. *Water Resources Research*, 6(4): 1148–1153.

Garstka, W. U., 1978: *Water Resources and the National Welfare*. Fort Collins, CO: Water Resources Publications, 638 p.

Gaussen, H., 1955: Les climats analogues a l'echelle du monde. *Compte Rendu Acad. Agr., France*, Vol. 41.

Geiger, R., 1957: *The Climate Near the Ground*. Cambridge, MA: Harvard Univ. Press.

_____, 1965: *The Atmosphere of the Earth*. Darmstadt, Germany: Justus Perthes.

Gloyna, E. F., 1966: "Major research problems in water quality," in Kneese, A. V. and S. C. Smith (eds.), *Water Research*. Baltimore: The Johns Hopkins Press, 479–494.

Goldfarb, W., 1980: Wild and scenic rivers. *Water Resources Bulletin*, 16(3): 560–561.

Gray, D. M. (ed.), 1970: *Handbook on the Principles of Hydrology*. Ottawa, Canada: Canadian National IHD Committee.

Gray, J. H., 1967: *Men Against the Desert*. Saskatoon, Canada: Western Producer Prairie Book, 250 p.

Grin, A. M., 1967: "Issledovanie vodnogo balansa estestvennykh i sel'skokhoziaist-vennykh ugodii lesostepi," in *Geogizika Landshafta* (D. L. Armand, ed.), 67–73. Moscow: Nauka.

Groen, C. L. and J. C. Schmulbach, 1978: The sport fishery of the unchannelized and channelized middle Missouri River. *Transactions of the American Fish Society*, 107(3): 412–418.

Hamel, L. and D. Nixon, 1978: Field control replaces design conservatism at world's largest underground powerhouse. *Civil Engineering–ASCE*, 48(2): 42–44.

Hanke, S. H. and J. E. Flack, 1968: Effects of metering urban water. *Journal of American Water Works Association*, 60: 1359–1366.

Hansen, V. E., O. W. Israelsen and G. E. Stringham, 1980: *Irrigation Principles and Practices*. New York: John Wiley & Sons, 417 p.

Hanson, M. and R. Bajwa, 1980: National Resources, Economics Division, U.S. Dept. of Agriculture (personal communication).

Harrold, L. L., D. L. Brakensiek, J. L. McGuinness, C. R. Amerman and F. R. Dreibelbis, 1962: Influence of land use and treatment on the hydrology of small watersheds at Coshocton, Ohio, 1938–1957. *USDA Technical Bulletin No. 1256*, Washington, DC.

Haynes, J. L., 1940: Ground rainfall under vegetative canopy of crops. *Journal of the American Society of Agronomy*, 32: 176–184.

Heaton, R. D., 1978: "Potable reuse: the U.S. experience," in Holtz, D. and S. Sebastian (eds.), *Municipal Water Systems*. Bloomington: Indiana University Press, 167–180.

_____, K. D. Linstedt, E. R. Bennett and L. G. Suhr, 1974: "Progress toward successive water use in Denver." Paper presented at the Water Pollution Control Federation Annual Meeting, Denver, CO. (mimeographed)

Hewlett, J. D. and W. L. Nutter, 1969: *An Outline of Forest Hydrology*. Athens: University of Georgia Press.

Hibbert, A. R., 1967: "Forest treatment effects on water yield," in Sopper, W. E. and H. W. Lull (eds.), *Forest Hydrology*. Proceedings of the National Science Foundation Advanced Science Seminar, Pennsylvania State University, Aug. 29–Sept. 10, 1965. Oxford: Pergamon Press.

Hidore, J. J., 1971: The effects of accidental weather modification on the flow of the Kankakee River. *Bulletin of the American Meteorological Society*, 52: 99–103.

Holmes, Beatrice, 1972: A history of Federal water resources programs 1800–1960. *U.S. Dept. of Agriculture, Economic Research Service, Misc. Pub. No. 1233*. Washington, DC: Government Printing Office, 51 p.

Holtan, H. N. and M. H. Kirkpatrick, Jr., 1950: Rainfall, infiltration, and hydraulics of flow in run-off computations. *Transactions of the American Geophysical Union*, 31: 771–779.

Holzman, B., 1937: Sources of moisture for precipitation in the United States. *U.S. Dept. of Agriculture, Tech. Bull. 589*, 41 p.

———— and H. C. S. Thom, 1970: The LaPorte precipitation anomaly. *Bulletin of the American Meteorological Society*, 51: 335–337.

Hoover, M. D., 1944: Effect of removal of forest vegetation upon water yields. *Transactions of the American Geophysical Union*, 25: 969–975.

————, 1962: Water action and water movement in the forest. Food and Agriculture Organization of the United Nations. *FAO Forestry and Forest Products Studies*, 15: 31, 33–80.

Hoppe, E., 1896: Regenmessungen unter Baumkronen. *Mitt. a. d. forstl. Vers. wesen Österr. 21*, Wien.

Houser, E. W., 1970: Santee project continues to show the way. *Water and Wastes Engineering*, 8: 41.

Howe, C. W. and K. W. Easter, 1971: *Interbasin Transfers of Water: Economic Issues and Impacts*. Baltimore, MD: The Johns Hopkins Press for Resources for the Future, Inc.

———— and F. P. Linaweaver, Jr., 1967: The impact of price on residential water demand and its relation to system design and price structure. *Water Resources Research*, 3(1): 13–32.

Howe, E. D., 1974: *Fundamentals of Water Desalination*. New York: Marcel Dekker, 344 p.

Hubbert, M. K., 1974: World potential and developed waterpower capacity. *Water Resources Bulletin*, 10(2): 397.

Hult, J. C. and N. C. Ostrander, 1973: Applicability of ERTS to Antarctic iceberg resources. *Rand Corporation Paper P-5137*, Santa Monica, CA.

Humlum, J., 1969: *Water Development and Water Planning in the Southwestern United States*. Aarhus, Denmark: Kulturgeografisk Institut, Aarhus Universitet.

Hussain, S. N., 1978: "Iceberg protection by foamed insulation," in Husseiny, A. A. (ed.), *Iceberg Utilization*. New York: Pergamon Press.

Husseiny, A. A. (ed.), 1978: *Iceberg Utilization*. New York: Pergamon Press.

Intermediate Technology Development Group, Ltd., 1969: *The Introduction of Rainwater Catchment Tanks and Micro-irrigation to Botswana*. Intermediate Tech. Devel. Group, London, England, 74 p.

Jahn, L. R. and J. B. Trefethen, 1973: Placing channel modifications in perspective. *Proceedings of the National Symposium on Watersheds in Transition*, American Water Resources Assocaition, 15–21.

Jeffrey, W. W., 1970: "Hydrology of land use," Section 13 in Gray, D. M. (ed.), *Handbook on the Principles of Hydrology*. Ottawa: Canadian National Committee for the IHD.

Johns Hopkins Magazine (The), 1966: The water we use. *The Johns Hopkins Magazine*, 17(8): 10–11.

Johnson, J. F., 1971: Renovated waste water: an alternative source of municipal water supply in the United States. *Department of Geography Research Paper No. 135*, University of Chicago, Chicago, IL, 155 p.

Johnson, V., 1947: *Heaven's Tableland: The Dust Bowl Story.* New York: Farrar Straus and Co.

Kalnicky, R. A., 1976: Recreation use of small streams in Wisconsin. *Dept. of Natural Resources, Tech. Bull. 95*, Madison, WI, 20 p.

Kaplan, Rachel, 1977: Down by the riverside: informational factors in waterscape preference, in "River Recreation Management and Research Symposium Proceedings." *USDA Forest Service, Gen. Tech. Rep. NC-28*, 285–289. North Cent. For. Exp. Stn., St. Paul, MN.

Kasperson, R. E. and J. X. Kasperson (eds.), 1977: *Water Re-Use and the Cities.* Hanover, NH: Clark University Press, 238 p.

Kazmann, R. G., 1972: *Modern Hydrology* (2nd ed.). New York: Harper & Row, 365 p.

Kellsall, K. J., 1962: *Construction of Bituminous Surfaces for Water Supply Catchment Areas in Western Australia.* Hydraulic Engineers Branch, Public Works Department, Perth, Western Australia. (mimeographed notes)

Kemper, J. P., 1949: *Rebellious River.* Boston: Bruce Humphries, Inc. Publishers, 279 p.

Kilbourne, E. D. and W. G. Smillie, 1969: *Human Ecology and Public Health.* London: Macmillan.

Kittredge, J., 1948: *Forest Influences.* New York: McGraw-Hill Book Co.

Koschmieder, H., 1934: Methods and results of definite rain measurements. *Monthly Weather Review*, 62(1): 5–7.

Law, J. P., 1968: *Agricultural Utilization of Sewage Effluent and Sludge, an Annotated Bibliography.* Washington, DC: U.S. Federal Water Pollution Control Administration.

Leopold, Luna B., 1974: *Water: A Primer.* San Francisco: W. H. Freeman & Co., 172 p.

Linsley, R. K., M. A. Kohler and J. L. H. Paulhus, 1949: *Applied Hydrology.* New York: McGraw-Hill Book Co.

————, ———— and ————, 1958: *Hydrology for Engineers.* New York: McGraw-Hill Book Co.

Loane, E. S., 1957: Pumped-storage hydro attractive. *Electrical World*, 148(13): 73.

Lofting, E. M. and H. C. Davis, 1977: "Methods for estimating and projecting water demand for water resources planning," in *Climate, Climatic Change, and Water Supply.* Geophysical Research Board, National Research Council, Washington, DC.

Love, L. D. and B. C. Goodell, 1960: Watershed research on the Fraser Experimental Forest. *Journal of Forestry*, 58(4): 272–275.

Lull, H. W. and K. G. Reinhart, 1972: Forest and floods in the eastern United States. *Forest Service Research Paper NE-226.* Upper Darby, PA: U.S. Dept. of Agriculture.

L'vovich, M. I., 1973: The water balance of the world's continents and a balance estimate of the world's freshwater resources. *Soviet Geography*, 14(3): 135–152 (trans. from *Izvestiya Akademii Nauk SSR, seriya geograficheskaya*, 1972, No. 5, 5–20).

_____ and S. P. Ovtchinnikov, 1964: *Physical-Geographical Atlas of the World* (in Russian). Academy of Sciences, USSR and Department of Geodesy and Cartography, State Geodetic Commission, Moscow.

Maass, Arthur, 1951: *Muddy Waters; the Army Engineers and the Nation's Rivers.* Cambridge, MA: Harvard University Press, 306 p.

Maine Department of Conservation, 1974: *1973 Survey of Allagash Wilderness Waterway Visitor Use and Visitor Use Characteristics.* Bureau of Parks and Recreation, Augusta, ME, 64 p.

Manabe, S. and J. L. Holloway, Jr., 1975: Seasonal variation of hydrologic cycle as simulated by a global model of the atmosphere. *Journal Geophysical Research,* 80(12): 1617–1649.

Marsh, G. P., 1864: *Man and Nature; or Physical Geography as Modified by Human Action.* New York: Scribners, 560 p.

Martin, W. J. and E. L. Thackston, 1980: A retrospective benefit-cost analysis of water resource projects in the Cumberland River basin. *Water Resources Bulletin,* 16(6): 1006–1011.

Mather, J. R., 1954: The measurement of potential evapotranspiration. *Publications in Climatology,* Laboratory of Climatology, 7(1): 1–225.

_____, 1968: "Irrigation agriculture in humid areas," in Court, A. (ed.), *Eclectic Climatology.* Yearbook, Association of Pacific Coast Geographers, Vol. 30. Corvallis: Oregon State University Press, 107–122.

_____, 1969: The average annual water balance of the world. *Proceedings of the Symposium on the Water Balance in North America,* American Water Resources Association, Series 7, 29–40.

_____ and C. M. Rowe, Jr., 1979: The use of the climatic water budget to evaluate the validity of a precipitation record. *University of Delaware Water Resources Center Contribution No. 28,* Newark, 25 p.

McCauley, David, 1977: "Water reclamation and re-use in six cities," in Kasperson, R. E. and J. X. Kasperson (eds.), *Water Re-Use and the Cities.* Hanover, NH: Clark University Press, 49–73.

McDonald, J. E., 1962: The evaporation-precipitation fallacy. *Weather,* 17(5): 1–9.

Meiman, J. R., 1976: "Snow measurement," in Kunkle and Thames (eds.), *Hydrological Techniques for Upstream Conservation.* Food and Agriculture Organization of the United Nations, Rome, 103–115.

Miller, D. H., 1955: Snow cover and climate in the Sierra Nevada. *Publications in Geography,* University of California, Berkeley, 11: 1–218.

_____, 1977: *Water at the Surface of the Earth.* New York: Academic Press.

More, R. J., 1969: "Water and crops," in Chorley, R. J. (ed.), *Water, Earth, and Man.* London: Methuen and Co. Ltd., 197–208.

Muller, R. A., 1966: The effects of reforestation on water yield: a case study using energy and water balance models for the Allegheny Plateau, New York. *Publications in Climatology,* Laboratory of Climatology, 10(3): 251–304.

Murray, C. E. and E. B. Reeves, 1972: Estimated use of water in the U.S. in 1970. *U.S. Geological Survey Circular 676,* Washington, DC.

Musgrave, G. W., 1938: Field research offers significant new findings. *Soil Conservation,* 3: 210–214.

Nadeau, R. A., 1950: *The Water Seekers.* Garden City, NY: Doubleday.

Nanda, Ved P., (ed.), 1977: *Water Needs for the Future: Political, Economic, Legal, and Technological Issues in a National and International Framework.* Westview Special Studies in Natural Resources and Energy Management. Boulder, CO: Westview Press, 329 p.

National Academy of Sciences, 1966: *Weather and Climate: Modification Problems and Prospects*, Pub. 1350. Washington, DC: National Academy of Sciences/ National Research Council.

_____, 1972: *Water Quality Criteria 1972*, National Academy of Engineering. Washington, DC: Government Printing Office.

_____, 1974: *More Water for Arid Lands: Promising Technologies and Research Opportunities.* Report of an Ad Hoc Panel of the Advisory Committee on Technology Innovation, BOSTID, Commission on International Relations, Washington, DC, 153 p.

National Association of Manufacturers, 1965: *Water in Industry: A Survey of Water Use in Industry.* Nat. Assn. of Mfrs. and the Chamber of Commerce of the U.S., New York, and Washington.

National Water Commission, 1973: *Water Policies for the Future.* Final Report to the President and to the Congress of the United States. Washington, DC: Government Printing Office, 579 p.

Niskanen, W. A., Jr., 1971: *Bureaucracy and Representative Government.* Chicago: Aldine, Atherton.

North Central Regional Committee 40, 1979: *Water Infiltration into Representative Soils of the North Central Region.* Agricultural Experiment Station, University of Illinois, Bull. 760 and North Central Region Res. Pub. 259.

Office of Saline Water, 1972: *1971-1972 Saline Water Conversion Summary Report*, U.S. Dept. of Interior, Office of Saline Water. Washington, DC: Government Printing Office.

_____, 1973: *1972-1973 Saline Water Conversion Summary Report*, U.S. Dept. of Interior, Office of Saline Water. Washington, DC: Government Printing Office.

Office of Science and Technology Policy, 1978: *Scientific and Technological Aspects of Water Resources Policy.* Executive Office of the President, Washington, DC.

Office of Water Planning and Standards, 1975: *National Water Quality Inventory*, 1975 Report to Congress, Washington, DC.

Oliver, John E., 1977: *Perspectives on Applied Physical Geography.* North Scituate, MA: Duxbury Press, 315 p.

O'Riordan, T. and R. J. More, 1969: "Choice in water use," in Chorley, R. J. (ed.), *Water, Earth and Man.* London: Methuen and Co. Ltd., pp. 547-573.

Palmer, W. C., 1965: Meteorological drought: its measurement and classification. *Research Paper No. 45*, U.S. Weather Bureau, Washington, DC.

Penman, H. L., 1949: The dependence of transpiration on weather and soil conditions. *Journal of Soil Science*, 1: 74-89.

_____, 1956: Evaporation: an introductory survey. *Netherlands Journal of Agricultural Science*, 4(1): 9-29.

————, 1963: *Vegetation and Hydrology*. Technical Communication 53, Commonwealth Bureau of Soils, Farnham Royal. Bucks, England: Commonwealth Agricultural Bureaux.

Pereira, H. C., 1973: *Land Use and Water Resources in Temperate and Tropical Climates*. Cambridge, England: University Press.

Peterson, H. B. and J. C. Ballard, 1953: Effect of fertilizer and moisture on the growth and yield of sweet corn. *Utah Agricultural Experiment Station Bulletin 360*, Utah State University, Logan.

Probstein, R. F. and H. Gold, 1978: *Water in Synthetic Fuel Production: The Technology and Alternatives*. Cambridge, MA: The MIT Press.

Rainbird, A. F., 1970: *Methods of Estimating Areal Average Precipitation*. Report No. 3, World Meteorological Organization, International Hydrological Decade (IHD), Geneva, Switzerland, 45 p.

Ralph M. Parsons Company, 1964: *NAWAPA: North American Water and Power Alliance*, Brochure 606-2934-19, Los Angeles.

Reed, Jack W., 1973: Cloud seeding at Rapid City: a dissenting view. *Bulletin of the American Meteorological Society*, 54(7): 676–677.

Reid, G. W., 1971: "The macro-approach—urban water demand models," in Albertson, M. L., L. S. Tucker and D. C. Taylor (eds.), *Treatise on Urban Water Systems*. Fort Collins, CO: Colorado State University, 235–294.

Reifsnyder, W. E. and H. W. Lull, 1965: Radiant energy in relations to forest. *U.S. Dept. of Agriculture, Forest Service Tech. Bull 1344*, 111 p.

Reinhart, K. G., A. R. Eschner and G. R. Trimble, Jr., 1963: Effect on streamflow of four forest practices in the mountains of West Virginia. *Forest Service Research Paper NE-1*, U.S. Dept. of Agriculture, Upper Darby, PA.

Ricca, V. T., P. W. Simmons, J. L. McGuinness and E. P. Taiganides, 1970: Influences of land-use on runoff from agricultural watersheds. *Agricultural Engineering*, 53: 187–190.

Richardson, H. L., 1960: Increasing world food supplies through greater crop production. *Outlook on Agriculture*, 3: 9–25.

Robinson, F. L., 1976: *Floods in N.Y. 1972 with Special Reference to Tropical Storm Agnes*. Water Resources Investigations, U.S. Geological Survey, Albany, NY. Water Resource Div. WRI-34-75, WRD 76/011.

Romanelli, C. J. and W. M. Griffith, 1972: *The Wrath of Agnes: A Complete Pictorial and Written History of the June 1972 Flood in Wyoming Valley*. Wilkes-Barre, PA: Media Affiliates, Inc., 200 p.

Rosenberg, N. J., 1972: Frequency of potential evapotranspiration rates in central Great Plains. *Journal of the Irrigation and Drainage Division*, Proceedings ASCE, IR 2, 203–206.

————, H. E. Hart and K. W. Brown, 1968: *Evapotranspiration: Review of Research*, Pub. 20. Lincoln: University of Nebraska, College of Agriculture and Home Economics and Nebraska Water Resources Research Institute.

———— and S. B. Verma, 1978: Extreme evapotranspiration by irrigated alfalfa: a consequence of the 1976 midwestern drought. *Journal of Applied Meteorology*, 17(7): 934–941.

Rothacher, J., 1963: Net precipitation under a Douglas-fir forest. *Forest Science*, 9: 423–429.

Rouse, W. J., 1962: Moisture balance of Barbados and its influence on sugar cane yield. Master of science thesis, McGill University, Department of Geography.

Royce, W. F., 1967: Fish and fishing. *Bulletin of the Atomic Scientists*, 26–27.

Sartor, J. Doyne, 1969: Electricity and rain. *Physics Today*, 22(8): 45–51.

Savini, J. and J. C. Kammerer, 1961: Urban growth and the water regimen. *U.S. Geological Survey Water Supply Paper 1591-A*. Washington, DC: Government Printing Office.

Schmidt, C. J., R. F. Beardsley and E. V. Clements, III, 1973: "A survey of industrial use of municipal wastewater," in Cecil, L. K. (ed.), *Complete Water Reuse: Industry's Opportunity*. Papers presented at the National Conference on Complete Water Reuse, sponsored by American Institute of Chemical Engineers, Environmental Protection Agency-Technology Transfer, April 23–27.

Schomaker, C. E., 1968: Solar radiation measurements under a spruce and a birch canopy during May and June. *Forest Science*, 14(1): 31–38.

Schoof, Russell, 1980: Environmental impact of channel modification. *Water Resources Bulletin*, 16(4): 697–701.

Scharr, S. H. and B. C. Netschert, 1960: *Energy in the American Economy, 1850–1975*. Baltimore: The Johns Hopkins Press for Resources for the Future, Inc., 774 p.

Sebald, A. V., 1974: *Energy Intensity of Barge and Rail Freight Hauling*. Urbana: Center for Advanced Computation, University of Illinois.

Seidel, H. F. and E. R. Baumann, 1957: A statistical analysis of water works data for 1955 data. *Journal of American Water Works Association*, 49(12): 1531–1566.

Select Committee on National Water Resources, 1960: "Electric power in relation to the nation's water resources," in *Water Resources Activities in the United States*, Committee Print 10, Senate Resolution 48, 86th Congress. Washington, DC: Government Printing Office.

————, 1960a: "Fish and wildlife and water resources," in *Water Resources Activities in the United States*, Committee Print No. 18, Senate Resolution 48, 86th Congress. Washington, DC: Government Printing Office.

————, 1960b: "Future water requirements for municipal use," in *Water Resources Activities in the United States*, Committee Print 7, Senate Resolution 48, 86th Congress. Washington, DC: Government Printing Office.

————, 1960c: "Water recreation needs in the United States, 1960–2000," in *Water Resources Activities in the United States*, Committee Print 17, Senate Resolution 48, 86th Congress. Washington, DC: Government Printing Office.

Shaw, L. H. and D. D. Durost, 1962: *Measuring the Effects of Weather on Agricultural Output*. Farm Economics Division, Economic Research Service, ERS-72, U.S. Dept. of Agriculture, Washington, DC, 49 p.

Shipley, J. and C. Regier, 1975: Water response in the production of irrigated grain sorghum, High Plains of Texas. *Texas Agricultural Experimental Station Misc. Pub. 1202*, Texas A&M, College Station, TX.

Smith, D. D., D. M. Whitt, A. W. Zingg, A. G. McCall and F. G. Bell, 1945: Investigations in erosion control . . . at Bethany, Mo., 1930–1942. *U.S. Dept. of Agriculture, Tech. Bull. 883*, Washington, DC.

Smith, R., 1968: Cost of conventional and advanced treatment of wastewaters. *Journal of the Water Pollution Control Federation*, 40: 1546–1574.

————— and W. F. McMichael, 1969: *Cost and Performance Estimates for Tertiary Wastewater Treating Processes*, U.S. Federal Water Pollution Control Administration Report No. TWRC-9, Washington, DC.

Snider, D., 1972: "Hydrographs," Chap. 16 in *Hydrology*, Soil Conservation Service, National Engineering Handbook, Section 4, U.S. Department of Agriculture. Washington, DC: Government Printing Office.

Spence, C. C., 1980: *The Rainmakers: American "Pluviculture" to World War II*. Lincoln: University of Nebraska Press, 181 p.

Sprinkler Irrigation Association, 1955: *Sprinkler Irrigation*, Washington, DC, 466 p.

Staple, W.J. and J.J. Lehane, 1962: Variability in soil moisture sampling. *Canadian Journal of Soil Science*, 42(1): 157–164.

Stephan, D. G. and L. W. Weinberger, 1968: Wastewater reuse—has it arrived? *Journal of the Water Pollution Control Federation*, 40: 529–539.

Subrahmanyam, V. P., 1967: Incidence and spread of continental drought. *WMO/IHD Report No. 2*, Geneva.

Sutcliffe, R. C., 1956: Water balance and the general circulation of the atmosphere. *Quarterly Journal Royal Meteorological Society*, 82: 385–395.

Sutton, O. G., 1953: *Micrometeorology*. New York: McGraw-Hill Book Co.

Swenson, H. A. and H. L. Baldwin, 1965: *A Primer on Water Quality*. Washington, DC: Geological Survey, U.S. Dept. of Interior, 27 p.

Tarplee, W. H., Jr., D. E. Louder and A. J. Weber, 1971: *Evaluation of the Effects of Channelization on Fish Populations in North Carolina's Coastal Plain Streams*. Raleigh, NC: North Carolina Wildlife Resources Commission.

Taylor, S. A., 1955: Field determinations of soil moisture. *Agricultural Engineering*, 36: 654–659.

Thomas, H. E., 1969: Water laws and concepts. *Transactions of the American Geophysical Union*, 50(2): 40–50.

Thompson, Herbert J., 1972: The Black Hills flood. *Weatherwise*, 25(4): 162–167, 173.

Thornthwaite, C. W., 1937: The hydrologic cycle reexamined. *Soil Conservation*, 3(4): 2–8.

—————, 1947: Climate and moisture conservation. *Annals of the Association of American Geographers*, 37(2): 87–100.

—————, 1948: An approach toward a rational classification of climate. *Geographical Review*, 38(1): 55–94.

————— and J. R. Mather, 1955: The water balance. *Publications in Climatology*, Laboratory of Climatology, 8(1): 1–104.

————— and —————, 1957: Instructions and tables for computing potential evapotranspiration and the water balance. *Publications in Climatology*, Laboratory of Climatology, 10(3): 185–311.

Todd, D. K. (ed.), 1970: *The Water Encyclopedia*. Syosset, NY: Water Information Center, 559 p.

Toumey, J. A., 1903: The relations of forests to streamflow. *Yearbook of Agriculture*. Washington, DC: U.S. Dept. of Agriculture.

U.S. Bureau of the Census, 1975: *Statistical Abstract of the United States*. Washington, DC: Government Printing Office.

_____, 1975a: *Water Use in Manufacturing, 1972 Census of Manufacturers*. Special Report Series MC 72(SR) 4. Washington, DC: Government Printing Office.

U.S. Department of the Army, Corps of Engineers, 1979: *Waterborne Commerce of the United States, Calendar Year, 1979, Part 5, National Summaries*. Washington, DC: Government Printing Office.

U.S. Department of Commerce, 1972: *Hurricane Agnes, June 14–23, 1972. Preliminary Reports on Hurricanes and Tropical Storms*. Silver Spring, MD: National Oceanic and Atmospheric Administration, National Weather Service, Office of Meteorological Operations.

U.S. Federal Water Pollution Control Agency, 1969: *The Cost of Clean Water and Its Economic Impact*. Washington, DC: Government Printing Office.

U.S. Water Resources Council, 1968: *The Nation's Water Resources, Part 1*, First National Water Assessment. Washington, DC: Government Printing Office.

_____, 1978: *The Nation's Water Resources 1975–2000, Vol. 1: Summary*. Second National Water Assessment. Washington, DC: Government Printing Office.

van der Leeden, Frits (ed.), 1975: *Water Resources of the World: Selected Statistics*. Syosset, NY: Water Information Center.

Wallis, James R., 1977: Climate, climatic change, and water supply. *EOS, Transactions of the American Geophysical Union*, 58(11): 1012–1024.

Walsh, R. G., 1977: "Recreational user benefits from water quality improvements," in *Economics of Outdoor Recreation Symposium Proceedings, USDA Forest Service General Technical Report WO-2*, 121–132, Northeast. Forest Exp. Sta., Upper Darby, Pa.

Walter, H., 1955: Die klimagramme als mittel zur deurteilung der klimäverhaltnisse für ökologische, vegetationskundliche und landwirtschafltiche zwecke. *Bericht Deutschen Botanischen Gesellschaft Jahrgang*, 68(8): 331–334.

Ward, R. C., 1975: *Principles of Hydrology* (2nd ed.). New York: McGraw-Hill Book Co., 367 p.

Ware, L. M. and W. A. Johnson, 1950: Value of irrigation with different fertility treatments for vegetable crops. *Alabama Exp. Sta. Bull. No. 276*, Auburn, AL, 69 p.

Warnick, C. C., 1969: "Historical background and philosophical basis of regional water transfer," in McGinnies, W. G. and B. J. Goldman (eds.), *Arid Lands in Perspective*. Tucson: University of Arizona Press and American Association for the Advancement of Science.

Weather Modification Advisory Board, 1978: *The Management of Weather Resources, Vol. I, Proposals for a National Policy and Program*, Report to the Secretary of Commerce. Washington, DC: Government Printing Office, 229 p.

Webster, Noah, Jr. (comp.), 1796: *A Collection of Papers on the Subject of Bilious Fevers.* New York: Hopkins, Webb and Co.

Weeks, W. F. and M. Mellor, 1978: "Some elements of iceberg technology," in Husseiny, A. A. (ed.), *Iceberg Utilization.* New York: Pergamon Press.

West, A. J. and K. R. Knoerr, 1959: Water losses in the Sierra Nevada. *Journal of the American Water Works Association,* 51: 481–488.

Whipkey, R. Z., 1965: Subsurface stormflow from forested slopes. *Bulletin of the International Association of Scientific Hydrology,* 10(2): 74–85.

White, G. F. (ed.), 1969: *Water, Health and Society: Selected Papers by Abel Wolman.* Bloomington: Indiana University Press.

Wiesner, C. J., 1970: *Climate, Irrigation and Agriculture.* Sydney, Australia: Angus and Robertson.

Willmott, C. J., 1977: WATBUG: A FORTRAN IV Algorithm for calculating the climatic water budget. *Publications in Climatology,* Laboratory of Climatology, 30(2): 1–55.

Winchester, James H., 1972: Night of terror in Rapid City. *The Reader's Digest,* November, 81–88.

Wittfogel, K. A., 1956: "The hydraulic civilization," in Thomas, W. L., Jr. (ed.), *Man's Role in Changing the Face of the Earth.* Chicago: The University of Chicago Press, 152–164.

Wollny, E., 1890: Untersuchungen über die beeinflussung der fruchtbarkeit der ackerkrume durch die thatigkeit der regenwürmer. *Forsch. Geb. Agric-Phys,* 13: 381–395.

Wolman, A., 1962: *Water Resources: A Report to the Committee on Natural Resources of the National Academy of Sciences-National Research Council,* NAS-NRC Publication 1000-B, Washington, DC.

Woodley, W. L., J. Simpson, R. Biondini and J. Jordan, 1977: NOAA's Florida area cumulus experiment, rainfall results 1970–1976. *Preprints, Sixth Conference on Planned and Inadvertent Weather Modification, Champaign-Urbana, IL.* Boston, MA: American Meteorological Society, 206–209.

Woodward, G. O., 1959: *Sprinkler Irrigation.* Washington, DC: Sprinkler Irrigation Association.

World Meteorological Organization, 1975: *Drought and Agriculture.* Technical Note No. 138. Report of the CAgM Working Group on the Assessment of Drought, WMO No. 392, Geneva.

Worstell, R. V., 1976: Estimating seepage losses from canal systems. *Journal, Irrigation and Drainage Division, Proceedings, ASCE,* March 137–147.

Zon, R., 1912: *Forest and Water in the Light of Scientific Investigation.* Senate Document 469, 62nd U.S. Congress, 2nd Session. Washington, DC: Government Printing Office.

————, 1945: Forests in relation to soil and water. *Proceedings of the American Philosophical Society,* 89(2): 399–402.

# SUBJECT INDEX

Hydrophylocia, 30

Icebergs
  to supply freshwater, 242-245
    encapsulation problems, 244
    feasibility of use, 243-244
    problems of towing, 243
    relative water volumes, 244
    size and shape for towing, 243
Iceberg utilization conference, 243
Ice crystal process
  in precipitation, 47
Industrial water use
  categories of demand, 111
  change over time, 12-16, 109
  effect of water cost, 16-17
  future demands, 19, 21-23
  by industry, 109-110
  localized geographically, 115-117
  per unit of product, 113-115
  recycle ratio, 109-110, 112
  reuse and recycling, 109-113
  techniques to control use, 118-119
Industry
  future locations, 118-119
  geographic distribution, 115-117
Infectious agents in effluent, 249
Infiltration
  definition, 77
  effect of ecosystem, 77-80
  factors affecting, 77
  rain exceeding infiltration rate, 76
  role in basin hydrologic cycle, 44-46
  variation over time, 77
  variation with soil type, 78-79
Inland Waterways Commission, 296
Inorganic chemicals in effluents, 250-251
Instream flow need
  for fish, wildlife, 175-179
  future needs, 180-181
  for recreation, 177-179
Interbasin water transfer, see Water transfer
    plans
Interception losses
  by agricultural crops, 59-60
  deciduous vs. coniferous vegetation, 56
  effect of crown density, 58-60
  effect of precipitation intensity, 55, 57
  effect of vegetation, 55-60, 232
  role in basin hydrologic cycle, 44-46
  tropical vs. temperate vegetation, 57
Interflow, see Throughflow
Interstate commerce
  Congressional right to regulate, 295

  constitutional provisions, 294
Ion exchange process, 207
  schematic cross-section of system, 220
Irrigation
  acreage, by country, 121-122
  in Bureau of Reclamation states, 123, 125-126
  change in acreage over time, 122-125
  control of salt problem, 134-135
  costs, 127
  critical water stress periods, 132-134
  efficiency, 127
  estimates of water losses, 404-405
  and fertilizer application, 129-132, 134
  funded by Reclamation funds, 297
  funds for development, 123, 125-126
  future, 139-140
  history, 120-121, 141
    in Anaheim, Calif., 121
    by Chinese, 120
    by Egyptians, 120
    by Hohokams, 120
    by Mormons, 121, 141
    by Spaniards, 120
  in humid areas, 128-135, 137
  humid vs. arid areas, 135, 137
  to increase crop yields, 129-134
  off-stream storage, 141
  program in humid area, 137
  purposes of, 121
  by states, 121-125
  techniques, 127
  water use, over time, U.S., 12-16

La Grande hydroelectric development, 157-159
  cost, 159
  location, 157-158
  Phase I construction, 157
  Phase II, III, IV construction, 159
La Porte precipitation anomaly, 54-55
Landuse-streamflow relations
  history of use, 224-225
  hydrometric studies, 223
  paired watershed technique, 224
    limitations, 224
  physical studies, 223, 228
  response to clearcutting, 225-226, 228
Latent evaporation, 70-71
Latrobe, Benjamin, 98
Leakage surveys, 106-107
Legislation, water resources, federal
  authorized by Constitution, 294
  benefit-cost analysis, 307-315

Pumped storage plants, 156-157
  environmental problems, 156

Qanat, 120

Radiation and snowmelt, 232-234
Radioactive pollution, 250
Rain gage
  network density, in U.S., 50-51
  standard Weather Bureau, 47-48
  tipping-bucket, 47, 49
  weighing, 47
  wind effect on measurements, 51, 53
  winterizing, 49
Rainmaking, see also Weather modification
  by chemical fumes, 186
  Colorado River Basin Pilot Project, 199
  by concussion, 185-186
  by convective activity, 194
  downwind effects, 198
  dry ice experiments, 189
  ecological effects, 201
  economic potential, 197-199
  effect of electrical charges, 189
  effect of large condensation nuclei, 189
  Hatfield, Charles, 186
  High Plains Cooperative Program (HIPLEX)
    199-200
  hydrologic effects, 201-202
  ice crystal process, 187-190
  necessary conditions, 188
  physical principles, 186-190
  Project Skywater, 198-200
  results of trials, 190-191, 197
  role in Rapid City flood, 345-347
  seeding of orographic clouds, 200
  Sierra Cooperative Pilot Project, 200
  use of silver iodide, 189
Rainwater harvesting, see Water harvesting
Ralph M. Parsons Company, 269-270, 272
Rapid City flood, 1972, 344-351
  action of volunteers, 350
  geographic setting, 345
  meteorological situation, 345, 348-349
  role of cloud seeding, 345-347
Rayleigh, Lord, 189
Reasonable use doctrine
  in groundwater law, 289-290
  in riparian law, 279-283
Reclamation Act, 1902, 297
Reclamation fund, 297
Recreation
  economic benefit, 177

expenditures for, 170
future water needs, 180-181
instream-flow needs, 177-179
need for water, 169, 176-179
uses of streams, 176-177
Recycle ratio, 109-110, 112
Red Cross
  activities in Canadian drought, 365
  shelters and camps in Mississippi River
    flood, 336, 340-342
  life in relief camps, 342-343
Renovation of waste water, see Waste water
  renovation
Reservoir, multipurpose
  development under Roosevelt, 298
  effect on fish, wildlife, 171-172
  filling with sediments, 145
  for flood control, 143-144
  history, 141-147
  for irrigation, 141-142
  loss of water, 145-147
  problems in operation, 143-144
  for recreation, 144
  seasonal vs. cyclical, 144-145
  storage of water, 10
  subterranean, in hydrologic cycle, 28-30
  for water power, 142-143
  Water Supply Act of 1958, 300-301
Reverse osmosis, 207, 217-218, 220, 222,
    see also Desalination
  costs, 222
  membrane thickness, 217
  pressures needed, 220
  schematic cross-section of system, 219
  use of synthetic membranes, 217
Riparian doctrine, 278-284, 288
  applied to Federal reserved water rights,
    288
  condemnation, 281
  continuous flow theory, 279
  development of permit system, 280-283
  individual court decisions, 281-282
  loss of rights, 282
  prescriptive rights, 282
  reasonable use theory, 279, 283
  rejected in California, 284
  rights to water, 280
  stream ownership, 279-280
Rivers
  increase in salt content, 135
  origin of, 28-31
River basin management
  Federal-interstate water compacts, 318-
    319

# AUTHOR INDEX